THE CHEMISTRY OF MILK AND MILK PRODUCTS

Physicochemical Properties, Therapeutic Characteristics, and Processing Methods

Innovations in Agricultural and Biological Engineering

THE CHEMISTRY OF MILK AND MILK PRODUCTS

Physicochemical Properties, Therapeutic Characteristics, and Processing Methods

Edited by

Megh R. Goyal, PhD, PE

Suvartan Ranvir, PhD

Junaid Ahmad Malik, PhD

AAP | APPLE ACADEMIC PRESS

First edition published 2024

Apple Academic Press Inc.
1265 Goldenrod Circle, NE,
Palm Bay, FL 32905 USA
760 Laurentian Drive, Unit 19,
Burlington, ON L7N 0A4, CANADA

CRC Press
6000 Broken Sound Parkway NW,
Suite 300, Boca Raton, FL 33487-2742 USA
4 Park Square, Milton Park,
Abingdon, Oxon, OX14 4RN UK

Library and Archives Canada Cataloguing in Publication

Title: The chemistry of milk and milk products : physicochemical properties, therapeutic characteristics, and processing methods / edited by Megh R. Goyal, PhD, PE, Suvartan Ranvir, PhD, Junaid Ahmad Malik, PhD.
Names: Goyal, Megh R., editor. | Ranvir, Suvartan, editor. | Malik, Junaid Ahmad, 1987- editor.
Series: Innovations in agricultural and biological engineering.
Description: First edition. | Series statement: Innovations in agricultural and biological engineering | Includes bibliographical references and index.
Identifiers: Canadiana (print) 20230189555 | Canadiana (ebook) 20230189628 | ISBN 9781774912249 (hardcover) | ISBN 9781774912256 (softcover) | ISBN 9781003340706 (PDF)
Subjects: LCSH: Milk—Composition. | LCSH: Dairy products—Composition. | LCSH: Dairy processing. | LCSH: Milk—Health aspects. | LCSH: Dairy products—Health aspects. | LCSH: Agricultural chemistry.
Classification: LCC SF251 .C47 2023 | DDC 637—dc23

Library of Congress Cataloging-in-Publication Data

CIP data on file with US Library of Congress

ISBN: 978-1-77491-224-9 (hbk)
ISBN: 978-1-77491-225-6 (pbk)
ISBN: 978-1-00334-070-6 (ebk)

ABOUT THE BOOK SERIES: INNOVATIONS IN AGRICULTURAL AND BIOLOGICAL ENGINEERING

Under this book series, Apple Academic Press Inc. is publishing book volumes over a span of 8–10 years in the specialty areas defined by the American Society of Agricultural and Biological Engineers (<asabe.org>). Apple Academic Press Inc. aims to be a principal source of books in agricultural and biological engineering. We welcome book proposals from readers in areas of their expertise.

The mission of this series is to provide knowledge and techniques for agricultural and biological engineers (ABEs). The book series offers high-quality reference and academic content on agricultural and biological engineering (ABE) that is accessible to academicians, researchers, scientists, university faculty and university-level students, and professionals around the world.

Agricultural and biological engineers ensure that the world has the necessities of life, including safe and plentiful food, clean air and water, renewable fuel and energy, safe working conditions, and a healthy environment by employing knowledge and expertise of the sciences, both pure and applied, and engineering principles. Biological engineering applies engineering practices to problems and opportunities presented by living things and the natural environment in agriculture.

ABE embraces a variety of the following specialty areas (<asabe.org>): aquaculture engineering, biological engineering, energy, farm machinery and power engineering, food, and process engineering, forest engineering, information, and electrical technologies, soil, and water conservation engineering, natural resources engineering, nursery, and greenhouse engineering, safety, and health, and structures and environment.

For this book series, we welcome chapters on the following specialty areas (but not limited to):

- Academia to industry to end-user loop in agricultural engineering;
- Agricultural mechanization;
- Aquaculture engineering;
- Biological engineering in agriculture;

- Biotechnology applications in agricultural engineering;
- Energy source engineering;
- Farm to fork technologies in agriculture;
- Food and bioprocess engineering;
- Forest engineering;
- GPS and remote sensing potential in agricultural engineering;
- Hill land agriculture;
- Human factors in engineering;
- Impact of global warming and climatic change on agriculture economy;
- Information and electrical technologies;
- Irrigation and drainage engineering;
- Nanotechnology applications in agricultural engineering;
- Natural resources engineering;
- Nursery and greenhouse engineering;
- Potential of phytochemicals from agricultural and wild plants for human health;
- Power systems and machinery design;
- Robot engineering and drones in agriculture;
- Rural electrification;
- Sanitary engineering;
- Simulation and computer modeling;
- Smart engineering applications in agriculture;
- Soil and water engineering;
- Micro-irrigation engineering;
- Structures and environment engineering;
- Waste management and recycling;
- Any other focus areas.

For more information on this series, readers may contact:

Megh R. Goyal, PhD, PE
Book Series Senior Editor-in-Chief: Innovations in Agricultural and Biological Engineering

OTHER BOOKS ON AGRICULTURAL AND BIOLOGICAL ENGINEERING BY APPLE ACADEMIC PRESS, INC.

Management of Drip/Trickle or Micro Irrigation
Megh R. Goyal, PhD, PE, Senior Editor-in-Chief
Evapotranspiration: Principles and Applications for Water Management
Megh R. Goyal, PhD, PE and Eric W. Harmsen, PhD Editors

Book Series: Research Advances in Sustainable Micro Irrigation
Senior Editor-in-Chief: Megh R. Goyal, PhD, PE

- **Volume 1:** Sustainable Micro Irrigation: Principles and Practices
- **Volume 2:** Sustainable Practices in Surface and Subsurface Micro Irrigation
- **Volume 3:** Sustainable Micro Irrigation Management for Trees and Vines
- **Volume 4:** Management, Performance, and Applications of Micro Irrigation Systems
- **Volume 5:** Applications of Furrow and Micro Irrigation in Arid and Semi-Arid Regions
- **Volume 6:** Best Management Practices for Drip Irrigated Crops
- **Volume 7:** Closed Circuit Micro Irrigation Design: Theory and Applications
- **Volume 8:** Wastewater Management for Irrigation: Principles and Practices
- **Volume 9:** Water and Fertigation Management in Micro Irrigation
- **Volume 10:** Innovation in Micro Irrigation Technology

Book Series: Innovations and Challenges in Micro Irrigation
Senior Editor-in-Chief: Megh R. Goyal, PhD, PE

- Engineering Interventions in Sustainable Trickle Irrigation: Water Requirements, Uniformity, Fertigation, and Crop Performance
- Management Strategies for Water Use Efficiency and Micro Irrigated Crops: Principles, Practices, and Performance

- Micro-Irrigation Engineering for Horticultural Crops: Policy Options, Scheduling, and Design
- Micro-Irrigation Management: Technological Advances and Their Applications
- Micro-Irrigation Scheduling and Practices
- Performance Evaluation of Micro-Irrigation Management: Principles and Practices
- Potential of Solar Energy and Emerging Technologies in Sustainable Micro-Irrigation
- Principles and Management of Clogging in Micro-Irrigation
- Sustainable Micro-Irrigation Design Systems for Agricultural Crops: Methods and Practices

Book Series: Innovations in Agricultural and Biological Engineering
Senior Editor-in-Chief: Megh R. Goyal, PhD, PE

- Advanced Research Methods in Food Processing Technologies
- Advances in Food Process Engineering: Novel Processing, Preservation and Decontamination of Foods
- Advances in Green and Sustainable Nanomaterials: Applications in Energy, Biomedicine, Agriculture, and Environmental Science
- Advances in Sustainable Food Packaging Technology
- Analytical Methods for Milk and Milk Products, 2-volume set:
    - o Volume 1: Sampling Methods, Chemical and Compositional Analysis
    - o Volume 2: Physicochemical Analysis of Concentrated, Coagulated and Fermented Products
- Biological and Chemical Hazards in Food and Food Products: Prevention, Practices, and Management
- Bioremediation and Phytoremediation Technologies in Sustainable Soil Management, 4-volume set:
    - o Volume 1: Fundamental Aspects and Contaminated Sites
    - o Volume 2: Microbial Approaches and Recent Trends
    - o Volume 3: Inventive Techniques, Research Methods, and Case Studies
    - o Volume 4: Degradation of Pesticides and Polychlorinated Biphenyls
- Dairy Engineering: Advanced Technologies and Their Applications
- Developing Technologies in Food Science: Status, Applications, and Challenges

- Emerging Technologies in Agricultural Engineering
- Engineering Interventions in Agricultural Processing
- Engineering Interventions in Foods and Plants
- Engineering Practices for Agricultural Production and Water Conservation: An Interdisciplinary Approach
- Engineering Practices for Management of Soil Salinity: Agricultural, Physiological, and Adaptive Approaches
- Engineering Practices for Milk Products: Dairyceuticals, Novel Technologies, and Quality
- Enzyme Inactivation in Food Processing: Technologies, Materials, and Applications
- Field Practices for Wastewater Use in Agriculture: Future Trends and Use of Biological Systems
- Flood Assessment: Modeling and Parameterization
- Food Engineering: Emerging Issues, Modeling, and Applications
- Food Process Engineering: Emerging Trends in Research and Their Applications
- Food Processing and Preservation Technology: Advances, Methods, and Applications
- Food Technology: Applied Research and Production Techniques
- Functional Dairy Ingredients and Nutraceuticals: Physicochemical, Technological, and Therapeutic Aspects
- Handbook of Research on Food Processing and Preservation Technologies, 5-volume set:
    - Volume 1: Nonthermal and Innovative Food Processing Methods
    - Volume 2: Nonthermal Food Preservation and Novel Processing Strategies
    - Volume 3: Computer-Aided Food Processing and Quality Evaluation Techniques
    - Volume 4: Design and Development of Specific Foods, Packaging Systems, and Food Safety
    - Volume 5: Emerging Techniques for Food Processing, Quality, and Safety Assurance
- Modeling Methods and Practices in Soil and Water Engineering
- Nanotechnology and Nanomaterial Applications in Food, Health, and Biomedical Sciences
- Nanotechnology Applications in Agricultural and Bioprocess Engineering: Farm to Table

- Nanotechnology Applications in Dairy Science: Packaging, Processing, and Preservation
- Nanotechnology Horizons in Food Process Engineering, 3-volume set:
    - Volume 1: Food Preservation, Food Packaging and Sustainable Agriculture
    - Volume 2: Scope, Biomaterials, and Human Health
    - Volume 3: Trends, Nanomaterials, and Food Delivery
- Novel and Alternative Methods in Food Processing: Biotechnological, Physicochemical, and Mathematical Approaches
- Novel Dairy Processing Technologies: Techniques, Management, and Energy Conservation
- Novel Processing Methods for Plant-Based Health Foods: Extraction, Encapsulation and Health Benefits of Bioactive Compounds
- Novel Strategies to Improve Shelf-Life and Quality of Foods: Quality, Safety, and Health Aspects
- Phytochemicals and Medicinal Plants in Food Design: Strategies and Technologies for Improved Healthcare
- Processing of Fruits and Vegetables: From Farm to Fork
- Processing Technologies for Milk and Milk Products: Methods, Applications, and Energy Usage
- Quality Control in Fruit and Vegetable Processing: Methods and Strategies
- Scientific and Technical Terms in Bioengineering and Biological Engineering
- Soil and Water Engineering: Principles and Applications of Modeling
- Soil Salinity Management in Agriculture: Technological Advances and Applications
- State-of-the-Art Technologies in Food Science: Human Health, Emerging Issues and Specialty Topics
- Sustainable and Functional Foods from Plants
- Sustainable Biological Systems for Agriculture: Emerging Issues in Nanotechnology, Biofertilizers, Wastewater, and Farm Machines
- Sustainable Nanomaterials for Biomedical Engineering: Impacts, Challenges, and Future Prospects
- Sustainable Nanomaterials for Biosystems Engineering: Trends in Renewable Energy, Environment, and Agriculture
- Technological Interventions in Dairy Science: Innovative Approaches in Processing, Preservation, and Analysis of Milk Products

- Technological Interventions in Management of Irrigated Agriculture
- Technological Interventions in the Processing of Fruits and Vegetables
- Technological Processes for Marine Foods, From Water to Fork: Bioactive Compounds, Industrial Applications, and Genomics
- The Chemistry of Milk and Milk Products: Physicochemical Properties, Therapeutic Characteristics, and Processing Methods

ABOUT SENIOR-EDITOR-IN-CHIEF

Megh R. Goyal, PhD, PE

Retired Professor in Agricultural and Biomedical Engineering, University of Puerto Rico, Mayaguez Campus; Senior Acquisitions Editor, Biomedical Engineering and Agricultural Science, Apple Academic Press, Inc.

Megh R. Goyal, PhD, PE, is, currently a retired professor of agricultural and biomedical engineering from the General Engineering Department at the College of Engineering at the University of Puerto Rico–Mayaguez Campus (UPRM); and Senior Acquisitions Editor and Senior Technical Editor-in-Chief for Agricultural and Biomedical Engineering for Apple Academic Press Inc. During his professional career, he has worked as a Soil Conservation Inspector; Research Assistant at Haryana Agricultural University and Ohio State University; Research Agricultural Engineer/Professor at the Department of Agricultural Engineering of UPRM; and Professor of Agricultural and Biomedical Engineering in the General Engineering Department of UPRM. He spent a one-year sabbatical leave in 2002–2003 at the Biomedical Engineering Department of Florida International University, Miami, USA.

Dr. Goyal was the first agricultural engineer to receive the professional license in agricultural engineering from the College of Engineers and Surveyors of Puerto Rico. In 2005, he was proclaimed the "Father of Irrigation Engineering in Puerto Rico for the Twentieth Century" by the American Society of Agricultural and Biological Engineers, Puerto Rico Section, for his pioneering work on micro irrigation, evapotranspiration, agroclimatology, and soil and water engineering.

During his professional career of 52 years, he has received many awards, including Scientist of the Year, Membership Grand Prize for the American Society of Agricultural Engineers Campaign, Felix Castro Rodriguez Academic Excellence Award, Man of Drip Irrigation by the Mayor of Municipalities of Mayaguez/Caguas/Ponce and Senate/Secretary of Agriculture of ELA, Puerto Rico, and many others. He has been recognized as one of the experts "who rendered meritorious service for the development of [the] irrigation sector in India" by the Water Technology Centre of Tamil Nadu

Agricultural University in Coimbatore, India, and ASABE who bestowed on him the 2018 Netafim Microirrigation Award.

Dr. Goyal has authored more than 200 journal articles and edited more than 100 books. AAP has published many of his books, including *Management of Drip/Trickle or Micro Irrigation; Evapotranspiration: Principles and Applications for Water Management;* ten-volume set on *Research Advances in Sustainable Micro Irrigation.* He has also authored the textbooks *Elements of Agroclimatology* (Spanish) by UNISARC, Colombia, and two *Bibliographies on Drip Irrigation.* Dr. Goyal has also developed several book series with AAP, including Innovations in Agricultural & Biological Engineering (with over 60 titles in the series to date), Innovations and Challenges in Micro Irrigation; and Innovations in Plant Science for Better Health: From Soil to Fork.

Dr. Goyal received his BSc degree in Engineering from Punjab Agricultural University, Ludhiana, India, and his MSc and PhD degrees from the Ohio State University, Columbus, Ohio, USA. He also earned a Master of Divinity degree from the Puerto Rico Evangelical Seminary, Hato Rey, Puerto Rico, USA.

ABOUT THE COEDITORS

Suvartan G. Ranvir, PhD
Assistant Professor in Department of Dairy Chemistry, Warner College of Dairy Technology, Sam Higginbottom University of Agriculture, Technology and Sciences, Prayagraj, Uttar Pradesh, India.

Suvartan Gautam Ranvir, PhD, is working as Assistant Professor in the Department of Dairy Chemistry at Warner College of Dairy Technology, Sam Higginbottom University of Agriculture, Technology and Sciences, Prayagraj, Uttar Pradesh, India. Dr. Ranvir is currently working in ultra-high-temperature-treated milk (UHT), low glycemic index food, and extraction of micronutrients and their encapsulation in food. He has published several research papers in national and international journals, a competitive examination book on dairy science, book chapters, popular articles, conference papers, abstracts, and editorial opinions. He is advising several MTech scholars in Dairy Chemistry and has successfully guided five postgraduate students for their dissertation work. He also serves as an external examiner for various Indian state agricultural universities and as editor and reviewer for several journals. He also served as a senior executive in Mother Dairy Fruit and Private Limited, Delhi from 2013–2015. Dr. Ranvir has successfully completed ISO lead auditor and internal auditor certification. He is a life member of Association of Food Scientists & Technologists (AFSTI), India.

He holds a BTech (2011) degree in Dairy Technology from Maharashtra Animal and Fishery Sciences University (MAFSU) Nagpur, Maharashtra, India. He received his MTech (2013) and PhD (2019) degrees in Dairy Chemistry from ICAR-National Dairy Research Institute Karnal, Haryana, India. For his PhD research work, he has received best poster awards in national and international conferences four times. Also he has received a DuPont Nutri-Scholar Award, India (net cash prize 75000 Rs). He is a recipient of a Senior Research Fellowship (2015) from the Indian Council of Agricultural Research; Senior Research Fellowship from the Babasaheb Ambedkar Research and Training Institute; a Young Scientist Award (2020);

Young Research Achiever Award (2020); Indian Council of Agricultural Research-Senior Research-National Dairy Research Institute Fellowship for MTech and PhD programs.

Junaid Ahmad Malik, PhD
Lecturer, Department of Zoology, Government Degree College, Bijbehara, Kashmir, Jammu and Kashmir, India

Junaid Ahmad Malik, PhD, is a Lecturer with the Department of Zoology at Government Degree College, Bijbehara, Kashmir (J&K), India, and is actively involved with teaching and research activities. He has more than eight years of research experience. His areas of interest are ecology, soil macrofauna, wildlife biology, conservation biology, etc. Dr. Malik has published over 20 research articles and technical papers in international peer-reviewed journals and has authored and edited books, book chapters, and more than 10 popular editorial articles. He also serves as an editor and reviewer of several journals. He has participated in several state, national, and international conferences, seminars, workshops, and symposia and has more than 20 conference papers to his credit. He is a life member of the Society for Bioinformatics and Biological Sciences.

Dr. Malik received a BSc (2008) in Science from the University of Kashmir, Srinagar, J&K; MSc (2010) in Zoology from Barkatullah University, Bhopal, Madhya Pradesh, India; and PhD (2015) in Zoology from the same university. He completed his BEd program in 2017 at the University of Kashmir, Srinagar, J&K.

CONTENTS

CONTRIBUTORS

V. K. Ammu
Dairy Engineering Division, ICAR-National Dairy Research Institute, SRS Bengaluru, Adugodi, Karnataka, India

S. Athira
Department of Food Chemistry, College of Food Technology, Kerala Veterinary and Animal Sciences University, Thrissur, Kerala, India

Aarthy Bose
Department of Dairy Chemistry, Sam Higginbottom University of Agriculture, Technology and Sciences (SHUATS), Allahabad, Uttar Pradesh, India

Mitul R. Bumbadiya
College of Dairy Science, Kamdhenu University, Amreli-Rajkot highway, Near Shedubhar Village, Amreli, Gujarat, India

Ramesh F. Chavan
Department of Chemical Technology, Dr. Babasaheb Ambedkar Marathwada University, Jaisingpura, Aurangabad, Maharashtra, India

Nandita Das
Dairy Chemistry Division, National Dairy Research Institute (NDRI), Karnal, Haryana, India

Partha Pratim Debnath
Faculty of Dairy Technology, West Bengal University of Animal and Fishery Sciences (WBUAFS), Mohanpur, Nadia, West Bengal, India

Gajanan P. Deshmukh
Dairy Engineering Department, SRS of ICAR-National Dairy Research Institute (NDRI), SRS, Bengaluru, Karnataka

Jyotika Dhankhar
Maharshi Dayanand University (MDU), Rohtak, Haryana, India

Chandni Dularia
Dairy Technology Division, National Dairy Research Institute (NDRI), Karnal, Haryana, India

Megh R. Goyal
Senior Technical Editor-in-Chief for Apple Academic Press; Retired Professor, University of Puerto Rico-Mayaguez, Mayaguez, Puerto Rico

Subrota Hati
SMC College of Dairy Science, Kamdhenu University, Anand, Gujarat, India

Ankita Hooda
Dairy Technology Division, National Dairy Research Institute (NDRI), Karnal, Haryana, India

Shamim Hossain
Dairy Technology Division, National Dairy Research Institute (NDRI), Karnal, Haryana

Thejus Jacob
Sam Higginbottom University of Agriculture, Technology and Sciences (SHUATS), Allahabad, Uttar Pradesh, India

Naveen Jose
Dairy Engineering Department, SRS of ICAR- National Dairy Research Institute (NDRI), SRS, Bengaluru, Karnataka, India

Drishti Kadian
Dairy Technology Division, National Dairy Research Institute (NDRI), Karnal, Haryana, India

Payal Karmakar
Dairy Chemistry Division, ICAR - National Dairy Research Institute, Karnal, Haryana, India

Harish Kumar
Food Technology, Amity University, Amity House, Jaipur, Rajasthan, India

Preeti Kundu
Maharshi Dayanand University (MDU), Rohtak, Haryana, India

Soma Maji
Centurion University of Technology and Management, Paralakhemundi, Odisha, India

Junaid Ahmad Malik
Department of Zoology, Government Degree College, Kashmir, Jammu and Kashmir, India

Ronit Mandal
Faculty of Land & Food Systems, University of British Columbia, Vancouver, BC, Canada

Sunil Meena
Department of Dairy Science and Food Technology, Banaras Hindu University (BHU), Varanasi, India

Rita Mehla
National Dairy Research Institute (NDRI), Karnal, India

Chander Mohan
Dairy Technology Division, National Dairy Research Institute (NDRI), Karnal, Haryana, India

Laxmana Naik
Dairy Chemistry and Bacteriology Section, ICAR-National Dairy Research Institute, SRS, Bengaluru, Karnataka, India

Pranali Nikam
Department of Chemistry, National Dairy Research Institute (NDRI), Karnal, Haryana, India

Heena Parmar
SMC College of Dairy Science, Kamdhenu University, Gujarat, India

Sandeep G. M. Prasad
Sam Higginbottom University of Agriculture, Technology and Sciences (SHUATS), Allahabad, Uttar Pradesh, India

Dinesh Chandra Rai
Department of Dairy Science and Food Technology, Banaras Hindu University (BHU), Varanasi, India

Rekha Rani
Department of Dairy Technology, West Bengal University of Animal and Fishery Sciences (WBUAFS), Mohanpur, West Bengal, India

Soniya Ranveer
Department of Chemistry, National Dairy Research Institute (NDRI), Karnal, Haryana, India

Suvartan G. Ranvir
Department of Dairy Chemistry, Warner College of Dairy Technology, Sam Higginbottom University of Agriculture, Technology & Sciences (SHUATS), Allahabad, Uttar Pradesh, India

Priyanka Singh Rao
Dairy Chemistry and Bacteriology Section, ICAR-National Dairy Research Institute, SRS, Bengaluru, Karnataka, India

Menon Rekha Ravindra
Dairy Engineering Department, SRS of ICAR-National Dairy Research Institute (NDRI), SRS, Bengaluru, Karnataka, India

Kuntal Roy
Department of Dairy Technology, West Bengal University of Animal and Fishery Sciences (WBUAFS), Mohanpur, West Bengal, India

Sunil Sakhala
Dairy Technology Division, National Dairy Research Institute, Karnal, Haryana, India

Bhagwan K. Sakhale
Department of Chemical Technology, Dr. Babasaheb Ambedkar Marathwada University, Jaisingpura, Aurangabad, Maharashtra, India

Khusbhu Sao
National Dairy Research Institute (NDRI), Karnal, Haryana, India

Sonu K. Shivanna
Dairy Chemistry Division, ICAR-National Dairy Research Institute, SRS, Adugodi, Bengaluru, Karnataka, India

Rahul R. Sindhani
Department of Chemical Technology, Dr. Babasaheb Ambedkar Marathwada University, Jaisingpura, Aurangabad, Maharashtra, India

Manju Singh
Department of Chemistry, KVA DVA College for Women, Railway Road, Karnal, Haryana, India

Divyang Solanki
Sumul Dairy, Surat, Gujarat, India

M. A. Syama
Dairy Development Department, Government of Kerala, Pattom, Thiruvananthapuram, Kerala, India

Srishty Tyagi
Dairy Chemistry Division, National Dairy Research Institute, Karnal, Haryana, India

ABBREVIATIONS & SYMBOLS

α-La	α-lactalbumin
α-Lg	α-lactalbumin
α_{s1}-CN	α_{s1}-casein
α_{s2}-CN	α_{s2}-casein
β	beta
β-CN	β-casein
β-Lg	β-lactoglobulin
κ	Kappa
κ–CN	κ-casein
2,4,6-TNBS	trinitrobenzene sulphonic acid
A	alpha
ACE	angiotensin converting enzyme
ADI	acceptable daily intake
AGMARK	agricultural marketing
AOAC	Association of Official Agricultural Chemists
API	atmospheric pressure ionization
Arg	arginine
As	arsenic
ASE	accelerated solvent extraction
ATR-FTIR	attenuated total reflection-Fourier transformer infrared spectroscopy
a_w	water activity
B	boron
BALB	Bagg Albino
BD	bulk density
BDDT	Braunner Demming Demming Teller
BI	browning index
BP	boiling point
BSA	bovine serum albumin
Ca	calcium
$CaCl_2$	calcium chloride
CAGR	compounded annual growth rate
CCP	calcium carbonate precipitation/colloidal calcium phosphate
CCPR	Codex Committee on Pesticide Residues

CD Value	conjugated dienes value
CI	Carr index/chemical ionization
Cl	chloride
CMC	critical moisture content/carboxy methyl cellulose
CN	casein
Co	cobalt
CODEX	Codex Alimentarius
cP	centipoise
CPPs	casein phosphopeptides
Cr	chromium
CTX	clinical trials exemption
Cu	copper
DAD	diode array detectors
DDT	dichlorodiphenyltrichloroethane
DEA	diethylaminopropyl
DF	diafiltration
DI	direct immersion
DNA	deoxyribonucleic acid
DSC	differential scanning calorimetry
ECD	electron capture detector
ED	electrodialysis
EDTA	ethylenediamine tetraacetic acid
EI	electron ionization
EM	evaporated milk
EMC	equilibrium moisture content
EPS	exopolysaccharides
ERH	equilibrium relative humidity
ESI	electrospray
F	fluorite
FAO	Food Agriculture Organization
Fe	iron
FF	flow function
FFA	free fatty acids
FITC	fluorescein isothiocyanate
FP	freezing point
FPD	flame photometric detector/freezing point depression
FSSAI	Food Safety and Standards Authority of India
FSSR	Food Safety and Standards Regulation
GAB	Guggenheime – Andersone – de Boer

GACR	compound annual growth rate
GC	gas chromatography
GCB	graphitized carbon black
GI	glycemic index
GLC	gas liquid chromatography
GMP	good manufacturing practices
GRAS	generally recognized as safe
H^+	hydronium ion
HC	heat capacity
HCH	hexachlorocyclohexane
HCT	heat coagulation time/temperature
HR	Hausner ratio
HRMS	high-resolution mass spectrometry
HS	headspace/heat stability
HTST	high temperature short time
I	iodine
IDF	International Dairy Federation
IFN-γ	interferon gamma
IgE	immunoglobulin E
IGF-1	insulin-like growth factor
Igs	immunoglobulins
IL-1	interleukin-1
IL-6	interleukin-6
IPP	isoleucine-proline-proline
IR	infrared
IT	ion trap
JMPR	Joint Meeting on Pesticide Residues
K	thermal conductivity
K	potassium
Kg/cm^2	kilogram per centimeter square
LAB	lactic acid bacteria
LC–MS	liquid chromatography–mass spectrometry
LLE	liquid–liquid extraction
LTI	low-temperature inactivation
LTLT	low-temperature long time
Lys	lysine
M	molar
MAE	microwave-assisted extraction
Max	maximum

MDA	malonaldehyde
MFGM	milk fat globule membrane
Mg	magnesium
MHR	modified Hausner ratio
Min	minimum
mL	milliliter
Mn	manganese
Mo	molybdenum
MPC	milk protein concentrate
MRLs	maximum residue limit
MS	mass spectrometry
MSI	moisture sorption isotherms
MSNF	milk solid not fat
MUFA	monounsaturated fatty acids
N	normal
Na	sodium
Na_2SO_4	sodium sulphate
NaCl	sodium chloride
NDDB	National Dairy Development Board
NEM	N-ethylmaleimide
NF	nano filtration
NFDM	non-fat dry milk
NH_2	aminopropyl
Ni	nickel
NLT	not less than
NMR	nuclear magnetic resonance
NMT	not more than
NPD	nitrogen and phosphorus detector
NSLB	nonstarter lactic acid bacteria
O/W	oil-in-water
OCPs	organochlorine pesticides
OCs	organochlorine
OPs	organophosphate pesticide
P	phosphorus
p-AnV	p-anisidine value
PA	plasminogen activator/poly-acrilate
PD	particle density
PDMS	polydimethylsiloxane
PG	propylene glycol

Pi	inorganic phosphates
Po	organic phosphates
POPs	persistent organic pollutants
PP	Proteose peptone
PP3	Proteose peptone 3
PSA	primary secondary amine
Psi	pounds per square inch
PTH	parathyroid hormone
PUFA	polyunsaturated fatty acids
PV	peroxide value/Polenke value
Q	quadrupole
QqQ	triple quadrupole
QTrap	hybrid quadrupole ion trap
QuEChERS	quick, easy, cheap, effective, rugged, and safe
RH	relative humidity
RI	refractive index
RM Value	Reichert-Meissl value
RNA	ribonucleic acid
RO	reverse osmosis
RP-HPLC	reverse phase high-performance liquid chromatography
RSSC	residual skin surface components
SCM	sweetened condensed milk
SDS-PAGE	sodium dodecyl sulphate–polyacrylamide gel electrophoresis
Se	selenium
SEM	scanning electron microscopy
SH	sulfhydryl
SHMP	sodium hexametaphosphate
Si	silicon
SMP	skim milk powder
SNF	solid not fat
SPE	solid phase extraction
SPME	solid-phase microextraction
SPT	sticky point temperature
SSHE	Scraped Surface Heat Exchangers
ST	surface tension
TBA Test	thiobarbituric acid test
TCA	trichloroacetic acid
TD	tapped density

T_g	glass transition temperature
Th1	Type 1 T helper
Th2	Type 2 T helper
TNF- α	tumor necrosis factor- α
TOF	time of flight
tPA:	tissue-type plasminogen activator
TRAP 5b	tartrate-resistant acid phosphatase 5b
TS	total solids
TV	Totox value
UAE	United Arab Emirates
UF	ultrafiltration
UFA	unsaturated fatty acids
UHPLC	ultra-high pressure liquid chromatography
UHT	ultra-high-temperature
uPA	urokinase plasminogen activator
US	United States of America
VPP	valine-proline-proline
W/O	water-in-oil
WHO	World Health Organization
WI	whiteness index
WMP	whole milk powder
WPC	whey protein concentrate
YI	yellowness index
Zn	zinc

PREFACE

Milk is a complete food that principally comprises water, milk fat, proteins, carbohydrate (lactose), and salts (minerals). Apart from these primary components, milk contains trace amounts of pigments, enzymes, vitamins, phospholipids, and gases. It is a highly nutritious medium, and the low acidity in milk provides a favorable environment for growth of pathogenic and spoilage causing organisms. For inactivating or killing these organisms and elongating the shelf life of milk products, it is necessary to substitute various types of thermal and nonthermal treatments. However, these treatments not only guarantee the safety of milk and milk products microbiologically, but simultaneously also cause changes in the nutritional quality, physicochemistry composition reactions, and formation of numerous types of components.

This book consists of four parts; **Part 1:** Milk and Milk Products: Constituents and Chemistry; **Part 2:** Physicochemical Properties of Milk and Milk Products; **Part 3:** Therapeutic Characteristics of Milk and Milk Products; **Part 4:** Processing and Characterization of Milk and Milk Products.

This book assumes a sound knowledge of raw milk composition, mineral constituents in milk, and manufacturing processes of butter and ice cream. It covers the physicochemical changes and compositional changes occurring during manufacturing and storage of milk, concentrated milk, butter, butter oil, and ice cream. This volume also encompasses the therapeutic characteristics of fermented milk and milk products, milk-derived bioactive peptides, and potential aspects of whey proteins in dairy products. Specialized methods like proteolysis in ultra high temperature (UHT), heat and acid coagulation of milk products, processing and characteristics of dry dairy milk powders, and methods to monitor pesticide residues in milk and milk products have also been elaborated on.

The book is intended for undergraduate and junior postgraduate students and be valuable for dairy science teaching staff, researchers, and industrial personnel interested in theoretical and practical knowledge of changes in physicochemical and compositional properties and heat-induced changes that occur during manufacturing and storage of dairy products, not just in improving the quality and performance of dairy products but also in a much wider context.

We hope that the book will answer some of the concerns raised by the readers regarding the chemistry of milk and milk products as well as pique their interest in learning more about these topics.

—*Editors*

PART I

CONSTITUENTS AND CHEMISTRY OF MILK AND MILK PRODUCTS

CHAPTER 1

CHEMISTRY OF RAW MILK: COMPOSITION, DISTRIBUTION, AND FACTORS

SUVARTAN G. RANVIR, ANKITA HOODA, AARTHY BOSE, and HARISH KUMAR

ABSTRACT

Generally, milk consists of 85–89% water, 2.5–6.0% fat, 2.9–5.0% protein, 3.6–5.5% lactose, and 0.6–0.9% minerals. The composition of raw milk depends on the species, breed, feed of animals, stage of lactation, number of parturitions, age of animal, health, the interval of milking, physical environment, and season. The major constituents of milk are interdependent with each other. Lactose chlorides and sodium ions contribute together for maintaining the osmotic pressure. Fat is an independent parameter, which changes with animal species, genetic, period of lactation, feed, an interval of feed, etc.

1.1 INTRODUCTION

All female mammalian species can secrete milk after giving birth to the newborn. The basic purpose of milk secretion is to supply the newborn neonate with immunity and nutrition.[42] The constituents of milk whether present in large or small quantities are responsible for these functions. The physiological immunological and nutritional necessities of all species vary significantly and hence the composition of milk secreted is different for each species.[15,51] The milk that is consumed by the human being consists of milk by cow, buffalo, sheep, and goat. Milk provides nearly all nutritional

The Chemistry of Milk and Milk Products: Physicochemical Properties, Therapeutic Characteristics, and Processing Methods. Megh R. Goyal, Suvartan Ranvir, & Junaid Ahmad Malik (Eds.)

components, which are needed for growth and development. Milk is a good source of milk fat, carbohydrate (lactose), protein, minerals, vitamin, and enzymes. However, milk is poor source of vitamin C and vitamin D.[36] Generally, milk is defined as a lacteal and biological secretion with a purpose to meet nutritional and immunological necessities of the newborn infant.[38,48]

The composition of milk not only varies among the species but also between the species depending on the individuality of animal, breed, health, age, lactation stage, climatic conditions, milking interval, etc.[41] A linear correlation among the protein concentration of milk and rate of growth of neonates has been reviewed by Benhart,[8] who reported that logarithm of days to double birth weight newborn is linearly related to the calories derived from protein. For example, it takes a human baby 120–180 days to double the birth weight and has 7% of calories coming from milk protein only; on the other hand, carnivores that take 7 days to double the body weight have >30% of energy coming from protein. Mother milk for human infants is the ideal food. Although cow, buffalo, and goat milk can be suitable replacement to feed newborns. Even the diet of human adults can be supplemented by milk of these species.[22,36]

This chapter focuses on milk constituents and various factors affecting the gross composition of milk.

1.2 GROSS COMPOSITION OF MILK

There are four broad categories of constituents[22] in the milk:

- Specific for organ and species: lipids and proteins.
- Specific for organ but not for species: milk carbohydrate.
- Specific for species but not for organ: some of the milk proteins.
- Neither species nor organ specific: vitamin, minerals, and water.

Milk comprises on an average of 87% water content, 3.9% fat content, 4.9% lactose content, 3.5% protein content, and 0.7% of minerals and miscellaneous components.[22,24,36] The gross composition of milk among species is presented in Table 1.1.

The advent of calving is concomitant and stimulus with biosynthesis and secretion of milk, a biological phenomenon triggered by lactogenic hormones. For the first few days after parturition, mammary glands secret, a fluid known as colostrum, which has a bitter taste, strong odor, faint reddish yellow color and contain a higher percentage of immunoglobulin. It is a rich source of all milk constituents except lactose, potassium, and pantothenic acid. Gradual changes from colostrum to normal milk are summarized in Table 1.2.[24,37]

TABLE 1.1 Gross Composition of Milk (%) of Different Species.

Species	Water	Protein	Fat	Ash	Lactose
Buffalo	82–84	3.3–3.6	7.0–11.5	0.8–0.9	4.5–5.0
Camel	86–88	3.0–3.9	2.9–5.4	0.6–0.9	3.3
Cow	85–87	3.2–3.8	3.7–4.4	0.7–0.8	4.8–4.9
Goat	87–88	2.9–3.7	4.0–4.5	0.8–0.9	3.6–4.2
Human	88–89	1.1–1.3	3.3–4.7	0.2–0.3	6.8–7.0
Sheep	79–82	5.6–6.7	6.9–8.6	0.9–0.1	4.3–4.8

Source: Jilo et al.[24]

TABLE 1.2 Transition from Colostrum to Normal Milk.

Time after calving (h)	Fat (%)	Total solids (%)	Total protein (%)	Albumin (%)	Casein (%)	Lactose (%)	Ash (%)
0	5.10	26.20	17.5	11.34	5.08	2.19	1.01
6	6.85	20.46	10.0	6.30	3.51	2.71	0.90
12	3.80	14.53	6.05	2.96	3.00	3.71	0.89
36	3.55	12.22	3.98	1.03	2.77	3.97	0.84
72	3.10	11.86	3.86	0.97	2.70	4.37	0.84
120	3.75	12.67	3.86	0.87	2.68	4.76	0.85

Source: Fundamental of Dairy Chemistry;[62] and Puppel et al.[50]

1.2.1 WATER

Water is called as silent nutrient.[19] Our body requires about 6–8 glasses of water daily for proper hydration of body. Water is the major constituent of milk in various species; and other various components are dissolved and emulsified into the water medium.[38] The solutes that are dissolved to a strength of 0.3 M contribute to depression in freezing point (by about 0.54°C).[18] Some amount of water is present in a bound form to protein and fat globule membranes and this water does not dissolve small ions and molecules. Milk contains about 87% of water and it is well source of water in diet. Water does not provide any energy as is done by fats, proteins, and carbohydrates; but it is most important for human metabolism. Water is also necessary for transporting many types of nutrients to organs and tissues.[22,36,62]

1.2.2 LIPIDS

The dietary fat provides energy for the newborn infant; and the amount of fat percentage secretion in milk by any species is dependent on the requirement of energy for survival of animals in cold environments.[41] Milk lipids are the source of essential fatty acids and fat-soluble vitamins (such as retinol, calciferol, tocopherol, and phytomenadione, and these are denoted by A, D, E, and K, respectively). Milk lipids principally contribute to improve the flavor and rheological characteristics of milk products.[30] Fatty acids are precursors of many types of flavoring compounds, such as, lactones and methyl ketones in some manufacturing processes.[4] On the other hand, lipids may also cause off-flavor because of hydrolysis or oxidation of lipid molecules (hydrolytic and oxidative rancidity).[4]

The milk fat consists of substances extractable by different methods, such as, simple extraction by a nonpolar solvent like chloroform, ether, and efficient method Rose–Gottlieb,[40] which uses solvents, i.e., ethanol, ammonium hydroxide, petroleum ether, and diethyl ether. The most common laboratory method to determine milk fat content is that of Gerber and Babcock,[26] in which sulfuric acid is used to release the fat. Milk fat mostly comprises of triglycerides (98%) and small amounts of mono and di-glycerides, fatty acids, phospholipids, sterols, and hydrocarbons.[3] The milk fat exists in the form of globules with diameter from 0.2 to 22 μm. This size distribution is affected by species, breed, age, stage of lactation, etc.[34]

1.2.3 CARBOHYDRATES

The major carbohydrate in milk is lactose and trace quantities of other carbohydrates (such as, glucose, fructose, galactosamine, glucosamine), and *N*-acetyl neuraminic acid.[6] The concentration of lactose varies with animal breed, individual animal, stage of lactation and udder disease, etc.[41] During lactation, the content of lactose significantly decreases.[17] In contrast during early lactation, protein and lipids decrease, while there is an increase during the second half of lactation. The content of lipids and proteins in milk can be inversely proportional to the amount of lactose content.[21,23]

Lactose and lipids are chief sources of energy in milk, where more energy is provided by lipids as compared with lactose content.[41] When animal is suffering from mastitis, the level of NaCl in milk is increased, which causes a depressed secretion of lactose.[47] The osmotic pressure in mammary system is maintained by lactose together with potassium (K), sodium (Na), and chloride (Cl) ions,[37] and any variation in lactose percentage. The milks of breeds and species that have higher lactose content would therefore have lower ash concentration and vice versa. The percentage of lactose and chloride in milk are also inversely related to milk secreted by udder-infected animals. The Koestler's chloride lactose test is used for identification of udder-infected milk[45] using following equation:

$$\text{Koestler Number} = 100 \times [(\text{Chloride }\%)/(\text{Lactose }\%)] \quad (1)$$

Kosteler number of <2 and >3 indicates normal and abnormal milk, respectively.

1.2.4 PROTEINS

Bovine milk consists of 3.5% protein, which varies throughout the lactation. The milk proteins provide energy, essential amino acids, antimicrobial, and immune-boosting milk proteins (i.e., lactoferrin and immunoglobulin) to the newborn neonate.[11] The young ones have different requirements according to stage of maturity and hence this phenomenon is reflected in protein percentage of milk, which varies from 1 to 20%. The higher is the protein content, more is the rate of body growth.[57]

The milk protein consists of several classes of polypeptides. The major group is called caseins, which comprise of four kinds of polypeptides (namely: αs_1, αs_2, β, and K-casein), which further have their genetic variants, post-transitional modificants, and proteolysis products.[5] Casein exists in milk in the form of colloidal-calcium-phosphate, and in the form of micelles of 20–300 nm size. The next significant milk proteins are α-lactalbumin, β-lactoglobulin, blood serum albumin, and immunoglobulin and they are collectively known as whey protein.[61]

The classic method of milk protein determination is Kjeldahl analysis for Nitrogen determination.[54] The observed value is multiplied by 6.38 (as 15.67% of N is in milk protein) to get the value of total protein. The classical method of milk fractionation includes precipitation at pH of 4.6 to separate casein in precipitate, whereas whey protein in supernatant. All proteins can be precipitated by using an aliquot of 12% (w/v) trichloroacetic acid concentration.[13]

1.2.5 MINERALS

The amount of noncombustible matter in milk is designated as the ash content of milk. The normal value for ash content in milk is about 0.7%. The value of ash content is slightly increased in abnormal milk, such as, mastitis. Important minerals present in milk (Table 1.3) are calcium, magnesium, potassium, sodium, phosphorus, chloride, sulfur, and citrate.[13,16,22,36]

TABLE 1.3 Minerals Composition of Milk.

Constituent	Concentration	
	Range (mg/L)	Mean value (mg/L)
Calcium (Ca)	1000–1400	1000–1400
Carbonate	–	200
Chloride (Cl)	800-1400	1000
Citrate		1750
Inorganic phosphorus	–	750
Magnesium (Mg)	100–150	130
Phosphorus total	750–1100	950
Potassium (K)	1350–1550	1450
Sulfate (SO_4)	–	100

Source: Dairy Chemistry and Biochemistry;[13] Guetouache et al.;[16] and Tamime et al.[59]

1.2.6 VITAMINS

Vitamins are classified as fat-soluble and water-soluble. Fat-soluble vitamins are retinol, calciferol, tocopherol, and phytomenadione, whereas water-soluble vitamins are B complex and ascorbic acid. Milk is a good source of vitamins except vitamin K (phytomenadione) and vitamin C. Vitamin E is the natural antioxidant, while vitamin B acts as coenzyme and is important in many enzymes catalyzed reaction. Vitamin A is required for development and growth, deficiency of vitamin A causes serious problem such as night blindness.[13,22,36,43] Comparative vitamin content in cow, buffalo, goat, and sheep milk is summarized in Table 1.4.

1.3 FACTORS AFFECTING COMPOSITION OF MILK

Composition of milk is not constant, and it varies depending on two major classes of factors: (1) animal factors (e.g., species, breed, different quarters of the udders, lactation period feed and nutritional level, disease, age of the animal, gestation, and hormones); and (2) environmental factors (e.g., milking interval, milking efficiency, season, exercise and excitement).

TABLE 1.4 Vitamin Content in Buffalo, Cow, Goat, and Sheep Milk (mg/100 g).

Vitamins	Buffalo milk	Cow milk	Sheep milk	Goat milk
Ascorbic Acid	2.5	0.09	4.16	1.29
Biotin	13	2.0	0.93	1.5
Calciferol	2.0	2.0	1.18	133
Cyanocobalamin	0.40	0.45	0.712	0.665
Niacin	0.17	0.09	0.416	0.27
Pantothenic acid	0.15	0.37	0.408	0.31
Pyridoxal	0.33	0.04	0.08	0.046
Retinol	69	46	146	185
Riboflavin	0.11	0.17	0.37	0.21
Thiamine	0.05	0.05	0.08	0.068
Tocopherol	0.19	0.21	-	0.03

Source: Khan et al.[30]

1.3.1 SPECIES

The changes in composition of milk in different species depend on so many other factors so that it is difficult to identify a specific factor liable for changes in gross composition of milk. Although every species milk composition is naturally planned to supply enough nutrition for natural rate of growth, which is an inherited trait of newborn mammals for the species.[22,36]

Rapid growth rate of animal is required more the concentrated milk components for this growth. Milk from all mammals is very different especially based on protein, fat, and lactose contents. Protein constituents principally contribute for doubling the birth weight of newborn mammals.[22,39] Protein content of rabbit milk contains very high amount of protein percentage, so that rapid newborn babies take very less time for doubling the birth weight. While human milk contains very low amount of protein because human baby doubling birth weight is very slow as compared with other mammals.[36] Apart from this, human milk looks as waterier (thinner) as compared with cow and buffalo milks because it contains less amount of fat and solid nonfat (SNF). While buffalo milk looks thicker as compared with human milk and cow milk because it contains higher amount of fat and SNF content.[22,36] Comparison of protein contents and time required for doubling body weight by newborn of mammals is given in Table 1.5.

TABLE 1.5 Development of the Newborn Babies in Relation to Protein (%) in Mother Milk.

Species	Protein % in milk	Time (days) required to double the birth weight
Cat	7	9.2
Cow	3.5	47
Human	1.6	180
Pig	5.21	14
Rabbit	10.38	6
Sheep	4.80	15

Source: Textbook of Dairy Chemistry[36]

1.3.2 BREED

The compositional variation in milk is strongly dependent on the type of breed. The change in fat is more pronounced among different breeds of animals, while other constituents are also slightly changed. It has been reported that as the fat percentage is increased then SNF content is also increased and vice versa.[10,12,41] Many comparative research studies on variation in gross composition milk have been carried out for different breeds. The available data is very wide because it is very difficult to get all the information that could lead to correct appreciation of the precise extent of impact by breeds. The data on the composition of milk for different breed milks are summarized in Tables 1.6–1.8.

TABLE 1.6 Differences in Milk Composition (%) among Different Breeds.

Component of milk	Animal breed					
	Jersey	Friesian	Short Horn	Ayrshire	Guernsey	Red Poll
Fat	5.14	3.45	3.03	3.85	4.96	4.24
Lactose	5.04	4.65	4.89	5.02	4.98	4.77
Mineral	0.75	0.68	0.73	0.69	0.75	0.72
Protein	3.80	3.15	3.32	3.34	3.84	3.70
Solid nonfat	9.59	8,48	8.94	9.05	9.57	13.28
Water	85.27	88.01	87.43	87.10	85.47	86.72

Source: Textbook of Dairy Chemistry;[36] Dairy Chemistry and Animal Nutrition[52]; Alphonsus et al.[2]

TABLE 1.7 Difference of Milk Composition (%) in Different Indian Cow Breeds.

Component of milk	Indian cow breeds				
	Tharparkar	Gir	Sahiwal	Red Sindhi	Cross breed
Fat	4.55	4.73	4.55	4.90	4.50
Lactose	4.83	4.85	5.04	4.91	4.92
Mineral	0.68	0.66	0.66	0.70	0.67
Protein	3.36	3.32	3.33	3.42	3.37
Solid nonfat	8.70	8.63	8.82	8.76	8.63
Total solids	13.25	13.30	13.37	13.66	13.13

Source: Textbook of Dairy Chemistry[36] and Dairy Chemistry and Animal Nutrition[52]

TABLE 1.8 Difference of Milk Composition (%) among Different Indian Buffalo Breeds.

	Murrah	Jaffarabadi	Surti	Bhadawari	Mehsana
Fat	6.8	7.3	8.4	7.43	6.46
Protein	3.87	–	3.93	3.92	3.87
SNF	10.1	10.1	10.3	8.99	9.13

Source: Indian Dairy Products[53]; Boro et al.[9]

1.3.3 DIFFERENT QUARTERS OF UDDER

The bovine udder is comparted into two distinct halves, which are divided by suspensory ligaments and a thin septum of connective tissue that divides the front and rear quarters with no direct connections. Each mammary gland quarter is considered physiologically and anatomically independent.[1,44,53] The quarters milked first yield milk have the fatter percentage, while lowest fat content in the milk is from the last quarters being milked. Generally, the difference in composition of milk from different quarters has reported that yield of milk from right-hand quarters is marginally higher than the rest of the quarters. When a mixing machine is used to milk from different quarters, it will be very similar in fat percentage.[31,41] Variation in fat (%) in different parts of the udder is presented in Table 1.9.

1.3.4 LACTATION PERIOD

The lactation period significantly affects gross composition of milk. After parturition of mammals secreted colostrum, in comparison with normal milk, is rich in protein especially lactoferrin and immunoglobulin, higher concentration of minerals and lower concentration of ash content.[7,36,37,56] After parturition, fat percentage is reported higher, whereas fat percentage is decreased to low level during the reminder of the lactation period. Furthermore, composition of fat is also changed by lactation, while the lactose is only attributable to the stage of lactation.[20,41] In colostrum, minerals content (such as, calcium, sodium, magnesium, and phosphorus) is higher; and only K is decreased. The natural acidity commences from a high level and decreases to a normal value during the first 3 months and remain steady until the last month when it declines sharply.[35,37] The changes in principle components in first milking period and after parturition are summarized in Table 1.10.

TABLE 1.9 Variation in Fat (%) in Different Parts of the Udder.

Animal #	Left fore	Right fore	Average left side	Left hand	Right hand	Average right side
1	4.36	4.33	**4.39**	4.42	4.20	**4.29**
2	4.42	4.68	**4.37**	4.31	4.38	**4.52**
3	3.19	4.12	**4.28**	4.32	4.27	**4.29**

Source: Indian Dairy Products[53]

TABLE 1.10 Variation in Composition in Milk in First Milking and After Parturition.

	Milking							
	1	**2**	**3**	**4**	**5+6**	**7+8**	**15+16**	**27+28**
Ash	1.16	1.03	0.92	0.87	0.85	0.85	0.81	0.78
Casein	6.4	4.9	3.8	3.4	3.3	3.2	2.8	2.6
Fat	5.3	5.4	4.4	4.5	4.5	4.8	4.8	4.6
Lactose	2.6	3.5	4.3	4.6	4.8	4.9	4.9	5.1
Protein	16.5	10.3	5.9	4.6	4.1	4.0	3.5	3.2
Whey protein	10.1	5.5	2.2	1.2	0.8	0.8	0.7	0.6

Source: Fundamental of Dairy Chemistry[62]; Parrish et al.[46]

1.3.5 AGE

Age of the animal significantly affects the fat percentage of milk. Many reports have observed that as the age of cow increases, there is slight decrease in fat content. Also, the same effect has been observed in solid not fat, the decline in solid not fat content is about twice the magnitude of the declined fat content. Solid not fat components especially casein and lactose are most affected by the aging of the cow. It was reported that during 7 years of lactation period, there is about 0.21–0.45% decrease in solid-not-fat content.[22,36,41]

1.3.6 UDDER INFECTION

Infection of udder significantly affects the gross composition of milk. The major infection of udder disease is mastitis, which causes the increase in contents of immunoglobulin, chloride, blood serum albumin, and soluble nitrogen. However, there is a decrease in percentage of lactose, potassium, and whey protein especially β-lactoglobulin, α-lactalbumin, and casein.[28,56] Acidity of mastitis milk is low as compared with normal milk.[36] Freezing point of mastitis milk is unchanged because of normal osmotic pressure.[22,36] With onset of foot and mouth diseases, there is marked reduction in the milk yield, whether the udder is infected or not. If the udder is not infected, moderate increase has been observed in the content of fat, protein, and ash, while lactose content is decreased.[22,36,63] If the udder is infected, then the changes are more pronounced and resemble with those of severe mastitis.

1.3.7 FEED AND NUTRITIONAL QUALITY

Composition of milk varies with the feeding material type and plan of feeding. Increase in the amount of feed or feeding additional concentrates, as compared with normal feeding, has been reported compared with the slight increase in solid not fat content.[32] Animal feeding additional protein-rich diet does not show any noticeable change in the protein percentage of milk, while the nonprotein nitrogen content may be improved. If animals are allowed to graze for fresh pasturage, then there was a significant increase in solid not fat content.[22,36]

Animal feeding with high amount of butter oil, palm oil, lard, and coconut oil has shown increase in milk fat, whereas feeding herring oil or cod liver oil produced reverse results.[25,29,60] The unsaturated fats usually have the depressed action; and the saturated fat increases the fat percentage.[41] The specific effect of cod liver oil is largely removed either by hydrogenation, which renders the fat less saturated or by feeding the fat in a form in which it is protected from degradation and bio-hydrogenation in rumen. Food fats modify the composition of milk fat to a limited extent. However, the technology to prevent rumen degradation and bio-hydrogenation of the fat has enabled to modify the fatty acid profile tailored to research requirements. In several studies, the content of essential fatty acids was increased by 35%.[22,33,36,41]

As increase in the proportion of certain group of fatty acids in food fat has a similar impact, but slightly modified action on composition of the milk fat. This is particularly true of fatty acids of the oleic type, which produce soft fat. Similarly, grass feeding did increase the unsaturation of milk fat.[27,36]

1.3.8 SEASONAL VARIATIONS

Variations in composition of milk with change in season are commonly noticed. In summer, fat and solid not fat percentage are low as compared with winter. Animal secreting milk in May showed a higher value of the fat, while minimum fat value was in November. Solid not fat content was lowest in milk secreted by animal in July and September, while highest solid not fat content of milk was secreted in October (Red Sindhi). Other season-related factors also affected the composition of milk, such as, rainfall, drought, thunderstorm, temperature, length of day, and sunshine.[27,28]

1.3.9 OESTRUM AND HEAT

Effect of heat and oestrum on gross composition of milk has been studied. Yield of the milk is decreased. These changes were due to the increased excitability and nervousness, which caused the cow to either hold up some of the milk or secrete less milk.[41,49]

1.3.10 HORMONES

The principal hormones responsible for maintenance of lactation period are prolactin hormones. It contains 199 amino acid residues and generally present in milk at concentration of about 50 μg per liter.[14,55] Another hormone is thyroxin (thyroid gland hormones) to control metabolic rate of the body.[58] Additionally, feeding of hormones has depressing effect on production. Oestrogen of ovaries and the female sex organs have twofold effect on lactation: stimulating and depressing.[22,36] Stimulating effect causes a decrease in yield of milk but increases fat and nonfatty solids. Casein number is normal, while the protein content is increased.[41]

1.3.11 MILKING INTERVAL

Different intervals of milking do not significantly affect the gross composition except fat content. When milk is drawn from shorter period, it results in low yield of milk but higher percentage of the fat; while milking carried out at longer interval of time resulted in higher yield, but fat percentage is decreased.[36] A significant change in fat content was observed when milking done at morning and evening (Table 1.11).

TABLE 1.11 Effect of Milking Time on Fat Content (%) in Cow and Buffalo Milks.

Time of milking	Gir cow	Red sindhi cow	Buffalo
Evening	6.2	6.3	7.9
Morning	3.2	6.0	7.1

Source: Indian Dairy Products[53]

1.4 SUMMARY

Milk from healthy animals should be free from colostrum. Generally, milk consists of 85–89% water, 2.5–6.0% fat, 2.9–5.0% protein, 3.6–5.5% lactose, and 0.6–0.9% minerals. Fat, protein, and lactose provide energy around 9.1, 4.1, and 4.1 kcal per g, respectively. Vitamins, minerals, and enzymes do not provide energy, but they are important for various metabolism and synthesis of essential compounds in body. In milk, fat is present in the emulsion form, protein is present in colloidal form and lactose is present in true solution form. However gross composition of milk can be affected by species, breed, different quarters of the udders, lactation period feed and nutritional level, disease, age of the animal, gestation, hormones, milking interval, milking efficiency, season, exercise, and excitement.

KEYWORDS

- fat
- lactose
- milk
- minerals
- protein
- vitamin

REFERENCES

1. Akers, R. M.; Nickerson, S. C. Mastitis and Its Impact on Structure and Function in the Ruminant Mammary Glands. *J. Mammary Gland Biol. Neoplasia* **2011,** *16* (4), 275–289.
2. Alphonsus, C.; Akpa, G. N.; Nwagu, B. I.; Barje, P. P.; Orunmuyi, M.; Yashim, S. M.; Opoola, E. Evaluation of Nutritional Status of Friesian x Bunaji Dairy Herd Based on Milk Composition Analysis. *J. Anim.Sci. Adv.* **2013,** *3* (5), 219–225.
3. Aurand, L. W.; Woods, A. E.; Wells M. R. *Lipids, Food Composition and Analysis.* Springer Science & Business Media: New York, 2013; pp 178–231.
4. Azzara, C. D.; Campbell, L. B. Off-Flavors of Dairy Products. Off-Flavors in Foods and Beverages; In *Book Series on Developments in Food Science*; Charalambous G. Ed.; Elsevier: Amsterdam, 1992; 28; pp 329–374.
5. Balthazar, C. F.; Pimentel, T. C.; Ferrão, L. L.; Almada, C. N.; Santillo, A.; Albenzio, M.; Mollakhalili N.; Mortazavian A. M.; Nascimento J. S.; Silva, M. C.; Freitas, M. Q.; Sant'Ana, A. S.; Granato, D.; Cruz A. G. Sheep Milk: Physicochemical Characteristics and Relevance for Functional Food Development. *Compr. Rev. Food Sci. Food Saf.* **2017,** *16* (2), 247–262.
6. Berg, J.; Tymoczko, J.; Stryer, L. (Eds.) Carbohydrates can be Attached to Proteins to form Glycoproteins. Section 11.3. In *Biochemistry*, 5th ed. (e-book); WH Freeman: New York, 2002; p 1100.
7. Bernabucci, U.; Basiricò, L.; Morera, P. Impact of Hot Environment on Colostrum and Milk Composition. *Cell. Mol. Biol.* **2013,** *59* (1), 67–83.
8. Bernhardt, F. W. Correlation Between Growth Rate of the Suckling and Percentage of Total Calories from Protein in the Milk. *Nature* **1961,** *191*, 358–360.
9. Boro, P.; Debnath, J.; Kumar Das, T.; Naha, B. C.; Debarma, N.; Deabbarma, P.; Devi, T. G. Milk Composition and Factors Affecting it in Dairy Buffaloes: A Review. *J. Entomol. Zool. Stud.* **2018,** *340*, 140–153.
10. Cheruiyot, E. K.; Bett, R. C.; Amimo, J. O.; Mujibi, F. D. Milk Composition for Admixed Dairy Cattle in Tanzania. *Front. Genet.* **2018,** *9*, 142–148.
11. Davoodi, S. H.; Shahbazi, R.; Esmaeili, S.; Sohrabvandi, S.; Mortazavian, A.; Jazayeri, S.; Taslimi, A. Health-Related Aspects of Milk Proteins. *Iran. J. Pharm. Res.* **2016,** *15* (3), 573–580.

12. Domínguez-Salas, P.; Galiè, A.; Omore, A. O.; Omosa, E. B.; Ouma, E. A. Contribution of Milk Production to Food and Nutrition Security. *Encyclopedia of Food Security and Sustainability*, **2019,** *3*, 278–291.
13. Fox, P. F.; Unicke-Lowe, T.; McSweeney, P. L. H.; Mahony, J. A. O. (Eds.) *Dairy Chemistry and Biochemistry*, 2nd ed.; Springer International Publishing AG: Switzerland, **2015**; p 604.
14. Freeman, M. E.; Kanyicska, B.; Lerant, A.; Nagy, G. Prolactin: Structure, Function, and Regulation of Secretion. *Physiol. Rev*. **2000,** *80* (4), 1523–1531.
15. Gantner, V.; Mijić, P.; Baban, M.; Skrtic, Z.; Turalija, A. The Overall and Fat Composition of Milk of Various Species. *Mljekarstvo/Dairy*. **2015,** *65* (4), 223–231.
16. Guetouache, M.; Guessas, B.; Medjekal, S. Composition and Nutritional Value of Raw Milk. *Issues Biol. Sci. Pharm. Res.* **2014,** *2* (10), 115–122.
17. Henao-Velásquez, A.F.; Múnera-Bedoya, O.D.; Herrera, A.C.; Agudelo-Trujillo, J. H.; Cerón-Muñoz, M. F. Lactose and Milk Urea Nitrogen: Fluctuations During Lactation in Holstein Cows. *Rev. Bras. de Zootec.* **2014,** *43* (9), 479–484.
18. Hinrichs, W.; Dreijer-vander G. S. Physical Chemistry. In *Practical Pharmaceutics*; Bouwman-Boer Y., Fenton-May V., Le Brun P. Eds.; Springer: Cham, 2015; pp 357–382
19. Hossain, M. Z. Water: The Most Precious Resource of Our Life. *Global J. Adv. Res.* **2015,** *2* (9), 1–11.
20. Institute of Medicine (US Committee on Nutritional Status during Pregnancy and Lactation Milk Composition). *Nutrition During Lactation*; Washington (D.C.): National Academies Press (US), 1991, 6, Available online from: https://www.ncbi.nlm.nih.gov/books/NBK235590/; (accessed on August 31, 2021).
21. Jennes, R.; Sloan, R. E. The Composition of Milks of Various Species: Review. *Dairy Sci. Abstr.* **1970,** *32*, 599–612.
22. Jenness, R. Composition of milk. Chapter 1; In *Fundamentals of Dairy Chemistry*; Springer: Boston, MA, 1988; pp 1–38.
23. Jenness, R.; Holt, C. Casein and Lactose Concentrations in Milk of 31 Species are Negatively Correlated. *Experientia.* **1987,** *43*, 1015–1017.
24. Jilo, K.; Tegegne, D. Chemical Composition and Medicinal Values of Camel Milk. *Int. J. Res. Stud. Biosci.* **2016,** *4* (4), 13–25.
25. Kadegowda, A. K. G.; Piperova, L. S.; Delmonte, P.; Erdman, R. A. Abomasal Infusion of Butterfat Increases Milk Fat in Lactating Dairy Cows. *J. Dairy Sci.* **2008,** *91* (6), 2370–2379.
26. Kala, R.; Samková, E.; Pecová, L.; Hanuš, O.; Sekmokas, K.; Riaukienė, D. An Overview of Determination of Milk Fat: Development, Quality Control Measures, and Application. *Acta Universitatis Agriculturae et Silviculturae Mendelianae Brunensis* (Byrno; Mendel University Press), **2018**, *66* (4), 1055–1064.
27. Kalac, P.; Samkova, E. The Effects of Feeding Various Forages on Fatty Acid Composition of Bovine Milk Fat: A Review. *Czech J. Anim. Sci.* **2010,** *55* (12), 521–537.
28. Kayo, G. Significance of Feed Supplementation on Milk Yield and Milk Composition of Dairy Cow. *J. Dairy Vet. Sci.* **2019,** *13* (2), e555860.
29. Keady, T. W.; Mayne, C. S.; Fitzpatric, D. A. Effects of Supplementation of Dairy Cattle with Fish Oil on Silage Intake, Milk Yield and Milk Composition. *J. Dairy Res.* **2000,** *67* (2), 137–153.
30. Khan, I. T.; Nadeem, M.; Imran, M.; Ullah, R.; Ajmal, M.; Jaspal, M. H. Antioxidant Properties of Milk and Dairy Products: A Comprehensive Review of the Current Knowledge. *Lipids Health Dis.* **2019,** *18* (1), 41–48.

31. Kuehnl, J. M.; Connelly, M. K.; Dzidic, A.; Lauber, M.; Fricke, H. P.; Klister, M.; Olstad, E.; Balbach, M.; Timlin, E.; Pszczolkowski, V.; Crump, P. M.; Reinemann, D. J.; Hernadez, L. L. The Effects of Incomplete Milking and Increased Milking Frequency on Milk Production Rate and Milk Composition. *J. Anim. Sci.* **2019,** *97* (6), 2424–2432.
32. Lawrence, D. C.; Odonovan, M.; Boland, T. M.; Lewis, E.; Kennedy, E. The Effect of Concentrate Feeding Amount and Feeding Strategy on Milk Production, Dry Matter Intake, and Energy Partitioning of Autumn-Calving Holstein-Friesian Cows. *J. Dairy Sci.* **2015,** *98* (1), 338–348.
33. Loor, J. J.; Herbein, J. H.; Jenkins, T. C. Nutrient Digestion, Biohydrogenation, and Fatty Acid Profiles in Blood Plasma and Milk Fat from Lactating Holstein Cows Fed Canola Oil. *Anim. Feed Sci. Technol.* **2002,** *97* (1–2), 65–82.
34. Martini, M.; Salari, F.; Altomonte, I. The Macrostructure of Milk Lipids: Fat Globules. *Crit. Rev. Food Sci. Nutr.* **2016,** 56 (7), 1209–1221.
35. Mastroeni, S. S.; Okada, I. A.; Rondó, P. H.; Duran, M. C.; Paiva, A. A.; Neto, J. M. Concentrations of Fe, K, Na, Ca, P, Zn and Mg in Maternal Colostrum and Mature Milk. *J. Trop. Pediatr.* **2006,** *52* (4), 272–275.
36. Mathur, M. P.; Roy, D. D.; Dinakar, P. Composition of Milk. In *Textbook of Dairy Chemistry*; Directorate of Information and Publication of Agriculture (ICAR): New Delhi, 2005; pp 1–13.
37. McGrath, B. A.; Fox, P. F.; McSweeney, P. L.; Kelly, A. L. Composition and Properties of Bovine Colostrum: A Review. *Dairy Sci. Technol.* **2016,** *96* (2), 133–158.
38. Mehta, B. M. Chemical Composition of Milk and Milk Products. In *Handbook of Food Chemistry*; Springer: Berlin, Heidelberg; 2015; pp 511–553.
39. Melnik, B. C. Milk: Nutrient System of Mammalian Evolution Promoting Mtorc1-Dependent Translation. *Int. J. Mol. Sci.* **2015,** *16* (8), 17048–17087.
40. Mulder, H.; Walstra, P. *The Milk Fat Globule*. Commonwealth Agricultural Bureaux: Farnham Royal, UK, 1974; p 179.
41. National Research Council (NRC). Factors Affecting the Composition of Milk from Dairy Cows. In *Designing Foods: Animal Product Options in the Marketplace*. National Academies Press (US): Washington, DC, 1988; pp 224–241.
42. Oftedal, O. T. The Evolution of Milk Secretion and its Ancient Origins. *Int J Animal.* **2012,** *6* (3), 355–360.
43. Oste, R.; Jägerstad, M.; Andersson, I. Vitamins in Milk and Milk Products. In *Advanced Dairy Chemistry*; Fox, P. F. Ed.; Springer: Boston, 1997; pp 347–402.
44. Paixao, M. G.; Abreu, L. R.; Richert, R.; Ruegg, P. L. Milk Composition and Health Status from Mammary Gland Quarters Adjacent to Glands Affected with Naturally Occurring Clinical Mastitis. *J. Dairy Sci.* **2017,** *100* (9), 7522–7533.
45. Park, Y.W. Production and Composition of Milk are Affected by Multivariate Factors. *J. Adv. Dairy Res.* **2016,** *4* (3), 1–5.
46. Parrish, D. B.; Wise, G. H.; Hughes, J. S.; Atkeson, F. W. Properties of the Colostrum of the Dairy Cow, Part V: Yield, Specific Gravity and Concentrations of Total Solids and its Various Components of Colostrum and Early Milk. *J. Dairy Sci.* **1950,** *33* (6), 457–465.
47. Petrovski, K. R.; Buneski, G.; Trajcev, M. Review of the Factors Affecting the Costs of Bovine Mastitis. *J. S. Afr. Vet. Assoc.* **2006,** *77* (2), 52–60.
48. Pietrzak-Fiećko, R.; Kamelska-Sadowska, A. M. The Comparison of Nutritional Value of Human Milk with Other Mammals' Milk. *Nutrients.* **2020,** *12* (5), 1404–1410.

49. Pragna, P.; Archana, P. R.; Aleena, J.; Sejian, V.; Krishnan, G.; Bagath, M.; Manimaran, A.; Beena, V.; Kurien, E. K.; Varma, G.; Bhatta, R. Heat Stress and Dairy Cow: Impact on Both Milk Yield and Composition. *Int. J. Dairy Sci.* **2017,** *12* (1), 1–11.
50. Puppel, K.; Gołębiewski, M.; Grodkowski, G.; Slósarz, J.; Kunowska-Slósarz, M.; Solarczyk, P.; Przysucha, T. Composition and Factors Affecting Quality of Bovine Colostrum: A Review. *Animals.* **2019,** *9* (12), e1070.
51. Quigley, L.; O'Sullivan, O.; Stanton, C.; Beresford, T. P.; Ross, R. P.; Fitzgerald, G. F.; Cotter, P. D. The Complex Microbiota of Raw Milk. *FEMS Microbiol. Rev.* **2013,** *37* (5), 664–698.
52. Rai, M. M. (Ed.) *Dairy Chemistry and Animal Nutrition.* Ram Prasad and Sons: Agra–India, 1980; p 179.
53. Rangappa, K. S.; Achaya, K. T. *Indian Dairy Products.* Ram Prasad and Sons: Agra–India, 1974; p 218.
54. Sáez-Plaza, P.; Navas, M. J.; Wybraniec, S.; Michałowski, T.; Asuero, A. G. An Overview of the Kjeldahl Method of Nitrogen Determination, Part II: Sample Preparation, Working Scale, Instrumental Finish, and Quality Control. *Crit. Rev. Anal. Chem.* **2013,** *43* (4), 224–272.
55. Saleem, M.; Martin, H.; Coates, P. Prolactin Biology and Laboratory Measurement: An Update on Physiology and Current Analytical Issues. *Clin. Biochem. Rev.* **2018,** *39* (1), 3–11.
56. Sharif, A. A. M. I. R.; Umer, M. U. H.; Muhammad, G. H. Mastitis Control in Dairy Production. *J. Agric. Soc. Sci.* **2009,** *5* (3), 102–105.
57. Soliman, A.; De Sanctis, V.; Elalaily, R. Nutrition and Pubertal Development. *Indian J. Endocrinol. Metab.* **2014,** *18* (l), S39–S47.
58. Song, Y.; Yao, X.; Ying, H. Thyroid Hormone Action in Metabolic Regulation. *Protein Cell.* **2011,** *2* (5), 358–368.
59. Tamime, A. Y.; Wszolek, M.; Božanić, R.; Özer, B. Popular Ovine and Caprine Fermented Milks. *Small Rumin. Res.* **2011,** *101* (1–3), 2–16.
60. Tomkins, T.; Drackley, J. K. Applications of Palm Oil in Animal Nutrition. *J. Oil Palm Res.* **2010,** *2010*, 835–845.
61. Wang, X.; Zhao, X.; Huang, D.; Pan, X.; Qi, Y.; Yang, Y.; Zhao, H.; Cheng, G. Proteomic Analysis and Cross Species Comparison of Casein Fractions from the Milk of Dairy Animals. *Sci. Rep.* **2017,** *7* (1), 1–9.
62. Webb, B. H.; Johnson, R. H.; Alford, J. A. (Eds.) *Fundamental of Dairy Chemistry*, 2nd ed.; AVI Publishing Co.: Westport, CT, 1974; p 767.
63. Wheelock, J. V.; Rook, J. A. F.; Neave, F. K.; Dodd, F. H. The Effect of Bacterial Infections of the Udder on the Yield and Composition of Cow's Milk. *J. Dairy Res.* **1966,** *33* (2), 199–215.

CHAPTER 2

MINERAL PROFILES OF MILK AND MILK PRODUCTS: THEIR INTERACTION AND THERAPEUTIC BENEFITS

MANJU SINGH, SUVARTAN G. RANVIR, THEJUS JACOB, and SANDEEP G.M. PRASAD

ABSTRACT

Minerals in milk and milk products are vital components for creating solid bones, teeth, and preserving the body's fluid ionic balance. These minerals conduct many essential body roles even in trace quantities, such as, acting as catalysts and activators of physiological functions. Deficiency of minerals in the body causes inhibited growth, skin rashes, hair loss, weakness, mobility disorder, lack of balance, susceptibility to illness, anemia, diarrhea, hormone imbalance, low and high blood pressure, hypocalcemia, osteomalacia, nausea, vomiting, headache, premature ageing, etc. Minerals are present in milk and milk products as inorganic ions and salts as well as part of organic molecules like carbohydrates, nucleic acids, proteins, and fats. Several types of processing can affect the equilibrium mineral content in milk and milk products by transforming its soluble and colloidal state.

2.1 INTRODUCTION

Overall body weight and part of all the tissues, liquid cells, and organs in the human body accounts for 4% of minerals.[74] Almost 20 minerals are present

The Chemistry of Milk and Milk Products: Physicochemical Properties, Therapeutic Characteristics, and Processing Methods. Megh R. Goyal, Suvartan Ranvir, & Junaid Ahmad Malik (Eds.)

in milk (such as Calcium (Ca), Magnesium (Mg), Sodium (Na), Potassium (K), Phosphorus (P), Chloride (Cl), Iron (Fe), Zinc (Zn), Copper (Cu), Manganese (Mn), Molybdenum (Mo), Silicon (Si), Arsenic (As), Fluorite (F), Nickel (Ni), Selenium (Se), Cobalt (Co), Iodine (I), Chromium (Cr), and Boron (B)), and these are important for human health.[1,22] Milk and milk products are good source of minerals (such as, calcium and phosphorus).[28]

In the human body, every mineral has a specific significance. For healthy bones and teeth, the proper ratio of calcium and phosphorus in diet is crucial; each of these nutrients helps avoid hypertension, lessen the risks of developing colon or breast cancer, boost weight management, and decrease the likelihood of developing kidney stones.[28,64] Minerals, based on their composition, are dispersed differently in the soluble and micellar stages of milk. A complex balance between the diffusible or soluble and nondiffusible or colloidal phases occurs. Na, K, and Cl are present in the soluble phase, while phosphorus, magnesium, and calcium are partially bounded to casein micelles.[26,49]

Assessment of soluble and colloidal form of milk containing minerals is important, as soluble fraction is responsible for the absorption and bio-use of minerals in the gastrointestinal tract.[40] The mineral content has a marked influence on technical properties of milk products as it causes its resistance to renneting, gelation, and sedimentation.[6] The evidence available to date has shown that the mineral content varies in different species and breed of milk and its variation depending on genetic, physical, and environmental aspects. In terms of composition, milk and milk products are most diversified natural foods. For human beings, all the 20 different minerals present in milk are important. Most of them, including magnesium, phosphorus, calcium, potassium sodium, manganese, iron, copper, and zinc, are necessary to human life and are very valuable.[8] Minerals constitute small portion of milk, but they play a major role in casein micelles formation and stability.

This chapter focuses on the content, distribution, interaction, and role of major and traces of minerals in milk,[71] and on the changes in minerals content during different processing methods. The importance of these minerals to human health and nutrition is given particular attention.

2.2 TYPES OF MINERALS AND DISTRIBUTION INTO SOLUBLE AND COLLOIDAL PHASES

The mineral fraction can be categorized as per body requirements per day in two categories: major minerals (> 50 mg per day) and minor minerals (< 50 mg per day).[60,71] There is another set of minerals that are needed in very trace amounts are known as "Trace elements," which include zinc, iron, copper, molybdenum, manganese, etc.[4] The major minerals that exist in the milk are phosphorus, magnesium, sodium, potassium, chlorides, calcium and iron, copper, zinc, and manganese. Buffalo milk is the richest than the other species of milk, with major constituents, such as, proteins, total solids, lactose, and ash.[4,7,60,71] Significance of minerals in milk is indicated in Table 2.1.

TABLE 2.1 Significance (Physiological and Biological) of the Mineral Elements in Milk.

Mineral	Significance
	Major mineral elements
Ca	Acquisition, repair, and maintenance of bones and teeth, blood pressure, contraction muscular, blood clotting, enzyme cofactor
Mg	Enzyme system cofactor, phosphorylation; DNA transcription; protein synthesis; the transmission of neuro-muscles; contraction of the muscles
Na, K, Cl	Blood pressure; muscle contraction; ionic balance
P	Bone and teeth acquisition and maintenance, blood pH.
	Trace mineral elements
Cu	Enzyme pathway cofactor, feature in the breathing chain (cytochrome c oxidase); dopamine catabolism
Fe	The base of haeme in hemoglobin, myoglobin, cytochromes, enzyme system cofactor
Se	Enzyme system cofactor (glutathione peroxidase)
Zn	Enzyme framework factor that has functions in synthesis of proteins, RNA and DNA

Source: World Health Organization, 2004.[60]

2.3 MINERAL DISTRIBUTION IN MILK OF DIFFERENT SPECIES

Raising, species, lactation stage, and feed significantly affect the mineral content in milk.[7] Mineral content in milk of various species (mg/100 mL) is listed in Table 2.2. The Jersey cow's milk normally comprises higher calcium and phosphate content, but less sodium and chloride content than most of the bovine milk.[9,52] Absolute soluble sodium and chloride concentrations are small in the middle of the lactation cycle and highest in mastic milk. The distribution of minerals often varies with the lactation cycle and around parturition. During mastitis, mineral content in milk is often altered, especially refers to highly elevated levels of sodium and chloride ions. Owing to its high calcium content, buffalo milk having poor heat stability. The heat stability of buffalo milk can be improved by partly substituting Ca with K and sodium.[52]

TABLE 2.2 Mineral Content in Milk of Various Species (mg/100 mL).

Mineral content	Cow	Buffalo	Goat	Human
	(mg/100 mL)			
Ash (%)	0.7	0.8	0.79	0.2
Ca	112–123	188	85–198	28–34
Cu	0.01–0.08	0.035	0.02–0.05	0.02–0.06
Fe	0.03–0.10	0.161	0.05–0.10	0.02–0.20
K	106–163	92	140–242	53–62
Mg	7–12	18	10.36	3–4
Mn	0.02	0.027	0.03	0.07
Na	58	35	28–59	10–16
P	59–119	99	79–153	14–43
Zn	0.30–0.55	0.410	0.02–0.05	0.02–0.06

Source: Barreto et al., 2019;[7] Zamberlin et al., 2012.[74]

Distribution of minerals in cow milk is shown in Table 2.3. For each mineral, the distribution of large milk minerals into soluble and colloidal fractions is distinct (diffusible and nondiffusible) as shown in Table 2.4. The sodium, potassium, and chloride ions are highly soluble, while Mg, inorganic phosphates (Pi), and Ca are partially bound to the casein micelles.[16,26,59,60] The one-third part of Ca, two-third of Mg, half of Pi, and >90% of citrate are found in soluble phase of milk.[25,53] Major minerals are not bounded to fat globules and lactose; however, minor portion of calcium bound to α-lactalbumin (for one protein molecule there is one atom of calcium).[73] The Phosphorus and Calcium distribution into colloidal and soluble state varies considerably with the animal. In general, milk containing a high level of calcium or phosphorus (sheep or buffalo milk) often contains a high level of these minerals in the colloidal process.[2,26,48] Distribution of major minerals in Ewe's milk[72] is shown in Table 2.5. Trace mineral distribution of Ewe's milk[72] is shown in Table 2.6.

TABLE 2.3 Distribution of Minerals in Cow Milk.

Mineral	Colloidal (%)	Soluble (%)
Calcium	66.5	33.5
Chloride	0	100
Citrate	6	94
Magnesium	33	67
Phosphorus	57	43
Potassium	8	92
Sodium	8	92

Sources: De la Fuente et al., 1997;[17] Gaucheron, 2005.[25]

TABLE 2.4 Distribution of Calcium and Phosphorus Content in Different Species of Milks.

Milk	P		Ca	
	Soluble	Colloidal	Soluble	Colloidal
Buffalo	32	68	22	78
Cow	53	47	32	68
Goat	41	59	34	66
Sheep	39	61	20	80

Source: Singh et al., 2019;[61] De la Fuente et al., 1997;[17] Sahai, 1996[59]

TABLE 2.5 Distribution of Major Minerals in Ewe's Milk.

Ewe's milk	Ca	Mg	Na	K	P
Total (mg /100 g)	210.4	17.9	60.3	101.1	128.2
Soluble (mg/100 g)	54.7	10.56	56.08	93.01	52.53
Soluble (%age)	26%	59%	93%	92%	41%

Source: Yabrir et al., 2014.[72]

TABLE 2.6 Distribution of Trace Minerals in Ewe's Milk.

Ewe's milk	Fe	Zn	Cu	Mn
Total (mg /100 g)	0.83	0.679	0.44	0.05
Soluble (mg/100 g)	0.29	0.076	0.19	0.0085
Soluble (%)	35%	11%	43%	7%

Source: Yabrir et al., 2014.[72]

2.4 MINERALS CONTENT IN MILK AND MILK PRODUCTS

2.4.1 CALCIUM

Ca is chief mineral found in milk and its content is about 120 mg/100 mL in milk. Around 99% of calcium present in skim milk portion, which is dispersed in two phases: colloidal and soluble phases. In the colloidal phase, Ca is around 67%, whereas in the aqueous phase it is about 33%.[24] Calcium in ionic form is free in the soluble or aqueous fraction to form the related salts with inorganic phosphates and citrate.[31] The aqueous or soluble fraction is saturated with calcium phosphate. The clusters of Ca and P form granules called nanoclusters in the casein micelles. The diameter of these mineral structures is equivalent to 3 nm.[10]

The shifting of Ca and Pi from soluble phase to colloidal phase and vice-versa is controlled by various physicochemical properties like pH and temperature. Due to these transformations milk mineral equilibrium is maintained.[48] Biological acidification (lactose fermentation in lactic acid) is one of the most significant transformations widely carried out. Throughout this, calcium gets solubilized and passed to the soluble fraction of acidified milk.[26] The aqueous fraction is omitted in some cases, and in some milk products, like processed cheeses, have varying amounts of calcium.[15] This mineral plays an important part in changing various functional characteristics of milk products. In milk, calcium is found in three forms, i.e., soluble calcium, insoluble calcium, and free calcium in ionic form. The concentrations of these three forms in millimoles are 20, 10, and 1.5 mm, respectively.[28]

2.4.2 PHOSPHORUS

"P" is an essential part of milk and dairy products. It helps to preserve acid-base equilibrium, cell membrane composition, and protein-energy metabolism. In milk, total phosphorus is around 95 mg per 100 mL. The phosphorus occurs in milk as organic (Po) and inorganic phosphates (Pi) forms.[24,61] Organic phosphorus is bound to organic molecules likes "nucleosides, casein molecules, Ribonucleic acid (RNA), Deoxyribonucleic acid (DNA), sugar phosphates, phospholipids, and nucleotides."[25] The organic phosphorus is mostly related to casein molecules in milk, which are in the micellar phase. The aqueous phase includes other types of organic phosphorus like sugar phosphates, nucleotides, nucleosides, phospholipids, RNA, and DNA.

Organic phosphorus is divided into the aqueous and the micellar phase, responsible for milk's mineral balance.[38,48] Overall 50% of organic phosphorus is present in the soluble phase, whereas remaining 50% exists in micellar phase at pH 6.7 to form calcium phosphate nanoclusters. During acidification step of manufacturing cheese, the inorganic phosphorus could be pass into the whey. This is explaining the existence of varying phosphorus levels determined for different milk products. Through the process of melting in processed cheeses, salts corresponding to polyphosphates produce large quantities of phosphorus.[24,37,48,51,61]

2.4.3 CALCIUM AND PHOSPHORUS RATIO

Ca and P are the major important minerals for the good health of teeth and bones. Though, high consumption of phosphorus, simultaneously decreased with intake of calcium may have detrimental impact on the bone health.[8] Reasonable "Ca: P ratio" in human diets must occur to ensure optimum bone protection. Alone excessive consumption of dietary phosphorus can be detrimental to the bone by increased release of parathyroid hormone and negative impact on bone development when dietary consumption is less.[67]

2.4.4 MAGNESIUM

Magnesium stimulates around 100 enzymes and controls >300 enzymatic reactions in our body.[34] It is not plentiful in milk and milk products relative to calcium. Approximately, it is 12 mg/100 mL present in milk. Magnesium is divided into the micellar (5 mg/100 mL) and the soluble fractions (7 mg per 100 mL) in milk and is aligned with citrate and inorganic phosphate in the soluble fraction and as nanocluster in the casein micelles. The management of these distributions depends on physicochemical conditions particularly acidic pH.[24] Magnesium in the micellar phase gets soluble in an aqueous fraction during milk acidification. Its content differs in various milk products depending on processing treatments and methods.[26] Although the Mg concentration is comparatively low in milk and its products, still these can be treated as sources for Mg ion (100 mL milk yields 11 mg of magnesium, comprising approximately 15–16% of human RDA).[26,61]

2.4.5 SODIUM, POTASSIUM, AND CHLORIDE IONS

Monovalent ions (Sodium, Potassium, Chloride) are primarily found in the soluble fraction of milk and its products, where these are correlated freely with oppositely charged ions. The sodium chloride (NaCl) levels are improved in dairy products by salting process especially in product likes cheeses. The NaCl leads to curd drainage, improve the organoleptic qualities of different cheeses, enzyme selection, and microorganisms during the period of ripening. Different types of cheese have varying concentrations of NaCl. Sodium-ion holds the equilibrium between water and acid-base. The potassium has an important part in nerve conduction.[24,35,48]

2.4.6 IRON

Iron is a main part of hemoglobin and it is also important for a variety of oxidation reactions. It is roughly 0.03–0.10 mg/100 mL present in cow milk and 2 mg/mL present in human milk.[74] In cow's skim milk contains, iron mainly attached to casein fraction, inorganic P, whey protein, citrate, and Lactoferrin.[14,50] Milk and its products are known to be a poor source of iron with limited contribution to the overall iron consumption. Its concentration is influenced by presence of Lactoferrin (iron-binding protein, xanthine oxidase), which generates reactive oxygen species. Around 200 mL of milk provides, iron content approximately 2–3% of the RDA.[60,68]

2.4.7 COPPER

Like iron content, milk, and milk products also have poor source of Cu, in bovine milk its concentration around 0.01–0.08 mg/100 mL. In milk, around 2% of Cu is associated with fats, 8% associated with whey proteins, 47% associated with inorganic phosphate and citrate ions (low molecular-weight), and 44% associated to casein. This variance may be due to the various sample processing procedures and analytical approaches used for assessing the sample.[27] Milk and its products contribute only 5% of Cu content of RDA.[60,68]

2.4.8 ZINC

Zinc is helps in digestion, metabolism, reproduction, and wound healing. Its content in milk is around 0.3–0.4 mg/100 mL. In cow's milk, maximum Zn is present in skim milk portion (around 99%) and potentially in contact with Pi and Pi of casein micelles. About 95% of zinc is attached to casein micelles.[20,60,68] In aqueous phase, a small component is connected to citrate molecules. Milk and milk products are a good source of zinc (100 ml of milk provides 0.4 mg contributes to 20% of the RDA). Zinc is bound with ~95% with the casein micelles; this is the major reason for the lower soluble zinc content in all animal's milk.[16] Zinc is almost entirely linked to high molecular compounds in cow's milk.[43]

2.4.9 MANGANESE

The Mn in milk of different animals is scanty studied. The presence of manganese was reported partly in fat globular membrane. Lactose synthase is the only enzyme present in milk that depends on manganese. The oxidized flavor formation in bovine milk can be reduced by adding manganese as it can bind the salts containing Fe and Cu, their displacement by manganese leads to a reduction of fat oxidation.[44,61]

2.4.10 SELENIUM

Selenium has sulfur-like properties and is sometimes described as synonymous with antioxidant glutathione peroxidase.[75] Milk contains selenium concentration of about 30 µg/L.[56] Selenium is absent in milk fat and is majorly associated with casein and whey proteins.[66] The milk is considered as an essential selenium supplier (300 ml of milk supplies 10 µg of selenium); however, its concentration is not well-known in other dairy products.[63] The contribution of milk and milk products to a daily intake of Selenium ranges from 8 to 39% of RDA.[23]

2.5 EFFECTS OF VARIOUS PROCESSING TREATMENTS ON MINERAL COMPOSITION IN MILK

The different treatments that influence the minerals content in milk like as addition of additives, heating, cooling, freezing, dilution, concentration high pressure, and changes in pH due to processing.

2.5.1 ADDITION OF ADDITIVES (SALTS AND CHELATANTS)

The addition of NaCl to milk, slightly decreases the pH content as well as rise in content of calcium ion in the soluble fraction. Such modifications will lead to exchange by sodium ions of divalent cations or protons bound to phosphoryl casein micelles group.[26] As ionic strength increases, it reduces diffusion of ions, therefore ionic pair dissociation increases and more of casein micelles get hydrated.[29] Adding calcium to milk contributes to CCP (calcium carbonate precipitation) precipitation, an increase in ionized calcium, decreased soluble phosphate levels, and reduced pH. The addition of secondary sodium and potassium phosphate causes CCP precipitation with a decrease insoluble calcium and Ca ion concentration.[33] Chelates have a high cation affinity and can displace Calcium. The calcium chelating agents like EDTA, $C_2O_4^{-2}$, citrate, etc. and cation exchange resins contribute to an increase of Ca and Pi content in the milk's soluble fraction.[42] The milk fortified with $CaCl_2$, calcium gluconate, and calcium lactate contribute to marked rise in its viscosity. The milk fortified with calcium gluconate has highest viscosity, followed by calcium lactate and calcium chloride-fortified milk, while the lowest viscosity was for unfortified milk.

2.5.2 DILUTION AND CONCENTRATION

CCP solubilization occurs with milk dilution, while concentration has the opposite effect.[18,31] As the milk is saturated with Ca and phosphate, dilution declines their concentration and allows some CCP to dissolve, thus increasing alkalinity.[24,48,61]

2.5.3 FREEZING

Micellar calcium phosphate solubility improves with a decrease in temperature, while milk cooling causes micellar calcium phosphate dissolution.[54] Frozen liquid is more salt-saturated. Any soluble calcium phosphate is precipitated by release of H+ ions; and pH is decreased.[32]

2.5.4 HEAT TREATMENT

The soluble fraction of calcium phosphate is less diffused during thermal treatment and consequent decline in calcium or inorganic phosphate has been calculated following thermal treatment. During heat treatment, the interaction of these constituents with casein micelles was shown by "Nuclear Magnetic Resonance" of ^{43}Ca and ^{31}P (from 30°C to 64°C). Changes are reversible if the thermal treatment is <95°C for a few minutes, but serious heating at 120°C for 20 minutes produces permanent changes in casein micelles.[31] The stability of cow's milk is more as compared with sheep and goat milk during heat treatment. The temperature-related variations are similar across all animal species. The changes in micellar structure, soluble and colloidal phase's distribution, and the protein interaction are due to variations in temperature fluctuations. The colloidal solubility of caprine and ovine milk is therefore less than that of bovine milk. This thermal stability can be improved by pH changes, salt added like phosphate and citrate or using membrane technology.[58]

2.5.5 HIGH PRESSURE TREATMENT (HPP)

During HPP treatment, the mineral composition of milk and milk products is changed. HPP solubilizes CCPs of raw milk casein micelles, while a part of the colloidal phosphate dissociated is reorganized when stored for HP-treated milk.[60] An elevated level of soluble calcium, phosphorus, magnesium, and nonsedimentary caseins was found following 100 MPa treatment given to raw bovine milk at 20°C.[46] There are several factors affecting levels of minerals in the dairy product; however, the method used in the processing of the product is the most important. For example: ultra-filtrated concentrated milk has a lower mineral content than a reverse osmosis. The RO keeps all milk components except water. UF helps unbundles (Soluble) to flow through the membrane. Likewise, evaporation and drying goods preserve all parent milk's minerals.[69]

2.5.6 CHANGES IN PH VALUES

It is a physicochemical feature of milk that is directly associated with milk's salt balance. A difference in the milk salt balance induces CCP solubilization. The pH of the milk is elevated during cooling, dilution, or high-pressure handling. Changes that induce calcium phosphate precipitation, i.e., heat treatment or concentration, result in pH reduction. Increased in pH in milk because of CCP solubilization is usually due to elevation in amount of phosphate ion in the soluble fraction of milk as they can bind to hydrogen ion.[31,48,61]

With decrease in pH of milk, the various acid-basic classes of milk constituents (organic and inorganic phosphate, citrate, etc.) become more and more protonated. Therefore, the casein-associated micellar phosphate and minimal quantities of citrate and Mg in casein micelles are dissolved into a diffused milk fraction. Degree of dissociation is pH and temperature dependent. Milk's acidification is responsible for decreased in pH and increase in solubilization of CCP and solubilization is complete below pH 4.9, though alkali addition has opposite effect (pH 11), all soluble calcium phosphate occurs in colloidal process.[13,48] For buffalo milk, the milk's capacity to acidify was higher than for cow milk. In both milks, coagulation of casein at their respective isoelectric point, solubilization of Ca and Pi are significantly influenced by decrease in pH value. These molecular modifications were qualitatively identical but quantitatively distinct for both milks. These quantitative acidification variations were due to differences in milk composition.[3] Average content of major minerals in milk products (mg/100 g)[74] is given in Table 2.7. Average content of trace minerals in milk products (mg/100 g)74 is given in Table 2.8.

TABLE 2.7 Average Content of Major Minerals in Milk Products (mg/100 g).

Product	Na	K	Ca	P
Butter	750	15	15	24
Cheddar cheese	6670	77	720	490
Condensed milks	150	450	330	270
Evaporated milk	180	360	290	260
Fresh cream	37–49	65–120	50–99	50–82
Ice cream	69	160	130	110
Liquid milks	29–58	135–190	107–133	63–145
Pasteurized skimmed milk	55	150	120	95
Yogurt	80	280	200	170

Source: Zamberlin et al., 2012[74]

Milk salts especially calcium and phosphate have an essential part to play in moderating structural changes and in the rennet coagulation of cheese. Elevating concentration of calcium decreases duration of the milk's rennet coagulation because of neutralization of negatively charged residues of casein and increased aggregation of renneting micelles. Adding low amounts of calcium improves the density of the gel. During drainage of the whey, the acid content and the pH are the critical factors deciding the mineral content of the paneer and cheese. The concentration of undissolved milk salts and pH play important role in improving quality and texture of cheese.[65,74] Rubbery body and firmness in buffalo milk paneer mainly because of more content of casein molecules and minerals especially Ca and P. Calcium compounds contribute to formation of cross-linkages during curd formation, which leads to helps in increasing the yield, more recovery of milk solids and enhances the texture and body of paneers.[39,74]

TABLE 2.8 Average Content of Trace Minerals in Milk Products (mg/100 g).

Product	Fe	Cu	Zn	Mn	Se
Butter	0.2	0.03	0.1	Traces	Traces
Cheddar cheese	0.3	0.03	2.3	Traces	12
Condensed milk	0.23	Traces	1.0	Traces	3.0
Evaporated milk	0.26	0.02	0.9	Traces	3.0
Ice cream	0.1	0.02	0.3	Traces	1.5
Pasteurized skim milks	0.05	Traces	0.4	63–145	1.0
Sterilized cream	0.1–0.2	Traces	0.2–0.5	Traces	Traces
Yogurt	0.1	Traces	0.7	Traces	2.0

Source: Zamberlin et al., 2012[74]

The heat stability of condensed milk depends on composition of the salt. Less soluble salt concentrate prepared by ultrafiltration or evaporation of milk that was dialyzed prior to evaporation concentration has been found to be more stable than concentrate prepared by simply evaporating method.[30,74]

Yogurt may result in the redistribution of casein-related minerals. Changes can influence nutritional properties in this distribution. In consideration of the rigorous quality requirements that this substance must follow in terms of safety, the presence of minerals in yoghurt is of special concern. However, no experiments have been performed to determine magnitude and type of mineral redistribution during the production of yoghurt. Yoghurt is a probiotic, which helps to increase the immunity.[12,74] The mineral quantity in curd or *dahi* remains constant after fermentation, but the product obtained possibly increases the bioavailability of calcium and may facilitate absorption of these nutrients in gastrointestinal tract. Curd has many health benefits and is used as a beauty aid for fair skin. It has been said that curd is more valuable than milk in a variety of situations.[8,11,18,74]

Cheddar cheese can significantly alter the physical properties by supplementing with strontium chloride (SrCl2). The proportion of additional magnesium and strontium has been casein bound. The essence of binding appeared to be a consideration that determined capability of these ions to modification of physical properties of cheese. The addition of strontium forms CCP nanoclusters and increase density and strength of para casein matrix, while decreases melting and increasing stiffness.[18] Buffalo milk cheese has been concluded to be nutritionally better than cows because of superior compositional profile, especially minerals and lactic acid content. Also, the lactic acid drastically rises during ripening, although the mineral profile was not significantly changed.[61]

2.6 BIOAVAILABILITY OF MINERALS PRESENT IN MILK AND MILK PRODUCTS

A variety of compounds that are found in food affect the use of minerals. Vitamin D, lactose, and acidity are believed to increase calcium absorption in the intestine. Fibers like cellulose and hemicelluloses and acids like oxalic acid; phytic acid; saturated long-chain fatty acids; and uronic acid form, insoluble calcium, and calcium absorption complexes.[5] The availability of calcium in milk products is influenced by the presence of calcium complexes.[36] It is generally understood that absorption of calcium not only depending on food containing mineral but also on a several types of other factors such as solubility and ionization.[19]

The chemical forms of microelements can affect bioavailability; free form such as ions are well absorbed in soluble form, whereas bounded forms are poorly absorbed.[16] Due to their capacity to bind and retain calcium in a soluble state in lumen of the intestine, and thus usable for absorption, casein phosphatides produced during casein digestion have a positive effect on calcium absorption. During fermentation of milk, noticeable changes in amounts of nonsedimentable minerals favor the absorption of certain minerals (Ca, Mg, P, and Zn) in the gastrointestinal tract.[18]

Both macro and micronutrients are best consumed in the acidic atmosphere of the stomach as they are found as salts, which are relatively easy to solubilize: carbonates, sulfates, oxides. Compared with inorganic salts prescribed to animals as mineral supplements (blends, premixes) used in the compound feed, the mineral substances of vegetable feed have a lower assimilation coefficient, i.e., lower absorption.[45] The main sources of calcium are milk and milk products (cheese, yoghurt, etc.). Most of the dietary Ca comes from milk and milk products, as casein micelles are normal

Ca carrier.[24,61] The average absorption of calcium from cow's milk, cheese, yoghurt ranges between 21% and 45%.

The absorption of Ca depends on age of a person and vitamin D levels of body. It has also been shown that lactose inhibits absorption of calcium in intestine. The bioavailability of calcium compounds is higher in milk and milk products as compared with other products especially fruits and vegetables. This can be due to the presence of calcium in strongly phosphorylated casein fragments, known as casein phosphopeptides (CPPs). These peptides exist primarily when milk products, such as yoghurt or cheese, are processed by breakdown the casein using indigenous or microbial enzymes.[24,21] The ability to form CPP-metal ion complexes is a remarkable characteristic of CPPs, which will theoretically improve the bioavailability of Ca and Fe, likely by retaining metals soluble in the distal small intestine.[21,57]

Proteins also promote the absorption of calcium, and phosphopeptides derived from casein enzymatic hydrolysis (casein phosphopeptides) can form soluble calcium complexes. The phosphopeptides present in the gut are important because calcium precipitates owing to the phosphates found in intestinal tract.[41] Without consuming milk and milk products, the RDA for calcium is hard to achieve. Special focus should be given to bioavailability of milk calcium. The average calcium absorption of bovine milk in the ranges between 21% and 45%. The bioavailability of cheese and yoghurt calcium is almost equivalent to that of milk.[74] Absorption of iron from bovine milk by babies was also tested using isotopes of slightly lower absorption (10%).[47]

Infant formulations fortified with high doses of iron (12 mg/L) consume just 3–4% of iron. No explanation is available for extremely high bioavailability of Fe in human milk. Iron bioavailability in human milk varies between 49% and 70%, which is substantially higher than bovine milk (10–34%). There is no simple solution to this variance. But this may be due to the elevated levels in human milk of Lactoferrin (the glycoprotein that binds iron ions, or high amount of lactose and ascorbate content in human milk that promote the absorption of iron).[44]

Zinc bioavailability is more in human milk than cow milk because it is bound to the low molecular mass ligand (e.g., citrate) that facilitates absorption in human milk. Also zinc binding to casein is ten times lower in human milk than cow milk, it can block zinc formed in the abdomen inside casein curd, rendering it inaccessible for absorption.[55] Little research on copper absorption in human body is available. Experiments conducted on rats showed 83% human milk absorption, 76% cow milk absorption, and 86–87% dairy baby formula.[8] In the presence of soluble carbohydrates and proteins, organic acids and low molecular weight chelates, Cu absorption and bioavailability is improved by increasing its solubility. The intake of manganese from human milk is 8.2% and slightly greater than that of cow's milk (2.4%).[70]

2.7 SUMMARY

Minerals perform many important roles in the body, including the action of enzymes, the development of bones and the transport of oxygen. Milk is considering as good source of minerals especially "P, Mg, Ca, K, Zn, and Se." Several types of minerals found in milk are in the form of their salts, like $Ca_3(PO_4)_2$ (calcium phosphate). In milk around 44% of phosphate, 35% of Mg and 67% of Ca associated with casein micelle in colloidal phase and some are soluble in the serum phase. The fact that Ca and phosphate are correlated with protein binding salts does not affect the nutrient availability of any of them. Milk contains small quantities of copper, iron, sodium, and manganese and is not known to be a significant dietary source of these minerals. Temperature variation causes flow of calcium phosphate into and out of the casein micelle. The process is reversible at moderate temperatures and has a little impact on nutritional content of minerals found in milk. The calcium phosphate will precipitate out of solution at extremely high temperatures, creating permanent changes in composition of casein micelle. The distribution among soluble and micellar phases of main elements in milk not only contributes to the nutritional content of milk, but also to the consistency of colloidal dispersion. Various features of the technological processes to which milk can be exposed can influence the equilibrium between Ca and P salts. Milk and milk derivatives provide the body with important minerals and trace elements to satisfy the RDA.

KEYWORDS

- micelles
- milk derivatives
- minerals
- processing treatments
- trace elements

REFERENCES

1. Abbas, H. M.; Hassan, F. A.; El-Gawad, M. A.; Enab. A. K. Physicochemical Characteristics of Goat's Milk. *Life Sci. J.* **2014,** *11* (1), 11–17.
2. Abd El-Salam; El-Shibiny, S. Comprehensive Review on the Composition and Properties of Buffalo Milk. *Dairy Sci. Technol.* **2011,** *91* (6), 663–668.
3. Ahmad, S.; Gaucher, I.; Rousseau, F.; Beaucher, E.; Piot, M.; Grongnet, J. F.; Gaucheron, F. Effects of Acidification on Physicochemical Characteristics of Buffalo Milk: A Comparison with Cow's Milk. *Food Chem* **2004,** *106* (1), 11–17.
4. Aliasgharpour, M.; Rahnamaye F. M. Trace Elements in Human Nutrition: A Review. *Int. J. Med. Investig.* **2013,** *2* (3), 115–128.
5. Allen, D. G.; Eisner; D. A.; Morris; P. G.; Pirolo, J. S.; Smith, G. L. Metabolic Consequences of Increasing Intracellular Calcium and Force Production in Perfused Ferret Hearts. *J. Physiol.* **1986,** *376* (1), 121–141.
6. Bansal, B.; Chen, X. D. Critical Review of Milk Fouling in Heat Exchangers. *Compr. Rev. Food Sci. Food Saf.* **2006,** *5* (2), 27–33.
7. Barreto, Í. M. L. G.; Rangel, A. H. D. N.; Urbano, S. A.; Bezerra, J. D. S.; Oliveira, C. A. D. A. Equine Milk. and its Potential Use in the Human Diet. *Food Sci. Technol.* **2019,** *39*, 1–7.
8. Cashman, K. D. Milk Minerals (Including Trace Elements) and Bone Health. *Int. Dairy J.* **2006,** *16* (11), 1389–1398.
9. Cerbulis, J.; Farrell Jr, H. M. Composition of the Milks of Dairy Cattle. II. Ash, Calcium, Magnesium, and Phosphorus. *J. Dairy Sci.* **1976,** *59* (4), 589–593.
10. Choi, J.; Horne, D. S.; Lucey, J. A. Determination of Molecular Weight of a Purified Fraction of Colloidal Calcium Phosphate Derived from the Casein Micelles of Bovine Milk. *J. Dairy Sci.* **2011,** *94* (7), 3250–3261.
11. Closa, S. J.; de Landeta, M. C.; Anderica, D.; Pighín, A.; Cufré, J. A. Mineral Nutrient Content in Cow Milk and Dairy Products in Argentina. *Arch. Latinoam. Nutr.* **2003,** *53* (3), 320–324.
12. Cooke, D. R.; McSweeney, P. L. The Influence of Alkaline Earth Metal Equilibria On the Rheological, Melting and Textural Properties of Cheddar Cheese. *J. Dairy Res.* **2013,** *80* (04), 418–428.
13. Dalgleish, D. G.; Law, A. J. pH-induced Dissociation of Bovine Casein Micelles. II. Mineral Solubilization and its Relation to Casein Release. *J. Dairy Res.* **1989,** *56* (05), 727–735.
14. Davidsson, L.; Kastenmayer, P.; Yuen, M.; Lönnerdal, B. O.; Hurrell, R. F. Influence of lactoferrin on iron absorption from human milk in infants. *Pediatr. Res.* **1994,** *35* (1), 117–124.
15. Davoodi, H.; Esmaeili, S.; Mortazavian, A. M. Effects of Milk and Milk Products Consumption on Cancer: A Review. *Compr. Rev. Food Sci. Food Saf.* **2013,** *12* (3), 249–264.
16. De La Fuente, M. A. Changes in the Mineral Balance of Milk Submitted to Technological Treatments. *Trends Food Sci. Technol.* **1998,** *9* (7), 281–288.
17. De La Fuente, M. A.; Olano, A.; Juarez, M. Distribution of Calcium, Magnesium, Phosphorus, Zinc, Manganese, Copper and Iron Between the Soluble and Colloidal Phases of Ewe's and Goat's Milk. *Lait* **1997,** *77* (4), 515–520.

18. De La Fuente, M. A.; Montes, F.; Guerrero, G.; Juárez, M. Total and Soluble Contents of Calcium, Magnesium, Phosphorus and Zinc in Yoghurts. *Food Chem.* **2003,** *80* (4), 573–578.
19. Delisle, J.; Amiot, J.; Dore, F. Biological Availability of Calcium and Magnesium from Dairy Products. *Int. Dairy J.* **1995,** *5* (1), 87–96.
20. Drago, S. R.; Valencia, M. E. Influence of Components of Infant Formulas on *In Vitro* Iron, Zinc, and Calcium Availability. *J. Agric. Food Chem.* **2004,** *52* (10), 3202–3207.
21. Dupuis, N. F.; Wu, C.; Shea, J. E.; Bowers, M. T. Human Islet Amyloid Polypeptide Monomers Form Ordered β-hairpins: Possible Direct Amyloidogenic Precursor. *J. Am. Chem. Soc.* **2009,** *131* (51), 18283–18292.
22. Flynn, A. Minerals, and Trace Elements in Milk. *Adv. Food Nutr. Res.* **1992,** *36*, 209–252.
23. Foster, L. H.; Chaplin, M. F.; Sumar, S. The Effect of Heat Treatment on Intrinsic and Fortified Selenium Levels in Cow's Milk. *Food Chem.* **1998,** *62* (1), 21–25.
24. Fox, P. F.; Unicke-Lowe, T.; McSweeney, P. L. H.; Mahony, J. A. O. *Dairy Chemistry and Biochemistry*, 2nd edn; Springer International Publishing AG: Cham – Switzerland, 2015; p 241–269.
25. Gaucheron, F. The Minerals of Milk. *Reprod. Nutr. Dev.* **2005,** *45* (4), 473–483.
26. Gaucheron, F. Milk and Dairy Products: A Unique Micronutrient Combination. *J. Am. Coll. Nutr.* **2011,** *30* (suppl 1), 400S–409S.
27. Gooneratne, S. R.; Buckley, W. T.; Christensen, D. A. Review of Copper Deficiency and Metabolism in Ruminants. *Can. J. Anim. Sci.* **1989,** *69* (4), 819–845.
28. Górska-Warsewicz, H.; Rejman, K.; Laskowski, W.; Czeczotko, M. Milk and Dairy Products and Their Nutritional Contribution to the Average Polish Diet. *Nutrients.* **2019,** *11* (8), 1771.
29. Grufferty, M. B.; Fox, P. F. Effect of added NaCl on Some Physicochemical Properties of Milk. *Ir. J. Food Sci. Technol.* **1985,** *9* (10), 1–9.
30. Hardy, E. E.; Muir, D. D.; Sweetsur, A. M.; West, I. G. Changes of Calcium Phosphate Partition and Heat Stability During Manufacture of Sterilized Concentrated Milk. *J. Dairy Sci.* **1984,** *67* (8), 1666–1673.
31. Holt, C. Some Principles Determining Salt Composition and Partitioning of Ions in Milk. *J. Dairy Sci.* **1981,** *64* (10), 1958–1964.
32. Hu, Y. B.; Wolthers, M.; Wolf-Gladrow, D. A.; Nehrke, G. Effect of pH and Phosphate on Calcium Carbonate Polymorphs Precipitated at Near-Freezing Temperature. *Cryst. Growth Des.* **2015,** *15* (4), 1596–1601.
33. O'Mahony, J. A.; McSweeney, P. L. H.; Lucey, J. A. A Model System for Studying the Effects of Colloidal Calcium Phosphate Concentration on the Rheological Properties of Cheddar Cheese, *J. Dairy Sci.* **2006,** *89* (3), 892–904.
34. Jahnen-Dechent, W.; Ketteler, M. Magnesium Basics. *Clin. Kidney J.* **2012,** *5* (Suppl 1), i3–i14.
35. Kaminarides, S.; Moschopoulou, E.; Karali, F. Influence of Salting Method on the Chemical and Texture Characteristics of Ovine Halloumi Cheese. *Foods* **2019,** *8* (7), 232–239.
36. Kansal, V. K. Health Benefit Claims of Dairy Calcium: A Review. *Ind. J. Dairy Sci.* **2002,** *55* (3), 127–132.
37. Kapoor, R.; Metzger, L. E. Process Cheese: Scientific and Technological Aspects—A Review. *Compr. Rev. Food Sci. Food Saf.* **2008,** *7* (2), 194–214.
38. Schiffer, S.; Scheidler, E.; Kiefer, T.; Kulozik, U. Effect of Temperature, Added Calcium and pH on the Equilibrium of Caseins between Micellar State and Milk Serum. *Foods.* **2021,** *10* (4), 822–829.

39. Khan, S. U.; Pal, M. A. Paneer Production: A Review. *J. Food Sci. Technol.* **2011,** *48* (6), 645–660.
40. Lamichhane, P.; Kelly, A. L.; Sheehan, J. J. Symposium Review: Structure-Function Relationships in Cheese. *J. Dairy Sci.* **2018,** *101* (3), 2692–2709.
41. Lee, Y. S.; Noguchi, T.; Naito, H. Intestinal Absorption of Calcium in Rats Given Diets Containing Casein or Amino Acid Mixture: The Role of Casein Phospho-peptides. *Br. J. Nutr.* **1983,** *49* (1), 67–76.
42. Lin, M. J.; Lewis, M. J.; Grandison, A. S. Measurement of ionic calcium in milk. *Int. J. Dairy Technol.* **2006,** *59* (3), 192–199.
43. Lönnerdal, B. Effects of Milk and Milk Components on Calcium, Magnesium, and Trace Element Absorption During Infancy. *Physiol. Rev.* **1997,** *77* (3), 643–669.
44. Lönnerdal, B.; Keen, C. L.; Hurley, L. S. Iron, Copper, Zinc, and Manganese in Milk. *Annu. Rev. Nutr.* **1981,** *1* (1), 149–174.
45. López-Alonso, M. Trace Minerals and Livestock: Not Too Much Not Too Little. *Int. Sch. Res. Notices.* **2012,** *2012*, 1–18.
46. Lopez-Fandiño, R. High Pressure-Induced Changes in Milk Proteins and Possible Applications in Dairy Technology. *Int. Dairy J.* **2006,** *16* (10), 1119–1131.
47. McMillan, J. A.; Oski, F. A.; Lourie, G.; Tomarelli, R. M.; Landaw, S. A. Iron Absorption from Human Milk, Simulated Human Milk, and Proprietary Formulas. *Pediatrics.* **1977,** *60* (6), 896–900.
48. Mehta, B. M. Chemical Composition of Milk and Milk Products. In *Handbook of Food Chemistry*, Chaung, P. C. K.,Mehta, B. M. Eds.; Springer Nature: Switzerland, 2015; p 511–553.
49. Mekmene, O.; Le Graet, Y.; Gaucheron, F. A Model for Predicting Salt Equilibria in Milk and Mineral-Enriched Milks. *Food Chem.* **2009,** *116* (1), 233–239.
50. Datta, N.; Deeth, H.C. Heat Treatment, Alternatives to Other Nonthermal Technologies. In *Encyclopedia of Dairy Sciences* Roginski, H. Ed.; Elsevier: San Diego, 2002; p 1339–1346.
51. Nagyová, G.; Buňka, F.; Salek, R. N.; Cerníkova, M.; Mancík, P.; Grůber, T.; Kuchar, D. Use of Sodium Polyphosphates with Different Linear Lengths in the Production of Spreadable Processed Cheese. *J. Dairy Sci.* **2014,** *97* (1), 111–122.
52. National Research Council (NRC). Factors Affecting the Composition of Milk from Dairy Cows. In *Designing Foods: Animal Product Options in The Marketplace.* National Academies Press (US); National Research Council (US) Committee on Technological Options to Improve the Nutritional Attributes of Animal Products: Washington –DC, 1988; online; https://www.ncbi.nlm.nih.gov/books/NBK218193/;
53. Oh, H. E.; Deeth, H. C. Magnesium in Milk. *Int. Dairy J.* **2017,** *71*, 89–97.
54. Ozcan, T. U. L. A. Y.; Horne, D.; Lucey, J. A. Effect of Increasing the Colloidal Calcium Phosphate of Milk on the Texture and Microstructure of Yogurt. *J. Dairy Sci.* **2011,** *94* (11), 5278–5288.
55. Pabon, M. L.; Lonnerdal, B. Bioavailability of Zinc and Its Binding to Casein in Milks and Formulas. *J. Trace Elem. Med. Biol.* **2000,** *14* (3), 146–153.
56. Pechova A.; Misurova L.; Pavlata L.; Dvorak R. Monitoring of Changes in Selenium Concentration in Goat Milk During Short-Term Supplementation of Various Forms of Selenium. *Biol. Trace Elem. Res.* **2008,** *121*, 180–191.
57. Peres, J. M.; Bouhallab, S.; Bureau, F.; Neuville, D.; Maubois, J. L.; Devroede, G.; Bouglé, D. Mechanisms of Absorption of Casein phosphor-peptide Bound Iron. *J. Nutr. Biochem.* **1999,** *10* (4), 215–222.

58. Raynal-Ljutovac, K.; Park, Y. W.; Gaucheron, F.; Bouhallab, S. Heat Stability and Enzymatic Modifications of Goat and Sheep Milk. *Small Rum. Res.* **2007,** *68* (1), 207–220.
59. Sahai, D. Compositional Profile of Buffalo Milk. Buffalo Milk. *Chem. Proc. Technol.* **1996,** *231*, 20–57.
60. Schrader, K.; Buchheim, W.; Morr, C. V. High Pressure Effects on the Colloidal Calcium Phosphate and The Structural Integrity of Micellar Casein in Milk, Part 1: High Pressure Dissolution of Colloidal Calcium Phosphate in Heated Milk Systems. *Nahrung.*, **1997,** *41* (3), 133–138.
61. Singh, M.; Sharma, R.; Ranvir, S.; Gandhi, K.; Mann, B. Profiling and Distribution of Minerals Content in Cow, Buffalo and Goat Milk. *Ind. J. Dairy Sci.* **2019,** *72*, 480–488.
62. Singh, T. K.; Drake, M. A.; Cadwallader, K. R. Flavor of Cheddar Cheese: A Chemical and Sensory Perspective. *Compr. Rev. Food Sci. Food Saf.* **2003,** *2* (4), 166–189.
63. Smirnova, A.; Konoplev, G.; Mukhin, N.; Stepanova, O.; Steinmann, U. Milk as a Complex Multiphase Polydisperse System: Approaches for the Quantitative and Qualitative Analysis. *J. Compos. Sci.* **2020,** *4* (4), 151–157.
64. Spencer, H.; Kramer, L.; Osis, D.; Wiatrowski, E.; Norris, C.; Lender, M. Effect of Calcium, Phosphorus, Magnesium, And Aluminum on Fluoride Metabolism in Man. *Ann. N. Y. Acad. Sci.* **1980,** *355*, 181–194.
65. Troch, T.; Lefébure, E.; Baeten, V.; Colinet, F.; Gengler, N.; Sindic, M. Cow Milk Coagulation: Process Description, Variation Factors and Evaluation Methodologies. A Review. *Biotechnol. Agron. Soc. Environ.* **2017,** *21* (4), 276–287.
66. Van Dael, P.; Shen, L.; Van Renterghem, R.; Deelstra, H. Selenium Content of Goat Milk and its Distribution in Protein Fractions. *Z. Lebensm. Unters. Forsch.* **1992,** *195* (1), 3–7.
67. Vorland, C. J.; Stremke, E. R.; Moorthi, R. N.; Gallant, K. M. H. Effects of Excessive Dietary Phosphorus Intake on Bone Health. *Curr. Osteoporos. Rep.* **2017,** *15* (5), 473–482.
68. Wabel, N. A. Mineral Contents of Milk of Cattle, Camels, Goats and Sheep in the Central Region of Saudi Arabia. *Asian J. Biochem.* **2008,** *3* (6), 373–375.
69. Walstra, P.; Wouters, J. T. M.; Geurts, T. J. Milk: Main characteristics. *Dairy Sci. Technol.* **2006,** *2016*, 1–8.
70. Wapnir, R. A. Copper Absorption and Bioavailability. *Am. J.Clin. Nutr.* **1998,** *67* (5), S1054–S1060.
71. World Health Organization (WHO); Food and Agricultural Organization (FAO) of the United Nations. *Vitamin and Mineral Requirements in Human Nutrition*, 2nd ed.; World Health Organization: Geneva, 2005; p 341.
72. Yabrir, B.; Hakem, A.; Mostefaoui, A.; Titouche, Y.; Bouzidi, A.; Mati, A. Nutritional Value of Algerian Breed Ewe's Milk Related to its Mineral Content. *Pak. J. Nutr.* **2014,** *13* (3), 176.
73. Younes, E. Structural Properties of Casein Micelles in Milk: The Effect of Salt, Temperature, and pH. *IJBB*. **2017,** *36*, 202–215.
74. Zamberlin, S.; Antunac, N.; Havranek, J.; Samarzija, D. Mineral Elements in Milk and Dairy Products. *Mljekarstvo*. **2012,** *62* (2), 111–125.
75. Zoidis, E.; Seremelis, I.; Kontopoulos, N.; Danezis, G. P. Selenium-Dependent Antioxidant Enzymes: Actions and Properties of Selenoproteins. *Antioxidants (Basel)*. **2018,** *7* (5), 66–70.

CHAPTER 3

CHEMISTRY OF BUTTER: CLASSIFICATION, PROCESSING EFFECTS ON CONSTITUENTS, AND DEFECTS

NANDITA DAS, SRISHTY TYAGI, PAYAL KARMAKAR, and SUNIL SAKHALA

ABSTRACT

Butter is a water-in-oil (W/O) emulsion form of a fat-rich dairy product, containing predominantly milk fat (~80%) along with little quantity milk solids-not-fat (MSNF), and added ingredients, such as color and salt. Generally, its color varies from pale yellow to deep yellow. Churning of cream or agitation of milk or curd leads to rupturing of milk fat globule membrane protective layer of milk fat globules and due to coalescence of fat globules large granules are formed, which further leads to separation of butter granules and serum. These butter granules are further worked out and butter is produced. The serum portion is called the buttermilk, which is natural rich source of phospholipids. Depending on the treatment given to the cream during the production, various types of butter are produced, such as: sweet cream butter, mildly acidified butter, sour cream butter, salted butter, unsalted butter, etc. The defects observed in butter are related majorly to its flavor, color, body, and texture. These defects are observed in final product mainly due to feed of the cow, unhygienic processing conditions, improper storage conditions, and temperature fluctuations during transportation, production, and storage. Butter is generally used for direct consumption

The Chemistry of Milk and Milk Products: Physicochemical Properties, Therapeutic Characteristics, and Processing Methods. Megh R. Goyal, Suvartan Ranvir, & Junaid Ahmad Malik (Eds.)

or can also be used as a cooking medium, in the confectionary, bakery, or ice cream industries as a raw material, and in the production of ghee, butter oil, and recombined milk.

3.1 INTRODUCTION

Butter is thought to have originated during 9000–8000 BC. In modern written scriptures, the name of butter is derived from the *Latin* word butrum or butyrum. It is believed that chemically this name originated from butyric acid. The only dietary source of butyric acid is milk fat. Butter has been a staple on international markets since the 14th century. It was produced on a small-scale utilizing the gravity separation of cream from milk. In 1877, mass production of butter was started with the development of mechanical separators. Butter was a costly product and even after extensive production, the price remained high. There have been taken several attempts to come up with the inexpensive butter replacement.

Mege-Mouries developed oleomargarine in 1870, which has a same taste as butter. It became very popular by the name of margarine, but it was noted as inferior and poor lifestyle food. Butter as a margarine is still known to be butter substitute.[54]

Butter is a water-in-oil (W/O) emulsion generally having not less than (NLT) 80% (m/m) fat, not more than (NMT) 16% (m/m) moisture, and NMT 2% (m/m) milk solids-not-fat (MSNF).[9] It is essentially milk fat separated via means of churning or agitation the milk, cream, or curd. The color of butter varies from pale yellow to deep yellow and white depending upon the source of milk and its beta-carotene content. Fat is present in milk as milk fat globules ranging in size from 0.1 to 10.0 μm. These globules are covered with a membrane made up of surface-active agents, i.e., phospholipids and lipoproteins and prevent different globules from coming together. This membrane is named as milk fat globule membrane. Agitation or churning the milk, cream or curd leads to rupturing this membrane and enables the fat globules to join together and coalesce to form large granules of fat and separate out from serum portions. This separated fat is called raw butter and serum portion as buttermilk. Most of the membrane part goes into the buttermilk and leads to an increased phospholipid content seven times compared to whole milk.[28,52]

This chapter focuses on chemistry and characteristics of butter (a fat-rich dairy product).

3.2 DEFINITION OF A BUTTER

Butter is a fatty product derived exclusively from milk of cow and/or buffalo or its products principally in the form of W/O emulsion. The product may be with or without added common salt and starter cultures of harmless lactic acid and/or flavor producing bacteria.

Table butter is obtained from pasteurized milk and/or other milk products, which have undergone adequate heat treatment to ensure microbial safety. It shall be free from animal, body fat, vegetable oil and fat, mineral oil, and added flavor. It shall have pleasant taste and flavor free from off flavor and rancidity. It may contain food additives permitted based on Food Safety and Standards Authority of India (FSSAI) Regulations. It shall conform to the microbiological requirements prescribed by FSSAI.[17] Table 3.1 indicates specifications of butter according to FSSAI. Specifications of table butter according to *Codex Alimentarius* are indicated in Table 3.2. Additive/color in butter according to *Codex Alimentarius*[9] is listed in Table 3.3. Acidity regulators in butter according to *Codex Alimentarius* are shown in Table 3.4.

TABLE 3.1 Specifications of Butter According to Food Safety and Standards Authority of India.[17]

Product	Milk fat	Milk solids not fat	Common salt	Moisture
Desi cooking butter	NLT 76% (m/m)	–	–	–
Table butter	NLT 80% (m/m)	NMT 1.5% (m/m)	NMT 3.0% (m/m)	NMT 16% (m/m)

NMT—Not more than; NLT—Not less than

TABLE 3.2 Specification of Butter According to *Codex Alimentarius*.[9]

Product	Milk fat	Milk solids not fat	Common salt	Moisture
Table butter	NLT 80% m/m	NMT 2.0% m/m	–	NMT 16% m/m

TABLE 3.3 Additive (Color) in Butter According to *Codex Alimentarius*.[9]

INS No.	Name of color	Maximum limit (mg/kg)
160a(i)	Carotene (synthetic)	25
160a(ii)	Carotenes (natural extracts)	600
160b(i)	Annatto extracts	20 (bixin basis)
160e	Carotenal, beta-apo-8	35
160f	apo-8′-carotenoic acid, methyl or ethyl ester	35

TABLE 3.4 Additive (Acidity Regulators) in Butter According to *Codex Alimentarius*.[9]

INS No.	Name of additive	Maximum limit (mg/kg)
339	Phosphates	880 mg/kg
500(i)	Sodium carbonate	Limited
500(ii)	Sodium hydrogen carbonate	by GMP
524	Sodium hydroxide	
526	Calcium hydroxide	

3.3 CLASSIFICATION OF BUTTER

Butter is majorly composed of milk fat (i.e., around 80%) and a small quantity of milk solids-not-fat besides added salt or other ingredients. Butter is classified under high calorie food as 700 calories that is obtained from 100 g of butter.[28] Nutritional composition of butter[8] is shown in Table 3.5. The ratio of unsaturated fatty acid to saturated fatty acid of butter is around 1:2. In butter, the palmitic acid is most abundant fatty acid, followed by oleic acid, myristic, and stearic acid.[1,45]

TABLE 3.5 Nutritional Composition of Butter.

Constituent	Per 100 g of butter
Carbohydrates (g)	0
Energy (calories)	720
Fat (g)	80
Minerals and added salt (g)	3
Protein (g)	0.5
Vitamins (g)	0.5
Water (g)	16

Butter can be classified based on various classes (such as: acidity content, salt content, end use, or manufacturing practices),[4,31,34,54] as follows:

I. Classification of butter based on acidity of cream

- Sour-cream butter is made up of ripened cream, i.e., having more than 0.2% acidity or pH less than 5.1.

- Mildly acidified butter is made using sweet cream, which is partially acidified having pH range of 5.2–6.3.
- Sweet cream butter is made of cream having pH > 6.4 or not exceeding acidity of 0.2%. No bacterial culture is added to this cream for enhance the diacetyl content in butter.

II. Classification of butter based on salt content

- Unsalted butter contains no salt, which is used for the production of ghee or butteroil.
- Salted butter is the butter in which salt is added to improve flavor of the end product along with its quality.

III. Classification of butter based on end use

- White butter produced from pasteurized cream (un-ripened and free from any preservative, flavor, or color.
- Table butter produced from pasteurized cream (un-ripened, but addition of lactic culture, salt, color, and diacetyl).

IV. Classification of butter based on manufacturing practices

- *Desi* butter is obtained by churning curd or cream to obtain product. This is the traditional method.
- Pasteurized cream butter produced from pasteurized sweet cream. This butter has milder flavor compared to the butter made from unpasteurized cream.

3.4 PREPARATION OF CREAM

Cream can be either sweet cream or sour (cultured) cream. In India and USA, mostly sweet cream is used in majority of the creameries.[20] This results in producing sweet butter milk, whose economic value is much better as compared to sour butter milk. Flow diagram of butter manufacturing is given in Figure 3.1.

Milk received	**Cooling & Aging**	**Addition of color & salt**
↓	↓	↓
Cream separation	**Churning**	**Final working**
↓	↓	↓
Neutralization	**Draining of butter milk**	**Butter formed**
↓	↓	↓
Standardization	**Butter grain obtained**	**Packaging**
↓	↓	↓
Pasteurization	**Washing of butter**	**Storage**
↓	↓	↓
Ripening	**Initial working**	**Distribution**
↓	↓	

FIGURE 3.1 Flow diagram of butter manufacturing.

3.4.1 NEUTRALIZATION OF CREAM

The neutralization of cream indicates the step, where acidity of the cream is reduced. This step is more important in case of sour cream, which has more chances of fat loss during churning (Figure 3.1). Thus, pasteurization will have much lesser chances of curdling resulting in a butter of better storage stability.[22] Neutralization can also help in getting rid of the undesirable (fishy) flavors.

For neutralization of cream, definite standard for churning acidity needs to be adopted. After estimating the acidity, calculated amount of neutralizer, be it soda neutralizer or lime neutralizer, needs to be added in the correct manner in diluted form and mixed thoroughly. Acidity should be rechecked by appropriate testing methods. The temperature should not be raised above 85 to 90° F and the agitation of cream should continue for about 5 to 10 min.[7]

3.4.2 STANDARDIZATION OF CREAM

The required quantity of fat in cream which is used for butter making should be 33 to 40%. Standardization means adjustment of fat level in the cream to get the desired level.[10] Both lower and higher values can cause fat losses in buttermilk. It is preferred to add skim milk or butter milk rather than water, otherwise a bland or flat flavor may be induced.[26]

3.4.3 PASTEURIZATION OF CREAM

This method is defined as the heating of every particle to a temperature of 71°C and then holding at that temperature for 20 min or any other time–temperature combinations employing appropriate equipments. The purpose of pasteurization is to get rid of pathogens and other harmful bacteria, yeasts, and mold and some gaseous particles.[60] For preparing ripened cream, the temperature employed is in the range of temperature at 90–95°C for 15 s or at 105–110°C with no holding. However, excessive heating can denature lactoglobulins and other whey proteins, exposing –SH groups.[13]

3.4.4 RIPENING OF CREAM

As the term suggests, ripening refers to the fermentation of the cream using starter culture at the rate of 0.5–2.0% of the weight of the cream. This step can be omitted in case of sweet cream. It improves in keeping quality of the butter.[5] Ripening helps in improving the diacetyl content of the butter. The starter cultures used are namely "*Streptococcus lactis, Streptococcus cremories* (acid producing) and *Streptococcus diacetylactis, Leuconostoc citrovorum,* and/or *Leuc. dextranicum* (flavor producing)." After addition of culture, incubation is maintained at 21°C followed by cooling at 5–10°C at a pH of 5.2, beyond which, biosynthesis of diacetyl is not possible.[61]

3.4.5 COOLING AND AGING OF CREAM

It prepares the cream for churning. While leaving the pasteurizer, the cream is in liquid form, but until crystallization starts, churning is not possible. Hence, cooling is required for fat crystallization, without which, there are high chances of fat losses. Temperature of 7–9°C in summer and 10–13°C in winter is maintained. If too high temperature is used, butter with a greasy body will be produced.[25] After proper cooling and aging, churning does not take more than 35–45 min. Rapid cooling produces a greater number of smaller fat crystals. However, slow cooling produces lesser number of bigger crystals. The fat is bound to the surface with the crystals by adsorption.

Vigorous cooling process will result in higher amount of fat crystallization into solid phase, thereby, less liquid fat separating out during working and churning.[37]

3.4.6 CHURNING OF CREAM

This is the most important step in the preparation where the cream is converted into butter. During churning, O/W emulsion form is transformed to W/O emulsion. The primary step includes agitation of the cream for incorporation of several air bubbles into cream thereby increasing its volume. Fat globules are disrupted, and milk fat is agglomerated.[43] The fat film causes the air bubble to burst by acting as foam depressant.

3.4.7 WASHING OF BUTTER

3.4.7.1 PURPOSE OF WASHING OF BUTTER

Washing of butter refers to the cleaning of the curd remnants adhered to the surface of butter grains. The main purpose of washing of butter is enhancement of shelf-life by washing out free butter milk from butter granules thus reducing content of curd in butter.[40] The bacterial deterioration is resisted, and fresh flavor is retained for a longer period of time. An optimum firmness is ensured by washing as proper working is facilitated by controlling the temperature of granules. It also helps in removal of unpleasant flavors in case where the cream is of poor quality.[59] However, washing acts as a bane, in case of cream having good quality as it may cause fading of the pleasant flavor. It can also result in development of a stale flavor which may even follow cheesy flavor.

3.4.7.2 METHOD OF WASHING OF BUTTER

The curd content of unwashed butter is in the range of 1.1–1.5% and that of washed butter is 0.6–1.0%.[2] The temperature of the wash water needs to be controlled as it has profound effect on the quality of the butter. The temperature should be kept slightly below the temperature of the buttermilk but not lower than that of the cream in the churn.[42] It is best advised to keep the temperature of wash water higher in winter and lower in summer since too high temperature softens and too low hardens the butter.

Washing of butter may not completely remove the curd. The reason underlying is, the buttermilk adhering to the surface of granules and entrapped between the same can be easily removed, but the part in the interior is very difficult to remove on account of being finely dispersed with them.[36] Anyhow, 25% of the curd content can be washed off.

Draining of water is also a critical step. If the moisture content still remains high, even after draining out, the churn again needs to be closed and rotated until the granules form a lump followed by stopping the churn and draining out of free water.[18]

3.4.8 ADDITION OF COLOR AND SALT

3.4.8.1 NEED FOR COLOR IN BUTTER

Carotene is the natural component responsible for coloring of butter. However, it may vary depending on several factors like breed, feed, season, and stage of lactation. Hence, to maintain uniformity in the color of butter throughout the year, coloring of butter is a very important step followed during manufacture of butter.[21] To overcome this influence of varying natural coloring component of milk and cream on the color of butter, calculated amount of color is added while butter making.

3.4.8.2 DESIRED PROPERTIES OF A BUTTER COLOR

- Should be soluble in oil.
- Should be devoid of unpleasant odors and flavors.
- Must not be injurious to health.
- Needed in little amount.
- Must be stable in emulsion and not settle out on standing.

3.4.8.3 ADDITION OF COLOR

The butter color is usually added to cream itself before being loaded to the churn. However, it can also be added during the final working of butter by mixing with salt. This method has the advantage of giving uniform color to the butter on account of being worked into it. Butter color should not exceed 250 g/100 kg of butter.[23]

3.4.8.4 TYPES OF COLORS

3.4.8.4.1 Colors for Table Butter

The most common example of vegetable butter colors is extracted from annatto plant (*Bixa orellana*).[49] The extraction involves boiling for several hours the annatto seed in oil, which gives ultimate vegetable butter color. During the initial stage, the temperature is kept low whereas at the later stage, temperature as high as 115°C is used which results in formation of annatto extract in permanent emulsion with oil. Then, gravity or pressure is employed to filter the mixture through heavy canvas. Other than annatto, other color sources include carotene and curcumin.

3.4.8.4.2 Colors for Mineral Butter

The source of this type of color is coal-tar dyes. They must be harmless and oil soluble. They are, at first, mixed with neutral oil, then, they are boiled and filtered. Few examples of coal-tar dyes, which have got USDA certification are yellow A and yellow B (benzeneazo-β-naphthlyamine) and yellow O B (ortho-tolueneazo-β-naphthylamine). As compared to vegetable butter colors, mineral butter colors are required in much lesser quantity and the permanency of the emulsion with the latter one is more than that of the former one.[58]

3.4.8.5 NEED FOR SALTING IN BUTTER

The butter, which is meant for direct consumption is usually added with salt. The main motto behind addition of salt is the improvement of flavor, thereby increasing consumer satisfaction. The salt takes out the loosely held water droplets together forming bigger aggregates. Salting also offers the advantage of increasing the overrun of the butter as well as enhancing the shelf-life and keeping quality of butter by retarding growth of bacteria, yeasts, and molds, especially in case if butter is made from the sweet cream. However, salting proves to be a disadvantage if the butter is made from sweet cream, in which case, chemical defects are accelerated. The light creamy color obtained in the butter is induced because of the scattering of light caused by finely dispersed water droplets.[39]

3.4.8.6 METHOD OF ADDITION OF SALT IN BUTTER

According to FSSR (2011), salt can be added to butter with maximum permissible limit of 3.0%. Addition of salt causes diffusion of water toward brine solution, as an effect of osmosis, which creates bigger moisture droplets, thereby making the butter leaky.[19] Thus, butter should be worked upon properly. In addition to this, high salt content also causes the curd to precipitate thereby reducing the water holding capacity of the butter which can again produce bigger water droplets. Hence, the butter needs to be worked completely to ensure sufficient distribution of brine to make the butter free from leakiness and to provide it with even color.[14]

There are three distinct methods for salting of butter:

3.4.8.6.1 Dry Salting Method

This is the most common method in case of mildly soft butter, where dry salt is being sprinkled on butter just before working. One major disadvantage, in case of abnormally soft butter, is that it takes a lot of time to dissolve the salt and soft fat may coat the salt crystals, due to which, the salt crystals fail to get enough moisture for them to dissolve, thereby making gritty kind of butter.[35]

3.4.8.6.2 Wet Salting Method

Unlike dry salting, in this case, the salt is spread over butter granules followed by complete wetting with water and then the butter is worked upon.[52] This helps in making butter free from undissolved crystals by providing rapid solution.

3.4.8.6.3 Brine Salting Method

This is the rarest method and not much into use because of its high equipment and maintenance cost. This type of salting refers to the addition of salt as saturated brine solution; hence, this mode of salting is used only in case of lightly salted butter.[41]

3.5 WORKING (KNEADING) OF BUTTER

Kneading of butter granules to obtain a compact and even mass of butter is called working of butter. It helps in complete dissolution and uniform distribution of salt in butter. The compact mass obtained will help in convenient packaging, handling, and storing of butter. Working of butter is done at two stages of butter manufacturing: Initial and Final working.[51]

Buttermilk and excess moisture are removed during initial working, after which salt and make up water is added and then the final working is done. It is always better to overwork butter than under work as under work will leave moisture on standing, which will promote mold growth in butter. However, over working of soft butter may lead to the production of greasy butter.[4]

The moisture in butter can be present in two forms: minute water droplets enmeshed within the butter granules and free moisture entrapped in butter granules. Due to working, butter loses its granular state and becomes plastic in nature. During working of butter granules start entrapping moisture and the amount of moisture assimilated in the butter granules depend on firmness of butter, amount of free water added, and the speed of the worker rolls. Besides water, working may also increase the air content which is undesirable as it may cause oxidative defects and microbial growth which will eventually reduce keeping quality. To reduce the extent of air incorporation in butter, vacuum working of butter is an alternative. Various indicator papers are used for checking proper dispersion of moisture based on the pH ranges, such as bromo phenol red (pH range 5.4–7.0), bromo thymol blue (pH range 6–7.6), and bromo phenol blue indicator papers (pH range 3–4.6). Worked butter is then removed either manually from churn, or by gravity, or by means of compressed air.[12,56]

3.6 BUTTER OVER-RUN

Butter overrun is the "difference between the weight of the butter made from that fat and the weight of fat churned." In addition to milk fat, butter also contains salt, moisture, curd, small amount of lactose, ash, etc. These nonfatty constituents contribute to the butter overrun. Maximum overrun possible in butter is around 25%. Factors influencing overrun of butter are fat losses in buttermilk, fat content of cream, mechanical losses of fat, etc.[3,38]

3.7 CONTINUOUS BUTTER MAKING

To increase the efficiency and output of butter manufacturing, the batch method of butter manufacturing was evolved to continuous butter manufacturing process. The advantages of continuous butter manufacturing process are that it is comparatively quick, hygienic, and economical method.[23] Few of the methods of continuous butter manufacturing are Fritz-Eisenrich Process, Contimab process, Alfa process, and Cherry-Burrell Golden Flow process. Three principles govern the continuous butter production process.[16,30,53]

3.7.1 FROTHING OR CHURNING METHOD

In this process, butter grains are formed by aggregation of fat globules due to incorporation of air through beating. During churning, the air incorporated is dispersed into small air bubbles, the fat globules attach themselves at the air–water interface. Due to churning, collisions take place which leads to coalescence. This process eventually destabilizes the fat emulsion of cream and buttermilk is released which is then separated to obtain the butter granules. Screw-type kneader is used for the working of the granules to obtain smooth butter.

3.7.2 PHASE REVERSAL METHOD

Concentrated cream is subjected to joint impact of working and cooling, which helps in direct conversion of cream to butter; this method omits the butter grain formation stage.

3.7.3 EMULSIFICATION METHOD

In this method, liquid butter fat is emulsified with serum and the emulsion obtained is cooled and worked to form butter.

3.8 PACKAGING, STORAGE, AND DISTRIBUTION OF BUTTER

Packaging of the butter is done to prevent contamination, degradation, loss of weight during storage, and for identification of the content and to increase sales appeal. Butter should be packaged immediately after churning, otherwise a color defect is observed called *Primrose*, which occurs due to the evaporation of moisture from surface of butter. During the time lapse between churning and packaging, the butter should be stored in tempering room at 5°C covered with a wet muslin cloth. There are three methods of packaging: manual molding and wrapping, mechanical molding and wrapping, and fully automatic molding, patting, and wrapping.[11,15]

After molding and wrapping, before dispatching the butter it should be stored in cold room for 24 to 48 h at 4°C. This helps butter in obtaining a solid consistency, the change observed is called setting of butter. Later, as the temperature rises, it would not become as soft as it will be at the same temperature if it had not been chilled before. For a short time of storage, butter should be stored at 4°C, but for a long duration of storage it must be deep frozen to a temperature of −23°C. Best quality butter should be chosen for a long period storage as at such a low temperature salt crystallization may take place due to its low solubility at low temperature. Solubility of salt at 0°C is 35.7%. After thawing of butter, the salt is re-dissolved.[32]

During storage, evaporation of moisture from butter take place that leads to loss in weight of packed butter.[44] The causes of moisture evaporation are:

- High temperature storage.
- Low humidity of the storage room, however high humidity of the storage room is also avoided to avoid any fungal growth.
- Small pats of butter, having greater surface area per unit volume.
- Incomplete working and hence incomplete incorporation of moisture in butter.

The quality of butter is mainly attributed to its firmness and standing-up property of butter, which depends mainly on the temperature. During transportation of the butter after packaging and setting from manufacture to distributors to retailers, a cold chain should be maintained. Butter may be transported in two ways: chilled (1–4°C) or frozen (−16 to −18°C). Temperature deviation during transportation may take place which may lead to body defects. To prevent quality loss, butter must be flash frozen, rapid cooling causes development of small crystals, which does not impart any defect, on other side, slow cooling leads to large ice crystal formation which results in crumbly texture and hence leads to quality loss.[6,52]

3.9 QUALITY ATTRIBUTES OF BUTTER

3.9.1 JUDGING AND GRADING OF BUTTER

The inclusion of a grade mark label on butter permits seller and buyer of that butter to understand its properties and prevent dispute. There are several grades available for butter, depending upon the country of production. Body, flavor, color, and salt are the four key variables that are used to grade butter (Table 3.6). Several different scorecards for sensory assessment of butter have been created, with maximum scores assigned to various sensory qualities of butter. The below enlisted characteristics are used to assess grade of butter.[16,50]

3.9.1.1 FLAVOR CHARACTERISTICS

Acidity	Feed	Scorched
Agedness	Flat	Smothered
Bitterness	Malty	Storage
Coarseness	Musty	Utensil
Cooked	Neutralizer	Weed
Culture	Old cream	Whey

3.9.1.2 COLOR, SALT AND BODY ATTRIBUTES

Some of the common characteristics used in the assessment scale are cheese body, color, and amount of salt. For each attribute, there are several basic points that should be considered (Table 3.6).

3.9.1.3 PACKAGING

It should be sound attractive, neat, clean, and tidy, and have a good finish (smooth, attractive surface). Packages should be securely and neatly fastened.

TABLE 3.6 Body, Color, and Salt Attributes.

Body	Color	Salt
Crumbly	Mottled	Gritty
Gummy	Speckled	Sharp
Leaky	Streaked	
Mealy or Grainy	Wavy	
Ragged-boring		
Short		
Sticky		
Weak		

3.9.1.4 GRADES OF BUTTER

Different organizations have given grads and scorecards. The score attained after sensory assessment of butter using BIS-recommended card is used to grade butter (Tables 3.7 and 3.8).

TABLE 3.7 BIS Grading of Butter.

Grade	Quality	Score
A	Excellent	>90
B	Good	80–90
C	Fair	60–79
D	Poor	<60

TABLE 3.8 BIS Score Card and ADSA Score Card.

Score	Flavor	Body and texture	Color	Salt	Package	Total
BIS Score Card	50	30	15	Not prescribed	5	100
ADSA Score Card	45	30	10	10	5	100

3.9.1.5 USDA SCORE CARD

Butter grades in United States can be divided into three grades (Grade AA, Grade A, and Grade B) by USDA (1989), which can be sold commercially in the United States. It should comprise not less than 80% fat from milk, and grading will be based on four attributes, i.e., color, body flavor, and salt (Table 3.9). Grade score on packaging may be found.[8,28,47]

TABLE 3.9 USDA Score Card for Butter.

Grade	Score out of 100
US grade AA	93
US grade A	92
US grade B	90

TABLE 3.10 Characteristics for Different US Grades for Butter.

US grade AA	Fine and very pleasing butter flavor, as well as a very pleasing fragrance.
	Produced from low natural acid sweet cream.
	May or may not be culture added.
	A mild feed and a distinct cooked taste may be present.
US grade A	Possesses a pleasing and desirable butter flavor.
	May possess slight degree of aged, acid, coarse, bitter, flat, smothered and feed flavor to a definite degree.
	May possess slight degree of following in flavor: Acid, bitter, coarse, and flat.
US grade B	Pleasing flavor of butter, may have slight degree of malty, musty, neutralizer, scorched, utensil, weed and whey.
	Definite degree of acid, aged, bitter, smothered, and old cream.
	Feed flavor to pronounced degree.

3.9.2 *RHEOLOGY OF BUTTER*

Inversion of mechanical phase in cream results in production of butter in that O/W emulsion changes to W/O emulsion. Butter is a plastic material example that appears to produce or undergo permanent deformation after application of force. More precisely butter is continuous phase of fat which consist of water droplets along with network of fat crystals and fat globules are distributed. The spreadability, appearance, and mouthfeel of butter are all determined by the fat crystal network, which is closely linked to overall structure and butter composition. The solid-to-liquid fat ratio is crucial for rheological properties of butter and spreads: without solid fat, butter is completely liquid.[57] The butter would be hard and brittle if it does not have any liquid fat. Butter's textural properties are influenced by the presence of cream heat treatment, milk fat globules, and fluctuations of temperature during storage. The texture of butter is a key factor in assessing its usability and consumer acceptance. It influences taste, spreadability, mouthfeel, appearance, and acceptability for different applications.[45]

3.9.2.1 TESTS USED IN RHEOLOGY OF BUTTER

Most rheology tests of butter are empirical in nature, which are intended to mimic sensory perceptions for quality control purposes. These are dependents on main principles of extrusion, penetrometry, and sectility, and comprise of large deformations, which break down the material's structure.

- **Compression:** Butter is placed between two flat plates by which a uniform stress is applied to both upper and lower surfaces of butter. Compression tests are less complex to analyze than penetrometer. Compression samples should not be too long (to prevent bowing), nor too short.
- **Extrusion:** Extrusion devices have also been used to evaluate butter texture, albeit to a lesser extent than the other techniques. Fat sample is extruded through an orifice at a constant speed and the thrust (force required to sustain motion) is recorded.
- **Penetrometry:** Placing a steel rod just above a butter sample and loading it with weights until it rapidly penetrated the fat (specified length of time, or the rate at which it falls, is measured).
- **Sectility:** Involves forcing a stretched steel wire through a sample. The load required for the wire to cut through the sample at a constant rate can be used to provide a measure of firmness. Sectility correlates best with the sensory perception of hardness.
- **Small deformation testing of butter:** Butter exhibits viscoelastic behavior at small stresses and its viscoelasticity can be used for evaluating the relationships between stress, strain, and time, using small deformations.

3.9.3 DEFECTS IN BUTTER

Basic defects of butter can be classified into defects of color, consistency, structure, and flavor,[27] as shown in Table 3.11.

TABLE 3.11 Different Defects in Butter.

Flavor defect	Body & texture defect	Color defect
• **Enzymatic off flavors**	f. Crumbly, Brittle body	k. Bleached Butter
a. Heating-induced off-flavors	g. Greasy Texture	l. Green discoloration
b. Light-induced off-flavors	h. Gummy butter	• **Moldy butter**
c. Lipolysis induced	i. Leaky butter	m. Mottled
d. Microbial off-flavors	• **Mealy butter**	n. Pink color
e. Oxygen-induced off-flavors	j. Sticky Butter	o. White specks
• **Taint off-flavors**	• **Weak Body**	• **Yellow specks**

3.10 RECENT ADVANCES IN MANUFACTURING OF BUTTER

Global butter market is growing at a CAGR of 3.9%. Various advances have been observed in manufacturing process of butter. Recombined butter has better spreadability and has firmer texture than fresh butter. It is produced by mixing butter oil, or anhydrous milk fat, water, salt, skim milk powder, flavors, and colors. The ingredients are melted and mixed with continuous agitation and emulsifiers are heated and are mixed with liquid oil.[55] Butter can also be produced from ozone-treated cream. This helps in removal of bad odors from cream and also helps in production of microbiologically safe butter and also helps in development of taste and aroma in butter.[34,48]

Using ultrasonication butter separation was achieved from cream. This includes three stages, first breaking of cream, adherence of small globules to each other, and finally a big mass of butter is formed. This method is more efficient and hence can be used at a large commercial level to produce butter with minimal fat loss during production.[29,33] As a nonthermal processing method, high-pressure homogenization treated cream can also be used to produce butter for increasing its keeping quality.[46] A lot of scope of experimentation with the manufacturing process of butter is there, which may be explored in coming future.

3.11 SUMMARY

Butter is a fat concentrated dairy product made from churning cream and working granules to make a compact mass butter. Till mid of the nineteenth century, manufacturing of butter was done at household or farm or cottage level. After the development of cream separator and introduction of pasteurization and artificial freezing, commercial production of butter started around 1980. Prior to this period, mostly butter was manufactured in a batch process method. Later, continuous butter making method started to gain interest to increase the manufacturing efficiency during World War II. Despite of all the modernization and developments the basic principle used for the butter manufacturing is destabilization of emulsion of cream. The texture and body of the butter are the major concern for consumer acceptance. Proper working of butter granules, salting, and temperature maintenance during production, storage, and distribution eliminates the chance of defect and maintains the quality of the butter manufactured. There are many recent advances like ozone treatment, high pressure homogenization, etc., which are nonthermal techniques to produce microbiologically safe butter.

KEYWORDS

- **emulsification**
- **frothing**
- **neutralization**
- **overrun**
- **phase reversal**
- **ripening**
- **salting**
- **washing**

REFERENCES

1. Azouz, A. Nutritional Evaluation of Egyptian Children's School Meal, Part I: Fat Sources and Fatty Acids Profile. *Ann. Agric. Sci.* **2011,** *56* (2), 73–76.
2. Bird, E. W.; Derby, H. A. Chemistry of Butter and Butter Making, Part III: The Relationships Between Percentage Fat and Protein Percentage of Cream and the Churning Loss. Ames, Iowa: *Iowa Agriculture and Home Economics Experiment Station Research Bulletin*; 1937; *19* (May); article 1; p 9.
3. Bird, E. W.; Breazeale, D. F.; Bartle, E. R. Chemistry of Butter and Butter Making, Part IV: The Relationships Among the Cream Acidity, the Churning Loss, and the Churning Time. Ames, Iowa: *Iowa Agriculture and Home Economics Experiment Station Research Bulletin*; 1937, *20* (227, November); article 1; p 10.

4. Bodyfelt, F. W., Tobias, J.; Trout, G. M. Sensory Evaluation of Butter. In *The Sensory Evaluation of Dairy Products;* Chapter 9. Van Nostrand Reinhold: New York, NY, 1988; pp 376–417.
5. Bradley, R. L.; Smukowski, M. Butter. Chapter 6; In *The Sensory Evaluation of Dairy Products*; Clark, S.; Costello, M.; Drake, M.; Bodyfelt, F. Eds.; Springer Science & Business Media: Singapore, 2009; pp 135–165
6. Buldo, P.; Kirkensgaard, J. J.; Wiking, L. Crystallization Mechanisms in Cream During Ripening and Initial Butter Churning. *J. Dairy Sci.* **2013,** *96* (11), 6782–6791.
7. Büsser, S.; Jungbluth, N. The Role of Flexible Packaging in the Lifecycle of Coffee and Butter. *Int. J. Life Cycle Assess.* **2009,** *14* (1), 80–91.
8. Casas, J.; Valverde, M. T.; Marín-Iniesta, F.; Calvo, L. Inactivation of *Alicyclobacillus Acidoterrestris* Spores by High Pressure CO_2 in Apple Cream. *Int. J. Food Microbiol.* **2012,** *156* (1), 18–24.
9. CODEX. *Codex Standard for Butter*; Codex Standard CXS-279-1971; Revised 1–1999; Amended 2018; p 12.
10. Czank, C.; Simmer, K.; Hartmann, P. E. Method for Standardizing the Fat Content of Human Milk for Use in the Neonatal Intensive Care Unit. *Int. Breastfeed. J.* **2009,** *4* (1), 1–7
11. Farkye, N.Y.; ur-Rehman, S. Concentrated Fluid Milk Ingredients. Chapter 5; In *Dairy Ingredients for Food Processing*, Chandan R. C. Ed.; Blackwell Publishing Ltd.: London, 2011; pp 123–140.
12. Fearon, A. M. Butter and Butter Products. In *Dairy Ingredients for Food Processing*, Chandan R. C. Ed.; Blackwell Publishing Ltd.: London, 2011; pp 199–223.
13. Fenelon, M. A.; Hickey, R. M.; Buggy, A.; McCarthy, N.; Murphy, E. G. Whey Proteins in Infant Formula. In *Whey Proteins*; Academic Press: London, 2019; pp 439–494.
14. Fernandes, R. (Ed.). *Microbiology Handbook: Dairy Products*, 3rd ed; Royal Society of Chemistry: London, 2009; p 173.
15. Forster, T. L.; Browny, R. W. Comparative Studies in the Prevention of Wood Taint, Toppiness, and Primrose Color of Storage Butter. *Sci. Agric.* **1943,** *23* (6), 342–354.
16. Frede, E.; Buchheim, W. Butter Making and the Churning of Blended Fat Emulsions. *Int. J. Dairy Technol.* **1944,** *47* (1), 17–27.
17. FSSAI. *Food Safety and Standards: Food Products Standards and Food Additives, and Regulations*. Food Safety and Standards Authority of India (FSSAI), New Delhi, India: Ministry of Health and Family Welfare, Government of India; 2011; p 877.
18. Gassi, J. Y.; Blot, M.; Beaucher, E.; Robert, B.; Leconte, N.; Camier, B.; Gaucheron, F. Preparation and Characterization of a Milk Polar Lipids Enriched Ingredient from Fresh Industrial Liquid Butter Serum: Combination of Physicochemical Modifications and Technological Treatments. *Int. Dairy J.* **2016,** *52*, 26–34.
19. Gemechu, A. T.; Tola, Y. B. Traditional Butter and Ghee Production, Processing and Handling in Ethiopia: A Review. *Afr. J. Food Sci.* **2017,** *11* (4), 95–105.
20. Hemme, T.; Garcia, O.; Saha, A. (Eds.). *Review of Milk Production in India with Particular Emphasis on Small-Scale Producers*. FAO: Rome, 2003; p 61.
21. Hettinga, D.; Van Nguyen, L.; Shahidi, F. Butter. *Bailey's Indus. Oil Fat Prod.* **2005,** *2005*, 1–55.
22. Hillig, F.; Ahlmann, S. W.; & Jordan, R. C. Effect of Neutralization of Cream in the Manufacture of Butter on Water-Insoluble Fatty Acids and Butyric Acid. *J. Assoc. Off. Agric. Chem.* **1951,** *34*, 720–725.

23. Ip, C.; Banni, S.; Angioni, E.; Carta, G.; McGinley, J.; Thompson, H. J.; Barbano, D.; Bauman, D. Conjugated Linoleic Acid–Enriched Butter Fat Alters Mammary Gland Morphogenesis and Reduces Cancer Risk in Rats. *J. Nutr.* **1999,** *129* (12), 2135–2142.
24. Ip, S. Y.; Phillips, T. L.; Cox, D. C. Economic Aspects of Continuous Butter making: Cost Comparison Between Batch and Continuous Processes. *Aust. J. Dairy Technol.* **1969,** *24* (4), 168–172.
25. Jahurul, M. H. A.; Zaidul, I. S. M.; Norulaini, N. A. N.; Sahena, F.; Jinap, S.; Azmir, J.; Sharif, K. M.; Omar, A. M. Cocoa Butter Fats and Possibilities of Substitution in Food Products Concerning Cocoa Varieties, Alternative Sources, Extraction Methods, Composition, and Characteristics. *J. Food Eng.* **2013,** *117* (4), 467–476.
26. Johnson, M. E.; Kapoor, R.; McMahon, D. J.; McCoy, D. R.; Narasimmon, R. G. Reduction of Sodium and Fat Levels in Natural and Processed Cheeses: Scientific and Technological Aspects. *Compr. Rev. Food Sci. Food Saf.* **2009,** *8* (3), 252–268.
27. Krause, A. J.; R. E.; Miracle, T. H.; Sanders, L. L.; Dean, M. A. The Effect of Refrigerated and Frozen Storage on Butter Flavor and Texture. *J. Dairy Sci.* **2008,** *91* (2), 455–465.
28. Lee, C. L.; Liao, H. L.; Lee, W. C.; Hsu, C. K.; Hsueh, F. C.; Pan, J. Q.; Chen, M. J. Standards and Labeling of Milk Fat and Spread Products in Different Countries. *J. Food Drug Anal.* **2018,** *26* (2), 469–480.
29. Lee, J.; Martini, S. Modifying the Physical Properties of Butter using High-Intensity Ultrasound. *J. Dairy Sci.* **2019,** *102* (3), 1918–1926.
30. Li, M.; Li, Y.; Wang, R.; Wang, Y.; Li, Y.; Zhang, L. Effects of Triglycerol Monostearate on Physical Properties of Recombined Dairy Cream. *Int. Dairy J.* **2020,** *103*, e104622.
31. Long, H. F.; Hammer, B. W. *Classification of the Organism in Dairy Products, Part III: Pseudomonas Putrefaciens*. Research Bulletin 285; State College of Agriculture Experiment Station and Mechanic Arts, Ames - Iowa, 1941 (January); p 23.
32. Lozano, P. R.; Miracle, E. R.; Krause, A. J.; Drake, M.; Cadwallader, K. R. Effect of Cold Storage and Packaging Material on the Major Aroma Components of Sweet Cream Butter. *J. Agric. Food Chem.* **2007,** *55* (19), 7840–7846.
33. Maheshwari, B.; Saravanathamizhan, R.; Balasubramanian, N. Butter Separation from Cream Using Ultrasonication: Optimization of Parameters Using RSM. *J. Food Biosci. Technol.* **2018,** *8* (2), 1–10.
34. Mallia, S.; Felix, E.; Hedwig, S. Aroma-active Compounds of Butter - Review. *Eur. Food Res. Technol.* **2008,** *226*, 315–325.
35. Minifie, B. W. Science and Technology of Chocolate and Confectionery. In *Chocolate, Cocoa and Confectionery: Science and Technology*; Springer: Dordrecht, 1989; pp 639–674.
36. Morin, P.; Britten, M.; Jiménez-Flores, R.; Pouliot, Y. Microfiltration of Buttermilk and Washed Cream Buttermilk for Concentration of Milk Fat Globule Membrane Components. *J. Dairy Sci.* **2007,** *90* (5), 2132–2140.
37. Mortensen, B. K. Physical Properties and Modification of Milk Fat. In *Developments in Dairy Chemistry*; Springer: Dordrecht; 1983, pp 159–194.
38. Mortensen, M. Material-Balance Method for Determining Losses of Butterfat in the Creamery. *Iowa Agric. Home Econ. Exp. Station Res. Bull.* **1945,** *27* (337), 1–10.
39. Panchal, B.; Truong, T.; Prakash, S.; Bansal, N.; Bhandari, B. Influence of Fat Globule Size, Emulsifiers, and Cream-Aging on Microstructure and Physical Properties of Butter. *Int. Dairy J.* **2021,** *117*, e105003.

40. Patel, A. A.; Sharma, P.; Patel, H. Butter and Fat Spreads: Manufacture and Quality Assurance. Chapter 11; In *Dairy Processing and Quality Assurance*; Chandan R. C. Ed.; Blackwell Publishing Ltd.: London, 2015; pp 266–286.
41. Pittia, P.; Antonello, P. Safety by Control of Water Activity: Drying, Smoking, and Salt or Sugar Addition. In *Regulating Safety of Traditional and Ethnic Foods*; Academic Press: London, 2016; p 7–28.
42. Pont, E. G. The Washing of Butter and its Effect on Curd Content and Quality. *J. Counc. Sci. Indus. Res. Australia.* **1946,** *19*, 432–437.
43. Rios, R. V.; Pessanha, M. D.; Almeida, P. F.; Viana, C. L.; Lannes, S. C. Application of Fats in Some Food Products. *Food Sci. Technol.* **2014,** *34* (1), 3–15.
44. Rønholt, S.; Kirkensgaard, J. J. K.; Mortensen, K.; Knudsen, J. C. Effect of Cream Cooling Rate and Water Content on Butter Microstructure during Four Weeks of Storage. *Food Hydrocoll.* **2014,** *34*, 169–176.
45. Rønholt, S.; Kirkensgaard, J. J. K.; Pedersen, T. B.; Mortensen, K.; Knudsen, J. C. Polymorphism, Microstructure and Rheology of Butter. Effects of Cream Heat Treatment. *Food Chem.* **2012,** *135* (3), 1730–1739.
46. Sert, D.; Mercan, E. Microbiological, Physicochemical, Textural Characteristics and Oxidative Stability of Butter Produced from High-Pressure Homogenization Treated Cream at Different Pressures. *Int. Dairy J.* **2020,** *111*, e104825.
47. Sert, D.; Mercan, E. Characterization of Physicochemical, Microbiological, Thermal, Oxidation Properties and Fatty Acid Composition of Butter Produced from Thermosonicated Cream. *Int. Dairy J.* **2020,** *109*, e104777.
48. Sert, D.; Mercan, E.; Kara, Ü. Butter Production from Ozone-Treated Cream: Effects on Characteristics of Physicochemical, Microbiological, Thermal and Oxidative Stability. *Lebensm. Wissens. Technol.* **2020,** *131*, e109722.
49. Shen, C. L.; Kaur, G.; Wanders, D.; Sharma, S.; Tomison, M. D.; Ramalingam, L.; Dufour, J. M. Annatto-extracted Tocotrienols Improve Glucose Homeostasis and Bone Properties in High-Fat Diet-Induced Type-2 Diabetic Mice by Decreasing the Inflammatory Response. *Sci. Rep.* **2018,** *8* (1), 1–10.
50. Sistrunk, W. A. Peach Quality Assessment: Fresh and Processed. In *Evaluation of Quality of Fruits and Vegetables*; Springer: Singapore, 1985; pp 1–46.
51. Tola, A.; Bekele, Y. Traditional butter and ghee production, processing and handling in Ethiopia: A review. *Afr. J. Food Sci.* **2017,** *11* (4), 95–105
52. Truong, T.; Lopez, C.; Bhandari, B.; Prakash, S. (Eds.) *Dairy Fat Products and Functionality*; Springer International Publishing: Singapore, 2020; p 583.
53. Van Aken, G. A. Aeration of Emulsions by Whipping, *Colloids Surf. A: Physicochem. Eng. Asp.* **2001,** *190* (3), 333–354.
54. Varnam, A. H.; Sutherland, J. P. Butter, Margarine, and spreads. In *Milk and Milk Products*; Springer: Singapore, **1994**; pp 224–274.
55. Weeragul, K.; Pandolsook, S. Effects of Manufacturing Conditions on Physical Characteristics of Recombined Butter and Quality of Butter and Sponge Cakes. *J. Culin. Sci. Technol.* **2020,** *2020*, 1–16.
56. Wiseman, D. W.; Marth, E. H. Stability of Aflatoxin M1 During Manufacture and Storage of a Butter-like Spread, Non-fat Dried Milk and Dried Buttermilk. *J. Food Prot.* **1983,** *46* (7), 633–636.
57. Wright, A. J., Scanlon, M. G., Hartel, R. W., Marangoni, A. G. Rheological Properties of Milkfat and Butter. *J. Food Sci.* **2001,** *66* (8), 1056–1071.

58. Wright, P. G. (Ed.) *Tariff on Animal and Vegetable Oils, Institute of Economics*. Macmillan Company: New York, 1928; p 86.
59. Yeomans, M. R.; Chambers, L.; Blumenthal, H.; Blake, A. The Role of Expectancy in Sensory and Hedonic Evaluation: The Case of Smoked Salmon Ice-Cream. *Food Qual. Prefer.* **2008,** *19* (6), 565–573.
60. Yildiz, S.; Pokhrel, P. R.; Unluturk, S.; Barbosa, G. V. Identification of Equivalent Processing Conditions for Pasteurization of Strawberry Juice by High Pressure, Ultrasound, and Pulsed Electric Fields Processing. *Innov. Food Sci. Emerg. Technol.* **2019,** *57*, e102195.
61. Yusuf, A. O. Quality and Storage Stability of Yoghurt Produced from Pigeon Pea Milk Supplemented with *Propionibacterium Freudenreichii*. Doctoral dissertation; Durban University of Technology, Kwa Zulu Natal, South Africa, 2017; p 216.

CHAPTER 4

CHEMISTRY AND DIFFERENT ASPECTS OF ICE CREAM

NANDITA DAS and ANKITA HOODA

ABSTRACT

Ice cream is an emulsified frozen dairy product, in which milk fat globules are distributed all over the mixture of protein, sugar, water, and ice crystals, along with air bubbles. Based on its composition, several types of ice creams are available in the market, such as premium, super-premium, standard and economy ice creams. Also, there are various groups of ice creams based on production practices, such as plain ice cream, fruit ice cream, chocolate ice cream, nut ice cream, cassata, mousse, puddings, variegated ice cream, rainbow ice cream, fanciful ice cream. Moreover because of so many varieties available, composition of ice cream is dependent on legal requirements, trade demand, and type and quality of the product. The basic composition of ice cream contains milk fat, milk solids not fat, sweeteners, emulsifiers, and stabilizers. Additionally, color, flavor, eggs, fruits, and dried fruits can be used as desired. Also the ice-cream manufacturing involves ingredients selection followed by formulation, blending, pasteurization, homogenization, cooling, aging, freezing, packaging, and hardening of mix, which finally gives the desired texture of ice cream; and it is a stable four phase system: fat, ice crystal, and air suspended in serum phase. The physicochemical changes that are observed during storage majorly affect flavor (acid and storage flavor), appearance (lack of uniformity), melting quality (wheying—off), and texture (coarse, sandy, and gritty) of ice cream. Novel techniques in ice cream include introducing functional ingredients into the mix, such as bioactive

The Chemistry of Milk and Milk Products: Physicochemical Properties, Therapeutic Characteristics, and Processing Methods. Megh R. Goyal, Suvartan Ranvir, & Junaid Ahmad Malik (Eds.)

compounds, encapsulated essential oils, polyphenols, curcumin; and whey proteins and developing sugar, lactose or fat free ice cream.

4.1 INTRODUCTION

Ice cream is a frozen milk product that is obtained by agitating and freezing standardized pasteurized ice-cream mixture of dairy ingredients, sugars, and flavors. Ice cream is said to be an iced beverage and water ice, which was very famous during the medieval period.[6,18,33] Selected biblical references indicate that *kings Alexander—the Great and King Solomon enjoyed iced drinks and desserts*. Marco Polo returned to Italy from the Far East, with a recipe closely similar to one now referred to as sherbet. In the 16th century, this recipe has been evolved into ice cream. Ice cream was discovered in England around same period. The Chinese and Romans were the first people to report enjoying flavored ice desserts. These desserts were very popular among royal people of France in the 1500s. The process of making ice cream evolved through various stages in the next few decades.[18,33]

Ice cream includes broad range of frozen desserts comprising dairy ice cream, nondairy ice cream, frozen yogurt, milk ice, sorbet, sherbet, water ice, fruit ice. Ice cream is defined differently in different countries.[2]

According to FSSAI, ice cream is defined as "the frozen milk product obtained by freezing a pasteurized mix prepared from milk or other products derived from milk, or both, with or without addition of nutritive sweeteners and other permitted nondairy ingredients. The said product may or may not contain incorporated air and shall be frozen hard except in case of softy ice cream which is of soft consistency".[25]

This chapter focuses on different aspects of ice creams, such as chemistry, composition, classification, constituents, physicochemical properties, manufacturing processes, and defects. The chapter also includes discussions on recent research advances.

4.2 COMPOSITION OF ICE CREAM

Ice cream is a nutritious, well-balanced, comfortably digestible, and delightful dessert. Since ice cream is a nutrient-dense food, it is commonly served as a meal portion to hospital patients. Carbohydrate content in ice cream is about four times higher than the milk. Some of the factors to be considered in developing a formulation of an acceptable composition of ice cream are: customer demand for flavor, body and texture, and color characteristics of the finished product, i.e., natural or fortified flavor with artificial flavoring; chewy to heavy, higher overrun; or more cooling body and texture characteristics.[18,33,65]

Table 4.1 provides a comprehensive composition of ice cream. Ice cream consists of 15% more protein than milk and 3 to 4 times as much fat as milk. Ice cream also contains other ingredients, such as fruits, nuts, eggs, dry fruits, and sugar, which boost its nutritive value. Milk solids in ice cream are generally treated at quite high temperature as compared to pasteurized temperature of milk. Because of its increased fat content, it is an outstanding source of energy and 50% of its contents are sugar (sucrose, lactose, and corn syrup solid). Although these substances are almost fully assimilated, yet it makes ice cream suitable for children or people, who are required to put on weight. For the same purpose, its managed use makes its way into the diet of individuals, who have to choose to lose weight. Ice cream is not a significant source of trace elements and iron.[39] Table 4.2 indicates nutritional value of commercial ice cream.

TABLE 4.1 Composition of Plain Ice Cream (per 100 g of Edible Portion).

Components	Good/Average ice cream	For different fat (%)			Water ice	Ice-cream cone	Ice milk
		10%	12%	16%			
Food energy	196.7	193.0	207.0	222.0	78.0	377.0	152.0
Water	61.7	63.2	62.1	62.8	66.9	8.9	66.7
Fat	12.0	10.6	12.5	16.1	-	2.4	5.1
Protein	4.1	4.5	4.0	2.6	0.4	10.0	4.8
Total carbohydrate	20.7	20.8	20.6	18.0	32.6	77.9	22.4
Wt. (g)/100 cal. portion	50.8	51.7	48.3	45.0	128.4	26.5	65.6

Source: From Refs. [19, 63].

4.3 CLASSIFICATION OF ICE CREAM

Ice cream is classified into different types, which is based on the ingredients used for its manufacture. Selected types are mentioned[18,33,41] as follows:

- **Bisque** contains particles of bakery products with appropriate flavoring agent. For example, sponge cake.
- **Confection** ice cream contains particles, such as, butter crunch and chocolate chip with appropriate flavoring component.

TABLE 4.2 Nutritive Value (per 100 g of product) of Commercial Ice-cream and Related Products.

Product	Fat (g)	Carbohydrate (g)	Protein (g)	Total solids (g)	Calories
Chocolate ice cream	13.1	25.8	3.6	42.1	221
Diabetic ice cream	9.0	20.6	4.3	28.6	152
Ice milk	4.0	20.5	4.1	33.5	144
Low-calorie ice milk	2.0	23.5	4.2	28.9	91
Strawberry ice cream	8	21.2	3.6	40.8	194
Vanilla ice cream	12	20.7	4.0	38.3	204

- **Fancy molded ice cream** has combination of one or more color and flavor. For example, brick ice cream containing three layers in which the middle layer is of fruits and the rest two are of ice creams.
- **Frappe** is an ice drink having slushy consistency prepared by mixing of fruit juices.
- **Frozen custard** ice cream has one or more types of egg ingredients in such a way that egg yolk solids are less than 1.4% by weight of the finished frozen custard.
- **Fruit ice cream** comes with or without addition of flavoring (strawberry, apricot, apple, mango, banana, grape, or pineapple) or coloring material. The fruits added may be fresh, frozen, preserved, or canned. It contains comparatively higher percentage of sugar as compared to plain ice cream. It should be drawn at 1°F cooler from freezer.
- **Fruit salad** has large pieces of fruit that are blended with a frozen mix of mayonnaise and whipped cream that is frozen and is served as salad.
- **Granite:** When ice is frozen with very little agitation, it is called granite.
- **Ice milk** is similar to ice cream, but it contains <10% fat.
- **Mousse** is a frozen ice cream without agitation containing whipped cream, cream, sugar, and flavor.
- **Neapolitan ice cream** in a single package contains two or more distinct flavors; usually it is seen in brick ice creams.
- **Novelties** ice cream in a low-priced package contains individual serving, which is convenient for eating and is of special shape. For example, Eskimo pies (chocolate coated ice-cream bars), candy or chocolate coated ice-cream bars with or without sticks, fudge.
- **Nut ice cream** contains nuts, such as pistachio, almonds, walnuts, ground nuts, walnuts, with or without addition of flavor or color.

- **Plain ice cream** is one of most basic type of ice creams. In this total amount, coloring and flavoring agent comprises less than 5% of volume of unfrozen ice-cream mix. Also, only one flavor is used for its preparation, such as vanilla, chocolate, butterscotch, or strawberry.
- **Pudding** has a large quantity of nuts, fruits, raisins, with or without addition of spices, egg, or liquor.
- **Rainbow ice cream** from freezer is provided with six or more colors to give it a look of rainbow, when it is hardened.
- **Sherbet:** Contribution of milk solids is low in this ice cream.
- **Soft serve ice creams** are drawn directly from freezer without hardening.
- **Souffle:** Whole egg or egg yolk containing sherbet is called souffle.
- **Variegated ice cream:** This is similar to plain vanilla ice cream prepared in combination with chocolate syrup or butterscotch to provide a marbling effect in the finished ice-cream product

4.4 CONSTITUENTS OF ICE CREAM

Various ingredients for the manufacturing of ice cream are either dairy or nondairy origin. The constituents used in ice-cream mix play their specific role in the texture of ice cream.[18,19,33]

4.4.1 CONSTITUENTS OF DAIRY ORIGIN

- **Milk fats** are chief components that impart rich flavor to the ice cream.[40] It plays an important role to give structure to ice cream through partial coalescence and foam stabilization and also ensures desirable melting characteristics in the resultant product. It acts as a synergist and also acts as a good carrier of added flavor. The drawbacks of milk fat as a constituent in ice cream are: (1) it retards whipping capacity; and (2) it increases caloric value of the ice cream, which may lead to limit its consumption. Butter, cream, condensed milk and butter oil are major sources of milk fat.[47,51]
- **Milk solids-not-fat (MSNF):** For maintenance of a proper ice-cream mix balance, MSNF should vary inversely with fat content. It helps in providing texture, body, and storage properties to ice cream. It also improves palatability of ice cream. It increases food value and is economical. The major dairy origin of MSNF can be summarized

as skim milk powder, skim milk, condensed skim milk, buttermilk, caseinate, WPC, low lactose milk powder, etc. Skim milk provides sugar, mineral, and protein, which prevents weak body and coarse texture and makes ice cream more compact and smoother.[6]

4.4.2 NONDAIRY CONSTITUENTS

- **Air:** Mix is incorporated with air during the process of freezing. The internal structure as well as yield of the product depends on the quantity and form of the air present in product and hence it influences the profits.[35] Amount and quality of air influences quality of finished product and hence should meet the legal standards.[9]
- **Color:** All flavored ice creams are colored accordingly. Pale yellow color or off-white color is used for vanilla flavored ice cream. Chocolate ice creams do not need any external coloring agent. The other colors, which may be used are yellow, pink, green, etc.[18,33,46]
- **Egg-yolk solids:** In some ice creams, egg is also used. Egg-yolk solids increase the food value and help in improving the body and texture, but it also increases the cost of the product. The egg-yolk solids are responsible for increasing the whipping ability and the viscosity too. It has minimal effect on freezing point.[60,62]
- **Emulsifier** is present at the interface of two different phases (fat/plasma and fat/air) and reduces the surface tension of the system. It provides a uniform body to the mix and results into a smooth stiff-finished product with a decreased rate of melting in the end product. It improves the whipping properties by reduction of the time required for whipping and by production as well as uniform spreading of small air-cells; this leads to formation of the internal foam like structure of ice cream resulting in production of smoother ice cream. It also promotes superior drawing qualities and imparts excellent melt quality to the product. Excessive amount of emulsifier leads to fluffy and short body characteristic, which is undesirable. Monoglycerides and diglycerides improve whipping ability and fat dispersion; it has reasonable impact on melting property and stiffness of ice cream, whereas polyderivatives have more positive effect on stiffness and rate of melting of product. Ice-cream mix contains excess amount of protein because emulsifiers are not required for stabilization of fat. When emulsifiers are not present, the fat globules are covered with casein and whey protein after homogenization to create a new fat

globule membrane surrounding the fat globule.[17,18,33,65] However, if low molecular weight surfactants are used, then they form the new membrane to help in increasing the foam stability. During freezing, the low molecular weight surfactants formed membrane become weaker and due to the reduction in steric stabilization of the fat it gets adsorbed at the interface between air and plasma to help in the foam stability of ice cream.[30]

- **Flavor:** A wide range of flavors are used in ice cream, such as, vanilla, chocolate, mango, strawberry, lemon, banana, orange, pineapple. Vanilla flavor is derived from Vanillin, which is produced during processing of pods of the Mexican species, flat-leaved vanilla, *V. planifolia.*[34] After harvesting, pods undergo curing, fermentation, and sweating leading to the production of vanillin from the broken-down products from glucosides. This process takes about 4 weeks to 4 months, which ends with complete drying of the beans to prevent any mold or fungal growth.[49,50,56]

  Chocolate flavor is obtained from cocoa, which is derived from cacao beans. Cacao beans are harvested and are gathered in piles to ripen, which undergo fermentation for around ten days. The beans are then washed and the pulp is removed, the obtained seeds are then sorted, roasted, cooled and separated to obtain nibs. Nibs are then grounded, which leads to production of chocolate liquor due to the production of heat during grinding. Under high pressure, the chocolate liquor is segmented in two parts: cocoa butter comprising of major portion of fat (~40% fat) and dry cake (~20% fat); and the latter is further milled and sifted to obtain cocoa powder.[15,24,56]

- **Stabilizers** are water-soluble biopolymers that are used in ice cream in very low amounts. On the basis of their source, they can be classified as: animal origin (gelatin), plant origin (sodium alginate, carrageenan, agar, locust bean gum, pectin, guar gum, sodium carboxymethyl cellulose), and microbial origin (xanthan).[11,61] The most important function of stabilizers is to decrease free-water in the mix either by binding the water of hydration or immobilizing it inside the gel structure resulting rise in viscosity of mix. Hence, it has a crucial role in prevention of formation of large ice crystals in case of temperature fluctuations in retail cabinet to help in preventing the coarsening of the texture and providing the smooth texture.[12,27]

  These have high water holding capacity to provide a smooth finished body to the final ice cream and to impart desirable resistance to melting. Hence, it reduces the migration of water from the product to the package. It also plays role in preventing shrinkage of product volume during storage and transport. This is used in minute quantities to reduce its impact on flavor and food value of finished product to the minimum, it limits the whipping ability.[8]

Sodium alginate provides immediate stabilization and hence do not need any aging, whereas gelatin gives a thin mix and requires an aging period for its effect. Gelatin does not only form gel during aging but also during freezing (incorporation of air) and hardening. CMC produces chewy characteristic in finished product.[28,48]

- **Sweetener solids:** Sugar is a vital ingredient as it not only imparts sweetness to ice cream, but also enhances the pleasing creamy flavor. The sweetener solids are cheapest source of total solids and it improves ice-cream texture and body. Sweetening agents are cane sugar, beet sugar, corn sugar, corn syrup, invert sugar, saccharin, etc. Corn syrup can also be used to provide a firm and smooth body, to reduce heat shock effect, and to increase shelf-life. The freezing point of mix is mostly affected. The effect is correlated with the sweetener's concentration and molecular weight. Molecular weight of the sweetener and the freezing point are inversely related to each other. Therefore, the sweeteners affecting the freezing point can be arranged in the order: fructose ~ dextrose > sucrose > corn syrup solid.[59]

  The sucrose provides 100% relative sweetness. The sweetness of the sweeteners is related in the order: fructose > invert sugar > sucrose > glucose > galactose > maltose > lactose. In the final product, two sugars (lactose and sucrose) are super-saturated; and glassy state in crystals is formed. The addition of sugar amount in the ice cream depends on various factors, such as: desired concentration of sugar in final product, total solids of mix, and desired viscosity, whipping ability, and the properties of the mix. Higher concentration of sweetener leads to overshadow desirable flavor and low concentration results in flat taste of the product.[43]

4.5 PROCESSING METHODS FOR ICE CREAM

4.5.1 SELECTION OF INGREDIENTS, FORMULATION AND BLENDING OF MIX

Good quality raw material should be used for the manufacturing of ice cream to meet the legal requirements and customer expectations. Selection of the ingredients is based on fat%, TS%, and the type of ice cream. A properly balanced mix is the one, which has a correct TS to water ratio, balanced fat and SNF content, and a correct fat to sugar ratio. The ingredients of the mix should be mixed in such a way to meet the TS of 36–40% to result in organoleptically palatable ice cream. The relationship between fat and SNF should be inversely related. Ice cream needs to have high fat, low SNF and vice versa. Also to have a sugar to fat balance, the ice creams having high fat% should contain more sugar to prevent fatty mouth-feel.

For the calculation of the amount of ingredients used for the formulation of mix various methods can be used, such as: Algebraic method, Pearson Square method, Serum point method, Formula tables, or formulations developed by computer programs. The ingredients supposed to be utilized for formulation of ice-cream mix are divided into two parts: liquid ingredients and dry ingredients. The liquid ingredients are mixed in a vat, which is heated to a temperature of around 40–50°C for proper blending of the dry ingredients, which will be added to it gradually and slowly.[9,13] The ingredients used as the source of milk fat (such as: butter or cream) should be properly molten down and mixed with the mix prior to pasteurization. If sodium alginate is added as a stabilizer, then it is mixed with sugar and added to mix; whereas if gelatin is used, it is mixed with water and then is added to mix.[18,33]

4.5.2 HOMOGENIZATION AND PASTEURIZATION OF MIX

Homogenization of the mix is done to avoid separation of phases of the mix, as it decreases fat globules size to <2 μm.[64] It helps in providing a smooth texture. Homogenization also helps in the improvement of whipping property of the mix, shortens the aging period, prevents churning during freezing, and helps in reducing the quantity of the stabilizer to be utilized. Homogenization of ice-cream mix is generally done at temperature range of 65–75°C at a pressure of 2000–2500 psi (or by two stage homogenization process of pressure 2500–3000 psi followed by 500 psi).[38]

Pasteurization of mix is generally done at 68°C for 30 minutes (LTLT-Batch method), 80°C for 25s (HTST), 90°C for 1 to 3s (Vacreation), and 99–130°C for 0–40s (UHT). Relatively high heat is used for the pasteurization of mix to achieve greater bacterial kill in high TS environment. The high heat treatment facilitates uniform blending of ingredients in the mix. It also helps in provision of better texture and body of product to reduce the concentration of stabilizer in the mix, and it produces a uniform product and increases shelf-life of final product.[52]

4.5.3 COOLING AND AGING

After the completion of pasteurization stage of ice-cream mix, the mix is cooled down to about at 4°C temperature through plate heat exchanger. This helps in arresting the growth of bacteria in the pasteurized mix and also helps in preventing the mix to become highly viscous.[42,45]

The mix is then subjected for aging at 4°C for around 24 h. This is an optional step depending on the stabilizer used. If sodium alginate is used as a stabilizer, there is no need of aging of the mix, whereas if any other stabilizer is used, aging is a mandatory step to facilitate the action of the stabilizer. Due to the aging at low temperature, the fat solidifies (crystallizes) and swelling of the proteins and stabilizer takes place due to hydration to increase the viscosity of mix. Rearrangement of the fat membrane material takes place during this stage due to competitive replacement of protein from the fat surface by the low molecular weight surfactants. Longer period of aging should be avoided to prevent psychotropic microbial growth in the mix.[10]

4.5.4 FREEZING AND PACKAGING

Freezing of the mix is one of most prominent stage during manufacture of ice cream as it determines quality, yield and palatability of finished product. The key function of freezing is to reduce temperature of mix, to freeze free water of mix and to incorporation of air in the mix. Rapid agitation during freezing hastens the air incorporation and destroys gel-structure by disrupting fat globule clusters. Freezing may cause destabilization of ice-cream mix as it can affect fat emulsion and protein colloidal suspension.[55] The precipitating effect on the protein can be attributed to:

- Conversion of free water into ice crystals, which may lead to dehydration of protein and eventually destabilization of the system.
- Salting out effect on the protein due to increased concentration of salts in the remaining unfrozen water.
- Increased hydrogen ion concentration due to the precipitation of salts (di- and tri-calcium phosphate).

If the mix is quite stable, freezing does not cause any destabilization of the mix, but if mix is unstable (because of high acidity or salt balance) then the added effect of freezing may lead to precipitation. Hence, freezing should be rapid to minimize its destabilizing effects.[18,33]

Ice-cream mix prior to freezing contains around 60% water. During the freezing, water is frozen and ice crystals formation starts along with the continuous incorporation of air in the form of air-cells in the mix. The drawing temperature determines the amount of water frozen during freezing. When the drawing temperature of the mix is −5.6°C, around 50% of water of mix is frozen, which can be increased to 67% if the drawing temperature is decreased to −9.4°C. Temperature difference of even 1°C during drawing of ice cream from freezer can cause 1.9% of the ice being melted. These types of fluctuations in the temperature during freezing may lead to defects, such as iciness, crystallization. The freezing of the water and formation of resultant crystals cause increase in viscosity of the mix.[31]

Rapid freezing of ice-cream mix is desirable as compared to slow freezing; as the size of the crystals depend on rate of freezing. Slow freezing leads to formation of fine and uniform crystals smaller in size, which cannot be detected using tongue, whereas the crystals formed during slow freezing are larger in size, which can be detected by tongue, and is undesirable.

Therefore, it is desired that ice cream is freezed and drawn as quickly as possible from the freezer. It is desired to freeze ice cream as stiff as possible, but also flowy enough to draw it out from freezers. Freezing temperature and time are well-affected by type of freezer used for purpose.[21,53] There are two types of freezers, which are used for this purpose: Continuous freezer; and Batch freezer. Soft serve freezer is a freezer that is used for the drawing of soft-serve ice cream. which is consumed without hardening.[42]

Ice cream drawn from freezer is generally filled into cups or cones, molded into bars, or extruded into various shapes. Extruded ice cream is usually more viscous, stiffer, colder, and creamier. The filled molds are passed through brine solution of −40°C, which completely freezes the ice cream, before inserting the stick into molds. The molded ice-cream bars are then extracted from the molds by passing these through warm brine solution. Then, the bars are wrapped and moved to cold storage.[41]

4.5.5 HARDENING

After drawing the semisolid consistent ice cream, it is filled into containers or packages, which are then subjected to hardening, where the freezing of the ice cream continues exclusive of agitation till the temperature of core of the ice cream is achieved to 0°F. After drawing ice cream from the freezer, it should be transferred to hardening room as promptly as possible to avoid melting of the ice leading to coarse texture of ice cream.[18,33] The various factors affect hardening time, such as:

- Circulation of air in the cabinet—Forced circulated air requires less time compared to still air. The time may vary up to 60% depending on the type of air used.
- Dimension of the package—Larger is the size of the package, more is the time required to lower the temperature at the core of ice cream, and hence longer is the hardening time.
- Overrun—Increase in overrun will increase the hardening time.
- Temperature of air used—The temperature of the air generally in use ranges from −23 to −30°C. Lower is the temperature of air used, less is the time required for hardening.
- Temperature of ice cream—The temperature, when ice cream is drawn, also affects hardening time. Ice cream drawn at 1°F higher temperature will add to the hardening time by ~10–15%.

Soft serve ice cream is consumed after drawing them from freezer directly without the hardening process, which gives it a smooth, soft and fluffy body. Its overrun is around 30–50% and contains less sugar than the regular ice cream. Hardening can be done in the hardening room, hardening tunnel, or hardening cabinet.[6,18]

4.6 PHYSICOCHEMICAL PROPERTIES OF ICE CREAM

4.6.1 ICE-CREAM MIX

Ice-cream mix is an oil in water type emulsion, where the dispersed phase is the milk fat suspended in a continuous phase of aqueous serum consisting of protein, carbohydrate, minerals, salts, and sugars. All the constituents of the mix are present in such a way that it provides stability to the mix.[1,14]

Mix stability refers to the ability of a mixture to prevent the segregation of milk proteins from colloidal suspension or milk fat from an emulsion. Protein and fat are uniformly distributed in the mix, the instability of the mix leads to the curdling of the mix, whey separation, fat separation, or syrup separation during storage, aging, transportation, etc. The stability of the mix particle reduces in the following manner depending on the property of the particle: charged and hydrated > charged but not hydrated > not charged but hydrated > neither hydrated nor charged; indicating that the charged and hydrated particle will be stable in the mix compared to the least stable particle, which is neither hydrated nor charged.

Generally, alcohol coagulation test is conducted to determine the stability of the mix. The colloidal substances are more stable at lower temperature due to their high hydration at low temperature. On the other hand, salts (such as: calcium) decreases the hydration of the protein. Maximum hydration can be achieved at pH of 6.2; and homogenization also increases the hydration.[6]

The average pH of ice-cream mix is normally 6.3. High acidity of the mix may lead to excessive viscosity, decrease in whipping rate, and it may even lead to coagulation of the colloidal part of the mix, which will eventually lead to less stable mix.[16,20]

The viscosity is most important property of the ice-cream mix. The viscosity of mix affects whipping ability and retention of air in the structure of ice cream. Resultant higher viscosity reduced whipping ability. As viscosity of the ice cream increases, so does its resistance to melting also increases that imparts a smooth and stiff body to the final product. The apparent viscosity of a mix is the viscosity that disappears with agitation, whereas the basic viscosity remains even after agitation of mix. The ice-cream mix's basic viscosity varies from 50cP to 300 cP.[18,33]

The freezing point of ice cream mix is between −3.6 and −2.4°C. Presence of sugar in ice-cream mix causes depression of freezing point. After formation of ice crystals, the lowering of freezing point was observed because of concentration of soluble constituents.[5,8]

4.6.2 STRUCTURE OF ICE CREAM

Complex colloidal system of the ice cream can be divided into three discrete phases: partially coalesced fat globule, air bubbles bounded by their adsorbed interfacial material, and the ice crystals bordered by concentrated aqueous serum consisting of soluble sugar, salts, protein, polysaccharides, etc. Size of air bubble generally ranges from 20 to 50 μm.

Protein in ice cream has a major role to determine its structure as it helps in stabilization, emulsification (fat structure formation), foam formation, and stability. In the mix, protein is concentrated at the interface of the fat and plasma, which is then replaced by low molecular weight surfactants during aging. Protein is also found at the air and plasma interface, which helps in formation and stabilization of air cells. The proteins in the concentrated aqueous phase of ice cream helps in stabilization of ice cream with property of their water-holding capacity. This helps in increasing viscosity of ice cream to provide impressive body to the product, reduces dryness, and increases melt down time.[23,27]

4.6.3 WHIPPING ABILITY

Whipping ability of the mix is the ability of mix to incorporate air into its structure. It depends on the tensile strength of the wall or the lamellae surrounding the air cells. High whipping rate is desired as it incorporates finer air cells, which provides high over run. Sodium caseinate influences the whipping rate as it affects the air cells and ice-crystal formation and distribution. Fat globules of smaller diameter provide more whipping rate compared to clumped fat globules. The effect of sugar on whipping rate depends on the stage when it is added, it reduces whipping ability excepting when it is added after the homogenization.[60]

4.6.4 OVERRUN

The excess volume of ice cream obtained is represented in the form of overrun. It is a ratio of excess volume of ice cream obtained to mix.[18,33]

$$\%\text{Overrun} = \frac{\text{Volume of ice cream} - \text{Volume of ice cream mix}}{\text{Volume of ice cream mix}} \times 100 \quad (4.1)$$

The overrun in ice cream is observed because of incorporation of air in mix during freezing (Table 4.3). The quantity of air incorporated determines the body of ice cream obtained. More amount of air incorporation leads to fluffy, snowy texture of ice cream, which is undesirable. Too little amount of air incorporation also leads to development of heavy and soggy ice cream having coarse texture, which is also considered unpalatable. Overrun should be 2.5 times of the dry matter in ice-cream mix.[4] Overrun of ice cream to be achieved depends on the following factors:

- **Fruit and nuts:** Ice cream containing fruits and nuts need to have lower overrun for desirable body and stiffness.
- **Legal requirements** enforced by the regulatory body
- **Packaging:** Overrun of the bulk packaged ice cream can be increased up to 90–100%, whereas the carry home type packaged ice cream overrun should be limited to 70–80% overrun.[3,7]
- **TS content:** High TS ice cream permits higher overrun.

- **Yield and profit:** More is the overrun, more is the yield and eventually more profit will be obtained from the product.

TABLE 4.3 Overrun Observed in Various Types of Ice Cream.

Type of ice cream	Overrun (%)
Ice	25–30%
Ice milk	50–80%
Ice-cream bulk packaging	90–100%
Ice-cream carry home type packaging	70–80%
Sherbet	30–40%
Softy ice cream	30–50%

The factors, which suppresses overrun, are fat, MSNF, corn syrup, stabilizers, fruit, chocolate, cocoa, calcium salt, inefficient homogenization, insufficient refrigeration, high temperature of mix, and blunt freezer blades. On the other hand, the factors promoting the overrun are sodium caseinate, whey solids, egg yolk, emulsifiers, homogenization, aging and high temperature pasteurization.[32,57]

4.7 DEFECTS IN ICE CREAM[18,33]

4.7.1 FLAVOR DEFECT

- Acid flavor—Due to over production of lactic acid in product acid flavor is detected.
- Bitter flavor—This may be the result of using inferior products for the production.
- Cooked flavor—Due to the overheating of the mix, cooked flavor is detected in mix.
- Flat flavor—This may be due to the use of insufficient quantity of the flavor, sugar, or milk solids.
- Harsh flavor—It may be the result of using inferior flavor substances for flavoring or may be due to the use of excess quantity of flavoring material
- High flavor—Use of more than optimum quantity of flavoring agent used in the product may lead to high flavor defect in the finished product.

- Low flavor—Insufficient quantity of flavoring material used in mix can lead to low flavor in the ice cream.
- Metallic flavor—Contamination of the ice-cream mix with copper may lead to metallic, oxidized, tallowy, or cardboard flavor.[2,16,18,33]
- Unnatural flavor—This defect may be the result of use of synthetic flavor, which is not a perfect imitation of the true flavor, over ripened fruits, rancid nuts, fermented syrup, poor quality cream, etc.

4.7.2 BODY

- Crumbly/Brittle—The ice cream when breaks apart easily due to lack of cohesiveness, which is due to crumbly or brittle body. The reasons for such defect may be low TS, excess overrun, inefficient homogenization, or insufficient quantity of stabilizers used. The factors limiting the hydration of protein can also promote such defects in the finished product.
- Gummy—Excessive use of stabilizer or high TS may lead to production of gummy like body of the ice cream.
- Shrinkage defect—Due to change in altitude during transportation or shipping causes shrinkage of the product. The probable reasons for this defect are, high overrun, heat shock, small air cells, small ice crystals, temperature during storage or transportation, or inefficient mixing of the ingredients.[2,3,22,60]
- Soggy/ Heavy—Due to low overrun or high concentration of sugar soggy or heavy body of the ice cream can be seen.
- Weak—Because of low TS in ice cream, low cohesiveness can be seen resulting into rapid melting and lack of firmness or chewiness.

4.7.3 TEXTURE

- Buttery—Detection of lumps of butter in mouth is due to buttery texture, this is attributed to the large lumps of butter formation due to the churning during freezing or improper homogenization.
- Coarse/Icy—Formation of large ice crystals, which can be easily detected in mouth leads to coarse or icy texture. This may be due to slow freezing, insufficient quantity of stabilizer, insufficient aging, or insufficient hydration of proteins.

- Flaky—Large size of air cells may lead to flaky ice cream.
- Fluffy—Due to the incorporation of excess air an open texture is obtained, which leads to fluffiness of the ice cream.
- Sandy—Large sized lactose crystals may produce sandy texture. This may be the result of high lactose content, fluctuating storage temperature, presence of nuclei, unfrozen liquid phase, or low viscosity.[27,41,42,58]

4.7.4 *MELTING QUALITY DEFECT*

- Curdy—Protein destabilization due to excess acidity or improper salt balance leads to curdy ice cream.
- Foamy—Due to use of large amount of emulsifier in low solid ice-cream mix or because of incorporation of large quantity of air foamy defect is observed in the end product.
- Slow melting—Use of excess amount of stabilizer may resulting over stabilization and slow melting of ice cream.
- Whey leakage—Poor quality of mix, insufficient quantity of stabilizer, or improperly balanced mix may lead to whey leakage after melting.[6,18,33,41,42]

4.7.5 *COLOR*

- Uneven—Due to improper mixing of color, uneven coloring can be seen in the final product.[6,18,33]
- Unnatural—Excessive, insufficient, or use of uncharacteristic coloring agent can lead to production of unnatural color development in the product.

4.8 RECENT ADVANCES IN PROCESSING OF ICE CREAM

The recent advances are market driven as the health-conscious consumers want to indulge in the pleasure of consuming ice cream. Moreover in that list, the top rank is given to the low-calorie ice creams. Ice creams with added functional benefits are like treating one's tongue and health simultaneously. Hence, ice creams with added probiotics, fiber, CBD, minerals (such as calcium), vitamins, or protein rich do not just offer indulgence but adds sense of well-being. In addition to all these advances, industries are broadening their product profile day by day by introducing ice creams with layers, added textures, topping, fillings, etc..[26,29,30]

A lot of experimentation has been done with flavor by introducing flavor mix and mash ups using cross-category innovation (using confection or bakery into ice cream), or sophisticated flavor profiles (pairing ethnic flavors, such as alcohol, tea, or spices with familiar flavors, such as vanilla, chocolate, strawberry, butterscotch) targeting the consumers exploring various sensorial experiences. Also, now ice creams are served in various restaurants paired with potato chips, salty pretzels, French fries, blue cheese, etc., along with customization options to provide a new experience to the consumers.[45,54]

Innovations in shapes and sizes of the ice cream have made easier snacking on ice cream than ever. Many innovations have made ice cream a premium luxury product by exploring the various ranges of added ingredients, texture, color, flavor, technology used, etc.[37] During the forecast period of 2020 to 2025, the global ice-cream market is expected to grow at a CAGR of 4.9%. Along with the increase in demand, the innovations and the scope of introducing changes will increase; hence a lot of newer trends can be expected in the ice-cream processing industries.[36,44,66,67]

4.9 SUMMARY

Maintenance of quality, quantity of ingredients along with proper check on processing and storage conditions ensures a defect free ice cream. As, the consumer awareness begins to rise, the demand for functional ice cream is increasing. There has been introduction of probiotic, low calories, low sugar, bio ingredients, encapsulated essential oils, polyphenols, curcumin rich ice creams in the market. These products have the potential to replace the traditional ice-creams ad, these not only provide the same flavor profile but are rich in functional ingredients. Commercial examples of such ice creams are probiotic ice cream, diabetic ice cream, turmeric ice cream, etc. This also includes adding whey proteins and developing sugar, lactose or fat free ice cream. These techniques are not only the future of ice-cream industry but would also be healthier and hence will gain more consumer attention. There is huge scope in the market for more such variants.

KEYWORDS

- **aging**
- **emulsifiers**
- **freezing**
- **hardening**
- **ice crystals**
- **ice cream**
- **stabilizers**

REFERENCES

1. Adapa, S.; Dingeldein, H.; Schmidt, K. A.; Herald, T. J. Rheological Properties of Ice-cream Mixes and Frozen Ice-creams Containing Fat and Fat Replacers. *J. Dairy Sci.* **2000,** *83*, 2224–2229.
2. Ahadzi, P. F. Determination of a Suitable Sensory Protocols to Characterize Frozen Dairy-Based Products. PhD Thesis, University of Ghana, Ghana; **2019**; pages 218.
3. Aishwariya, M.; Geeta, P.; Aravindakshan, P. Development of Technology for Fruit-Based Frozen Sherbet. *IJRBAS.* **2019,** *1*, 13–24.
4. Akalın, A.S.; Erişir, D. Effects of Inulin and Oligofructose on the Rheological Characteristics and Probiotic Culture Survival in Low-Fat Probiotic Ice-cream. *J. Food Sci.* **2008,** *73*, M184–M188.
5. Alvarez, V. B.; Wolters, C. L.; Vodovotz, Y.; Ji, T. Physical Properties of Ice-cream Containing Milk Protein Concentrates. *J. Dairy Sci.* **2005,** *88*, 862–871.
6. Arbuckle, W.S. Defects, Scoring and Grading. In: *Ice-cream*; Boston, MA: Springer, **1986**; pp. 313–332.
7. Arifan, F.; Rizqiati, H.; Hintono, A.; Nurwantoro, N.; Susanti, S.; Sulistiyani, L. N. Effect of Sugar Substitution with Dates (*Phoenix Dactylifera* L.) Puree on the Physical and Organoleptic Characteristics of Kefir Ice-cream. *JITEK.* **2021,** *16*, 21–31.
8. Baer, R. J.; Keating, K. R. Determination of Ice-cream Mix Freezing Points: A Comparison of Methods. *J. Dairy Sci.* **1987,** *70*, 555–558.
9. Bahramparvar, M.; Mazaheri-Tehrani, M. Application and Functions of Stabilizers in Ice-cream. *Food Rev. Int.* **2011,** *27*, 389–407.
10. Barford, N. M.; Krog, N.; Larsen, G.; Buchheim, W. Effects of Emulsifiers on Protein-Fat Interaction in Ice-cream Mix during Ageing, Part I: Quantitative Analyses. *Lipid/Fett.* **1991,** *93*, 24–29.
11. Bisht, B.; Lohani, U. C.; Kumar, V.; Gururani, P.; Sinhmar, R. Edible Hydrocolloids as Sustainable Substitute for Non-Biodegradable Materials. *Crit. Rev. Food Sci. Nutr.* **2020,** *2020*, 1–33.

12. Bolliger, S.; Wildmoser, H.; Goff, H. D.; Tharp, B. W. Relationships between Ice-cream Mix Viscoelasticity and Ice Crystal Growth in Ice-cream. *Int. Dairy J.* **2000,** *10*, 791–797.
13. Cavender, G. A.; Kerr, W. L. Microfluidization of Full-Fat Ice-cream Mixes: Effects on Rheology and Microstructure. *J. Food Process Eng.* **2020,** *43*, e13350.
14. Cheng, J.; Ma, Y.; Li, X.; Yan, T.; Cui, J. Effects of Milk Protein-Polysaccharide Interactions on the Stability of Ice-cream Mix Model Systems. *Food Hydrocoll.* **2015,** *45*, 327–336.
15. Chin, E.; Miller, K. B.; Payne, M. J.; Hurst, W. J.; Stuart, D. A. Comparison of Antioxidant Activity and Flavanol Content of Cacao Beans Processed by Modern and Traditional Mesoamerican Methods. *Herit. Sci.* **2013,** *1*, 1–7.
16. Choo, S. Y.; Leong, S. K.; Henna-Lu, F. S. Physicochemical and Sensory Properties of Ice-Cream Formulated with Virgin Coconut Oil. *Food Sci. Technol. Int.* **2010,** *16*, 531–541.
17. Darling, D. F.; Butcher, D. W. Milk-Fat Globule Membrane in Homogenized Cream. *J. Dairy Res.* **1978,** *45*, 197–208.
18. De, S. (Ed.). *Outlines of Dairy Technology*, 44th ed.; Oxford University Press: New Delhi, India, 2018 (January); p 525.
19. Deosarkar, S. S.; Kalyankar, S. D.; Pawshe, R. D.; Khedkar, C. D. Ice-cream: Composition and Health Effects. In *Encyclopedia of Food and Health*, Caballero, B. Ed.; Oxford University Press: London, 2016 (December); pp 385–390.
20. Dervisoglu, M.; Yazici, F.; Aydemir, O. The Effect of Soy Protein Concentrate Addition on the Physical, Chemical, and Sensory Properties of Strawberry Flavored Ice-cream. *Eur. Food Res. Technol.* **2005,** *221*, 466–470.
21. Drewett, E. M.; Hartel, R. W. Ice Crystallization in a Scraped Surface Freezer. *J. Food Eng.* **2007,** *78*, 1060–1066.
22. Dubey, U. K.; White, C. H. Ice-cream Shrinkage: A Problem for the Ice-cream Industry. *J. Dairy Sci.* **1997,** *80*, 3439–3444.
23. Eisner, M. D.; Wildmoser, H.; Windhab, E. J. Air Cell Microstructuring in a High Viscous Ice-cream Matrix. *Colloids Surf. A: Physicochem. Eng. Asp.* **2005,** *263*, 390–399.
24. Engeseth, N. J.; Pangan, M. F. A. Current Context on Chocolate Flavor Development—a Review. *Curr. Opin. Food Sci.* **2018,** *21*, 84–91.
25. FSSAI (Food Safety and Standards Authority of India). Food Products Standards and Food Additives: Regulations 2011. *Gazette India* **2011**, *235*, 287–299.
26. Glicksman, M. Utilization of Seaweed Hydrocolloids in the Food Industry. In *Proceedings of the Twelfth international Seaweed Symposium*, Ragan, M. A. Ed.; Development in Hydrobiology – Volume 41; Dordrecht: Springer, 1987; pp 31–47.
27. Goff, H. D. Colloidal Aspects of Ice-cream—a Review. *Int. Dairy J.* **1997,** *7*, 363–373.
28. Goff, H. D.; Davidson, V. J.; Cappi, E. Viscosity of Ice-cream Mix at Pasteurization Temperatures. *J. Dairy Sci.* **1994,** *77*, 2207–2213.
29. Goff, H. D.; Hartel, R. W. Novelty Products and Ice-cream Cakes. In *Ice-cream*; Springer: Boston – MA, 2012; pp 261–287.
30. Goff, H. D.; Jordan, W. K. Action of Emulsifiers in Promoting Fat Destabilization during the Manufacture of Ice-cream. *J. Dairy Sci.* **1989,** *72*, 18–29.
31. Hartel, R. W. Ice Crystallization during the Manufacture of Ice-cream. *Trends Food Sci. Technol.* **1996,** *7*, 315–321.
32. Herald, T. J.; Aramouni, F. M.; Abu-Ghoush, M. H. Comparison Study of Egg Yolks and Egg Alternatives in French Vanilla Ice-cream. *J. Texture Stud.* **2008,** *39*, 284–295.

33. Jana, A.; Pinto, S.; Moorthy, P. R. S. (Eds.) Ice-cream and Frozen Desserts. [*Agrimoon.com*]; *India;* online; **2016**; Unpublished course modules.
34. Kanisawa, T.; Tokoro, K.; Kawahara, S. Flavor Development in the Beans of *Vanilla planifolia*. In *Olfaction and Taste XI*, Kurihara, K., Ed.; Springer: Tokyo, 1994; pp 268–270.
35. Keeney, P. G.; Maga, J. A. Factors Affecting Composition and Yield of a Foam Fraction Recovered from Ice-cream. *J. Dairy Sci.* **1965,** *48*, 1591–1596.
36. Khaira, N. M.; Abd Rahmana, N. A.; Baharuddina, A. S.; Hafidb, H. S.; Wakisakab, M. Capturing the Impact of Nanobubble Liquid in Enhancing the Physical Quality of Ice-cream. *J. Agric. Food Eng.* **2020,** *2*, e0012.
37. Konstantas, A.; Stamford, L.; Azapagic, A. Environmental Impacts of Ice-cream. *J. Clean. Prod.* **2019,** *209*, 259–272.
38. Koxholt, M. M.; Eisenmann, B.; Hinrichs, J. Effect of the Fat Globule Sizes on the Melt-down of Ice-cream. *J. Dairy Sci.* **2001,** *84*, 31–37.
39. Kyritsis, A.; Kanias, G.; Tzia, C. Nutritional Values of Trace Elements in Dried Desserts. *J. Radioanal. Nucl. Chem.* **1997,** *217*, 209–219.
40. Li, Z.; Marshall, R.; Heymann, H.; Fernando, L. Effect of Milk Fat Content on Flavor Perception of Vanilla Ice-cream. *J. Dairy Sci.* **1997,** *80*, 3133–3141.
41. Marshall, R. T.; Goff, H. D.; Hartel, R. W. (Eds.). Fancy Molded Ice-creams, Novelties and Specials. In *Ice-cream*; Springer Nature: Cham, 2003; pp 275–294.
42. Marshall, R. T.; Goff, H. D.; Hartel, R. W. (Eds.). Soft-frozen Dairy Desserts. In *Ice-cream*, Springer Nature: Cham, 2003; pp 253–263.
43. Miller, G. F.; Kropp, J. D.; Gupta, S.; Grogan, K. A.; Mathews, A. Do Elementary Students Substitute Ice-cream and Baked Goods for Healthier National School Lunch Program Meal Items? *Appl. Econ. Perspect. Policy.* **2017,** *39*, 41–64.
44. Mustafa, R.; He, Y.; Shim, Y. Y.; Reaney, M. J. Aquafaba, Wastewater from Chickpea Canning, Functions as an Egg Replacer in Sponge Cake. *Int. J. Food Sci. Technol.* **2018,** *53*, 2247–2255.
45. Patel, H.; Anema, S.; Holroyd, S.; Harjinder Singh; Creamer, L. Methods to Determine Denaturation and Aggregation of Proteins in Low-, Medium- and High-Heat Skim Milk Powders. *Le Lait.* **2007,** *87*, 251–268.
46. Pavlyuk, R.; Pogarska, V.; Pavlyuk, V.; Pogarskiy, A.; Kakadii, I.; Stukonozhenko, T.; Telenkov, O. The Development of New Method of Production of Healthy Ice-Cream-Sorbet of Fruits and Vegetables with a Record Bas Content. *EUREKA: Life Sci.* **2018,** 33–40.
47. Pelan, B. M. C.; Watts, K. M.; Campbell, I. J.; Lips, A. The Stability of Aerated Milk Protein Emulsions in the Presence of Small Molecule Surfactants. *J. Dairy Sci.* **1997,** *80*, 2631–2638.
48. Qadri, T.; Hussain, S. Z.; Rather, A. H.; Amin, T.; Naseer, B. Nutritional and Storage Stability of Wheat-Based Crackers Incorporated with Brown Rice Flour and Carboxymethyl Cellulose (CMC). *Int. J. Food Prop.* **2018,** *21*, 1117–1128.
49. Ramachandra Rao, S.; Ravishankar, G. A. Vanilla Flavor: Production by Conventional and Biotechnological Routes. *J. Sci. Food Agric.* **2000,** *80*, 289–304.
50. Romagnoli, L. G.; Knorr, D. Effects of Ferulic Acid Treatment on Growth and Flavor Development of Cultured *Vanilla Planifolia* Cells. *Food Biotechnol.* **1988,** *2*, 93–104.
51. Rønholt, S.; Mortensen, K.; Knudsen, J. C. The Effective Factors on the Structure of Butter and Other Milk Fat-Based Products. *Compr. Rev. Food Sci. Food Saf.* **2013,** *12*, 468–482.

52. Ruger, P. R.; Baer, R. J.; Kasperson, K. M. Effect of Double Homogenization and Whey Protein Concentrate on the Texture of Ice-cream. *J. Dairy Sci.* **2002,** *85*, 1684–1692.
53. Russell, A. B.; Cheney, P. E.; Wantling, S. D. Influence of Freezing Conditions on Ice Crystallisation in Ice-cream. *J. Food Eng.* **1999,** *39*, 179–191.
54. Saunders, L. *The Best of American Beer and Food: Pairing & Cooking with Craft Beer*, Illustrated Edition; Brewers Publications: Boulder–Colorado, 2007 (September); p 180.
55. Savaria, T. J. *Homogenized Cream as an Ingredient in Ice-cream Mix*. MSc Thesis, University of Massachusetts, Amherst–MA, 1937; p 40.
56. Schaller-Povolny, L. A.; Smith, D. E. Sensory Attributes and Storage Life of Reduced Fat Ice-cream as Related to Inulin Content. *J. Food Sci.* **1999,** *64*, 555–559.
57. Segall, K. I.; Goff, H. D. A Modified Ice-cream Processing Routine That Promotes Fat Destabilization in the Absence of Added Emulsifier. *Int. Dairy J.* **2002,** *12*, 1013–1018.
58. Singo, T. M.; Beswa, D. Effect of Roselle Extracts on the Selected Quality Characteristics of Ice-cream. *Int. J. Food Prop.* **2019,** *22*, 42–53.
59. Soukoulis, C.; Rontogianni, E.; Tzia, C. Contribution of Thermal, Rheological and Physical Measurements to the Determination of Sensorially Perceived Quality of Ice-cream Containing Bulk Sweeteners. *J. Food Eng.* **2010,** *100*, 634–641.
60. Syed, Q. A.; Anwar, S.; Shukat, R.; Zahoor, T. Effects of Different Ingredients on Texture of Ice-cream. *J. Nutr. Health Food Eng.* **2018,** *8*, 422–435.
61. ULM, M.; David, L. S.; Lavanya, N. *Study on The Ability of Selected Plant-Derived Stabilizers for Manufacturing Set Yoghurt*. M.Sc. Thesis, Colombo-Sri Lanka: Faculty of Agriculture, Research Gateway - Eastern University, 2009; p 103.
62. Varela, P.; Pintor, A.; Fiszman, S. How Hydrocolloids Affect the Temporal Oral Perception of Ice-cream. *Food Hydrocoll.* **2014,** *36*, 220–228.
63. Vivek, B. (Ed.). *Ice-cream: Manufacture and Technology*. Tata McGraw-Hill Book Publishing Co. Ltd.: New Delhi, 2001; p 341.
64. Vlassopoulos, A.; Masset, G.; Charles, V. R.; Hoover, C.; Chesneau-Guillemont, C.; Leroy, F.; Lehmann, U.; Spieldenner, J.; Tee, E.-S.; Gibney, M. Nutrient Profiling System for the (Re) Formulation of a Global Food and Beverage Portfolio. *Eur. J. Nutr.* **2017,** *56*, 1105–1122.
65. Walstra, P.; Oortwijn, H. The Membranes of Recombined Fat Globules: Modes of Formation (Milk). *Netherlands Milk Dairy J.* **1979,** *33* (2), 134–154

CHAPTER 5

FUNCTIONAL ICE CREAM: CHEMISTRY, CHARACTERISTICS, AND TECHNOLOGY

RAMESH F. CHAVAN, RAHUL R. SINDHANI, and
BHAGWAN K. SAKHALE

ABSTRACT

Ice-cream is one of the most popular desserts and the ice-cream industry is a fastest growing sector of food processing and dairy industry because ice-cream is famous among both children and adults. Artificially flavored ice-creams are very common in the current scenario due to its low cost of production. Demand for the functional ice-cream is rising rapidly in new food product development across the world because of its nutritional benefits and instant satisfaction to the consumer. The main aim of functional ice-cream is to introduce beneficial biocompounds into the body through daily dietary intake. It is a good carrier of all functional ingredients because of its low storage temperature, stabilization of ingredients, and popularity among consumers. This concept flips the idea of ice-cream being an unhealthy and turns ice-cream into a positive and an essential healthy treat. The ice-cream includes all supplements of functional nutrients, such as vitamins, minerals, antioxidants, fiber, probiotics, omega-3 fatty acids, and prebiotics.

5.1 INTRODUCTION

Over the last 20 years, the interest in food-health relationships and the quality of living for health welfare has increased.[29] Clarifying epidemiological and

The Chemistry of Milk and Milk Products: Physicochemical Properties, Therapeutic Characteristics, and Processing Methods. Megh R. Goyal, Suvartan Ranvir, & Junaid Ahmad Malik (Eds.)

clinical trials, it has now been well confirmed the relationship between the diet and human health. Recently, the correlation of the use of food added substances with expanded medical problems, just as expanded industry interest because of client inclinations for natural, organic, and free of synthetic additives foods, has led research into the study of bioactive components and their use in functional food production.[6,13]

Due to various critical factors, the interest in useful food sources has expanded, for example, familiarity with individual medical problems, exceptional way of life including hunger and insufficient active work, unregulated substance utilization, dietary instructions given by specialists and the media, logical advances in nutrition surveys, and the open furthermore serious food area.

Useful nourishments are characterized by the food sources that accomplish fundamental sustenance and decidedly influence soundness of a buyer.[3,28] Perhaps the main reasons that shoppers lean toward utilitarian food sources are that they can be devoured without changing dietary propensities, paying little heed to portion of utilization, and term not at all like prescriptions. Useful nourishments in such a manner incorporate engineered added substance free and sound normal food sources that are eaten in their food structure with day-by-day count calories and furthermore positively affect the health, bioactive food components, and different items enriched with these components.[3,16]

The size and growth of the market is measured by consumers' acceptance and attitude toward the functional foods. Japan, Korea, USA, and European countries make up the largest usable food market. Functional foods are eaten in greater proportions in Spain, Finland, Holland, and Sweden in the European Union. It usually must stick to some positioning for functional food to thrive.[22,33] In many human physiologies, utilitarian food consumption has significant capacities, for example, early turn of events and development, essential metabolic cycle's association, insurance against oxidative pressure, cardiovascular and gastrointestinal illnesses, psychological and mental execution, actual execution, and wellbeing.[2,11]

Dairy items are significant for the nourishment of local habitants since plenitude of supplements and heavenly taste of dairy items have made them famous around the world. Different functional food ingredients are naturally included in their structure and certain components are often used to enrich the dairy products, such as vitamins, minerals, antioxidants, and phenolics. Therefore, healthier, and more beneficial goods are obtained for customers by increasing functional characteristics. Due to their beneficial impact on health and their inclusion in the daily diet, the dairy and milk-based products are more favorable for developing functional products. In the functional properties of dairy products, functional ice-cream has an important position. Nutraceutical components widely used in functional ice-cream processing technologies are grouped in this analysis and different studies are discussed in this chapter.[10,27]

Functional ice-cream is a relatively nascent concept in the categories of the novel functional foods and trend of this functional ice-cream is increasing continuously because of reputation of ice-cream being very popular among all age groups.[30] Customary ice-cream is a frozen dessert made by freezing a purified blend of milk, sugar, corn syrup, enhancing, stabilizer, emulsifier, with or without eggs, and it is well known around the globe. The nature of frozen yogurt relies basically on the fixings utilized just as handling boundaries and capacity conditions. Frozen yogurt is regularly considered a fun food, which is undeserving thought, and even was considered lousy nourishment.[8,25]

Frozen dessert contains considerably higher amount of fat and almost more than 14–15% of the protein as compared with pure lacteal secretion. Similarly, it has abundance of other beneficial nutrients containing food items, such as organic products, nuts, eggs, dry organic products, and sugar, which support its nutritive worth. The milk solids in ice-cream are usually subjected to higher heat treatments than those of pasteurized milk that are also subjected to lower temperatures in the freezing process, and they are stored longer before consumption.[19] It is an excellent source of food energy due to its enhanced fat content than that of milk, and 50% of its total solids content is sugar, including lactose, sucrose, and corn syrup solids. The way that these constituents are totally absorbed makes ice-cream an attractive nourishment for developing youngsters and people, who need to gain weight.[20]

The present chapter focuses on the advances in novel area of the functional ice-cream with potential health benefits and commercial availability.

5.2 FUNCTIONAL ICE-CREAM AND HUMAN HEALTH

Research has shown that fortifications in ice-creams normally increase their emulsion stability[6] and have a higher rate of performance over other fortified foods. It is also a clever way for kids to get much of the goodness from these reinforced ice-creams. It is a fascinating way to produce a popular dessert with the combination of bioactive molecules with ice-creams. Recent advancements in the manufacturing of ice-cream have significantly improved the quality determining rheological, firmness, and organoleptic parameters of the dessert items.

Diet plays a very important role in human health. Today, many studies have reported that the diet will cause or stop diseases. In recent years, explicit attention has been dedicated to the assembly of practical foods. The most important role of functional foods is to introduce microorganisms or helpful compounds into the body through the daily dietary intake. A probiotic may be a live microbial food supplement that beneficially affects the host by improving its enteral microbic balance. Probiotics are viable microorganisms that are helpful to the host once consumed in applicable quantities. Advantages embody inhibition of microorganism pathogens; reduction of serum cholesterol levels; reduction in the incidence of constipation, diarrhea, and intestine cancer; improvement of disaccharide tolerance, Ca absorption, and vitamin synthesis; and stimulation of the immune system.

While designing new food products, certain minimum quality determining factors need to be considered for meeting the expectations of end-users and their desire toward a particular dessert item. Strategies must be defined based on consumer willingness about ice-cream in which the sensory properties of foods are combined health promoting contains of the products.[20] Fortification by the addition of essential nutrients can be known as food enrichment. Food fortification is also considered the most cost-effective long-term strategy.[26] Initiatives such as the aims of health departmental bodies should be providing adequate supply of proper nutritionally balanced and popular food items, such as ice-cream should be promoted among the nutritionally deprived section of the community, which should ensure proper nutritional supply to them is extremely important as nutrition, and it is a mandatory requirement for proper growth and nourishment of the body. Although the frozen desserts are not listed under the popular categories of the balanced nutrient providing items, yet their popularity is among the people of different ages and hence it has a great potential to fulfill the demand of nourishment of huge world population by providing nutrition along with satiety.[9]

Today, consumers are interested in low-cost healthy dairy products, such as functional dairy products in general and functional ice-cream to be specific. The scientists have considered the arrangement of utilitarian ice-cream and have referenced that the mash of the thorny pear organic product did improve the rheological qualities, cancer prevention agent action, and absolute acceptance of low-fat ice-cream. The use of avocado fruit pulp as functional ingredients rich in natural antioxidants and high in unsaturated fatty acids in ice-cream manufacture enhances its aesthetic and nutritional characteristics.[17]

5.3 SOURCES OF FUNCTIONAL INGREDIENTS

Various distinguished sources have been identified for isolation of different functional ingredients for incorporation in the ice-cream. Many of the functional ingredients are currently under-used for their possible usefulness in the preparation of functional ice-cream. However, the major sources, which are utilized for the extraction of the functional ingredients, include various sources that are discussed in this section.

5.3.1 NON-MILKFAT, NON-MILKFAT PORTIONS, AND HYDROGENATED STRUCTURES

In attempting to coordinate the practical and tangible properties of milkfat, it ought to be recollected that plant fats/oils are essentially not quite the same as milkfat and, much of the time, from one another while considering their impact on construction through fat agglomeration and softening point varieties.[21]

5.3.2 PLANT STEROLS

In some no-fat items, there stays a need to deliver fat/oil-dissolvable seasoning parts fundamental for the ideal portraying flavor. Limited quantities of plant sterols (Figure 5.1) can offer the capacity to give this essential usefulness.

Plant sterols are not fat for nourishment naming purpose; however, they do give calories. Therefore, how they are utilized, overseen, and pronounced should be deliberately thought of the sterols. Normally, application in close to without fat blends shows the most flavor improvement.[31]

5.3.3 LECITHIN

Lecithin is mother nature's liked emulsifier. Be that as it may, the usefulness of some random source can be required to shift significantly from different sources.

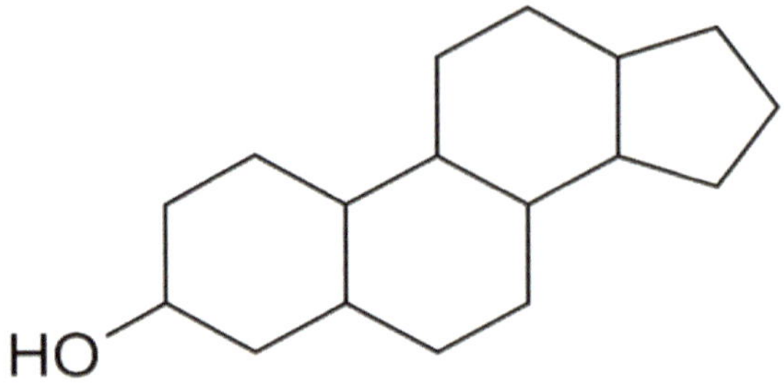

FIGURE 5.1 Structure of sterol.

Lecithin (Figure 5.2) contains unsaturated fats that are basic to its capacity yet additionally make affectability to oxidative rancidity.[3,11]

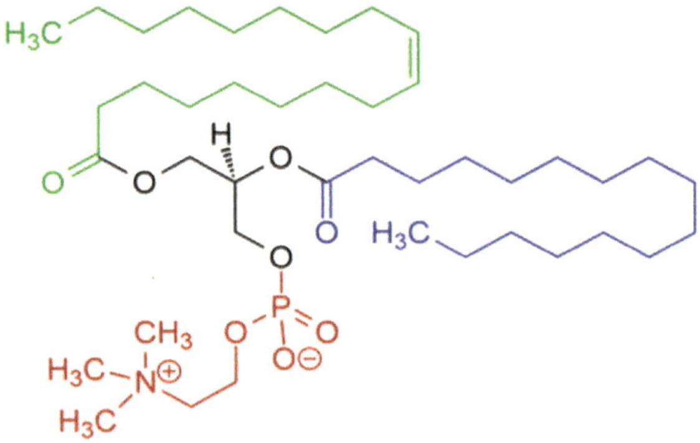

FIGURE 5.2 Structure of lecithin.

5.3.4 PROPYLENE GLYCOL (PG)

Propylene glycol is not like conventional emulsifiers, for example, mono-diglycerides, polysorbates, PG (Figure 5.3) capacities to cover little ice gems with a greasy film, keeping them from developing under warmth stun. In setting, PG basically makes a steady ice-in-water emulsion. Ice-cream is a complex colloidal food structure that in its frozen state contains ice diamonds, air cells, and generally combined fat beads scattered in a constant frozen and thought fluid (whey) stage containing polysaccharides, yet in addition sodium and potassium).[5]

Moreover, it is noteworthy in the ice-cream to observe ice recrystalliza-tion, Ostwald aging noticeable all-around cell, and lactose crystallization oversee organoleptic quality, including richness, unpleasantness, mouth feel, and tongue satiety.[12,29]

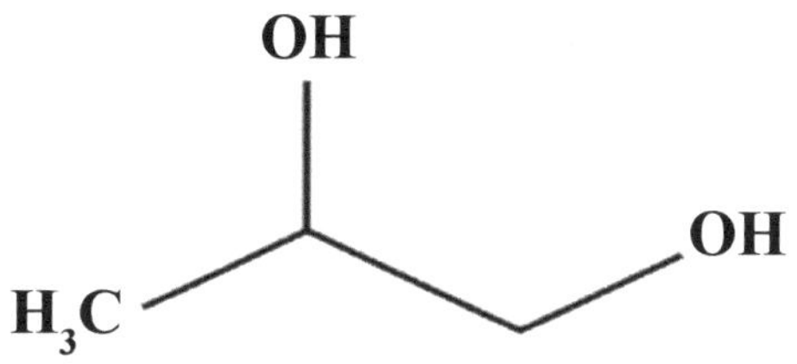

FIGURE 5.3 Chemical structure of propylene glycol (PG).

5.3.5 OIL OF FLAXSEED TO STANDARDIZE LOW FAT ICE-CREAM

The milk fat replacement in frozen dessert with edible oil might influence the utilitarian qualities of frozen dessert, contingent upon what quantity milk oil is supplanted. Additional extents of milk fat replacement with oil may cause less palatable and fewer firm desserts with high emergency rate. This might be due to the occurrence of distinctions within the liquefying temperature of fat or oil. Likewise, the degree of fat natural action happening throughout the attendant stirring and phase transition of the frozen dessert mix determines the immovableness.[7,15] The flocculated fat beads primarily settle the air entered yoghourt that offers firm construction of the dessert ice-cream. It has been discovered that if oil is consolidated in an exceedingly quantity in the ice-cream, then the final quality of the dessert is not influenced that much.[10,26]

Over the past 2 decades, the science and technology of ice-cream has seen a remarkable progress in exploring and understanding structure, texture, and storage. This has empowered food technologists to understand the joining of novel or utilitarian fixings into model or genuine ice-cream systems not only to provide customized techno-functionality (such as viscosity enhancement, cryo-protection, emulsification, union of water), but also to improve aspects related to health and nutrition, natural antioxidants, fat sources rich in polyunsaturated fatty acids,[11] minerals and sweeteners of low GI have been incorporated into ice-cream systems. This trend of using ice-cream as a vehicle for health-related compounds appears to be supported by consumer demand for healthier and more nutritious food products that lack food additives.[30] Due to its near global availability, high consumer acceptability, and attractive sensory attributes, resulting in a high sales rate, ice-cream can be considered a favorable vehicle for the delivery of bioactive compounds.[18]

In addition, generally low temperature manufacturing steps, including storage in the frozen state, typically in dark conditions, make ice-cream a good substrate for long-term maintenance of the functional characteristics of additional health. Promotion of compounds, including beneficial living cells (such as probiotics), reduces oxidative damage caused by light or heat. Recently, important advances in ice-cream technology in terms of structure-texture development and stabilization have been extensively studied.[15,31] The health-promoting antioxidant lycopene in food has limited natural bioavailability, but functional foods rich in lycopene can improve its bioavailability.

In a randomized hybrid study, 4-week dietary mediations with one or the other control or lycopene-invigorated ice-cream were explored. Samples of serum and residual skin surface components (RSSC) were taken from facial skin before the interventions at 2 weeks and at the end of the intervention.[18,33] Lycopene center, standard blood normal scientific markers, and biomarkers for oxidative pressing factor, which included combustible oxidative damage, and low thickness lipoprotein peroxidase proteins were assessed in the serum. Lycopene-related immunofluorescence, lipid drop size, corneocyte scaling, and microbial presence were assessed in the RSSC.

The results showed that lycopene concentrations in serum and skin were increased steadily during the consumption of lycopene-enriched ice-cream. Lippo-proteins having low density and oxidative inflammatory damages were decreased significantly in the traditional biochemical processes at the end of the lycopene-enriched ice-cream intervention, yet it stayed unaltered during control ice-cream utilization.[24] Control ice-cream essentially expanded coenocyte chipping and bacterial presence. These unfriendly impacts, which might incline customers to skin inflammation advancement, were found lacking when the volunteers under the study were given lycopene-rich frozen dessert. Along with the presence of lycopene in the frozen dessert, it can alleviate the pro-inflammatory action of ice-cream at the level of the facial skin, thus reducing the risk of acne associated with diet in young consumers.[16]

The transport passage of the probiotics is well facilitated by using the frozen dessert. The functional efficiency of the probiotics as a passenger of the probiotics in living being by stage change all through delivering and delay stockpiling are yet matter of thought. Probiotic societies add worth to the frozen treat and demonstrate higher delineation of being intentional food.[32] For ensuring useful properties, every strategy stage ought to be streamlined.

The fresh lacteal goes most likely as fermentative substrate, for keeping an important separation from detrimental changes throughout maturing and limit of food factor, ideal temperature, and gas molecule satiety levels square measures are preserved. Probiotic strains square measure unbelievably sensitive to chop down pH scale regards (4.0–4.5) that causes negative impact on material sufficiency of the factor. To overcome such issues with variable gas molecule centralization of fermentative substrate as outcomes of production of expendable damaging, the event system was rebuffed at ideal pH scale.[19]

5.4 ROLE OF SPECIFIC INGRDIENTS IN MANUFACTURING OF THE ICE-CREAM

Every component utilized for manufacturing of the ice-cream has potential nutritional properties. Different ingredients used in ice-cream preparation such as sugar, milk powder, fat, emulsifiers, stabilizers, color, and flavor play specific role in quality parameters of the ice-cream and hence govern the overall acceptability of the prepared ice-cream. Texture, taste, flavor, color, and overall acceptability of any ice-cream depend upon the quality of the used ingredients. In this section, specific functions of all such ingredients are elaborated with respect to their use and function.

5.4.1 FATS

Fats increment the extravagance of the frozen ice-cream flavor, produce a smooth surface, offer body to the frozen ice-cream, and produce great liquefying properties when the frozen ice-cream is eaten. Despite the fact that milk fats often do not produce ice cream, other vegetable fats, such as vegetable oil, lipids, or sans salt oleomargarine, are also more affordable and do not increase the cost of frozen ice cream.[3,27]

5.4.2 MILK SOLIDS-NOT-FAT

Milk SNF (solids-not-fat) is fused as milk powder or complete fat milk. These ingredients enhance the body texture and surface of frozen ice-cream, enable the following high invasion, and it produces very thick and delicious frozen treat.[14,18]

5.4.3 SUGARS

Sugars improve the flavor, surface, and satisfactoriness of ice-cream. They help to increase a lower point of cooling strong, with the aim that the reasonable frozen treat has some unfreeze water. Without lowering and achieving the desired lower freezing point, the ice-cream may not be eaten in a proper way. Smooth surface of the ice-cream can be achieved by reducing the oil content of ingredient mixtures. Granulated invert sugar syrup, and the dextrose powders may be added additionally in the ice-cream to achieve more desirable ice-cream texture in terms of its surface properties. Corn syrup crystallizes slowly and hence it creates tenderer and chewy texture of the ice-cream; therefore, these ingredients may be added as per the product specifications. It is accessible in various dextrose equivalents in ice-cream preparation. The pleasantness enhances with higher DE qualities depending upon the working of ice-cream mixture. Lower DE corn syrups have a more prominent settling impact than its higher quantities.[32]

5.4.4 STABILIZERS

Stabilizers are important constituents of the frozen ice-creams to assist the uniform mixing of the key ingredients of the ice-cream (such as fats, sugars, etc.), which build a stable texture for desirable properties thus increasing the consumer acceptance. Stabilizers lower the amount of the defrosted water inside ice-cream body and hence help to make stable emulsion of the fat and other ingredients.[23] Although gelatin is used nowadays as a stabilizer, yet CMC (carboxymethylcellulose) is still being the first choice of the ice-cream manufacturers because composition of CMC has guar gum, bean gum, and insect gum. The vegetable gums may likewise be utilized rather than the CMC. The measures of stabilizer must meet the requirements if the ice-cream manufacturers.[32]

5.4.5 EMULSIFIERS

Smooth surface of the ice-creams is attributable to the extent of the addition of emulsifiers. The conventional emulsifiers in the manufacturing of the frozen dessert were the yolk portion of the eggs, but recently it has been replaced by diglycerides and monoglycerides that are more convenient to use.[25,28]

5.4.6 FLAVORS AND COLORS

The popular flavors for the ice-creams include vanilla, butterscotch, strawberry, chocolate, and natural fruit flavors, etc.[4,23] For example, green tone with mint flavor or orange with mango is common.[21] The flavors and tones must be of food grade and are normally available in local stores and metropolitan networks or from bread shops. Vanilla flavor is as often as possible the most standard fixing upgrading, yet creators must find close tendencies before picking the flavor to offer.[24,31]

5.5 APPLICATIONS OF FUNCTIONAL INGREDIENTS IN PRODUCT DEVELOPMENT

While manufacturing of functional ice-creams, various conventional ingredients (such as minerals, probiotics, phytochemicals, bioactive peptides, supplements, dietary strands, prebiotics, whey, and its derivatives along with the unsaturated fats), vegetable oils, and flavors are utilized for providing acceptable quality of the frozen dessert with respect to organoleptic parameters.[14] Likewise, different components (such as organic products, wild organic products, vegetables, therapeutic sweet-smelling plants, honeybee items (such as nectar, dust, and propolis)), and different sugar substitutes generally consisting of vegetable sugars (which include sugar alcohols and stevia) can be utilized. Nutraceutical parts broadly utilized in utilitarian ice-cream creation innovation are assembled and introduced beneath.

In such a manner, valuable food sources embrace counterfeit added substance free and sustaining normal nourishments that are overwhelmed by daily diet in its food type and even decidedly affect individual's well-being, and bioactive components can be enhanced with these components.[17]

Accommodating food sources are the trademark sustenance alongside an important limit, for example, tomato-lycopene, fish pecan omega unsaturated fats, cancer prevention agents, anthocyanin, iodized salt, unsaturated fats, omega-3 unsaturated fats, changed supplements and minerals, phenolic tar substances, cell fortifications, dietary fibers, oligosaccharides, probiotics, prebiotics or disposing of a damaging part as sodium-reduced salt. Likewise, it is feasible to deliver helpful nourishments by altering a few mixes such as yogurt-protein-bioactive peptide, handled tomato-lycopene inside the food.[32]

5.5.1 PROBIOTICS, PREBIOTICS, AND SYMBIOTICS

Probiotics are important components giving helpful outcomes on the host by leveling microorganism vegetation inside the organ framework. To achieve desirable impact inside the body, food sources with probiotics should be devoured frequently and consequently the scope of probiotic microorganism in the item must be about 106–108 cfu/g. Prebiotics has tremendous potential to improve the functional characteristics of the ice-cream by incorporating wide range of the beneficial bioflora, which has potential to enhance the immunogenic response in our body.[26]

The most beneficial part of using prebiotics along with probiotics lies in its composition of oligosaccharides and polysaccharides, which can regulate the digestive mechanism.[13] Therefore, the digestive regulative mechanism of the prebiotics triggers the remedial action against some of the incurable illnesses relating to counteraction of pathology as a consequence of expanded Calcium admission, decrease of avoirdupois danger and type-2 diabetes, decrease of carcinoma hazard because of balance of hepatotoxic items, guideline of framework, and assurance of urogenital system.[15]

Symbiotics are probiotic bacteria that protect them while they travel through the stomach and small intestine. They do this by monitoring probiotic growth and imitating it in the gut.[21] Few frozen treat plans are created by exploitation of probiotics and prebiotics together and furthermore the product was dissected regarding the arranged quality highlights. Probiotic ice-creams are tested by utilizing *L. acidophilus* and *B. bifidum* to expand common sense and restorative outcome and discovered that the measure of probiotics is considered at an ideal level of>106 cfu/g for accomplishing remedial impact.[20]

5.5.2 PHENOLIC COMPOUNDS AND ANTIOXIDANTS

The life component for almost all living being is Oxygen. Free revolutionaries are particles that have in any event one unpaired electron. In living frameworks, free revolutionaries are vital in various metabolic capacities and free radical oxidation happens in natural particles because of high measure of these mixes, causing different infections, such as tissue harm in the body, cell demise, early maturing, malignancy, paleness, heart-related sicknesses, and neural complexities. Antioxidants play major role in circulating the adequate quantity of oxygen throughout the body or engrossing dynamic oxygen molecules.[12]

The vitamins E and C along with carotenoids and phenolic tar mixes are beneficial to us because of their inhibiting qualities. The recent investigations confirm that organic products, wild organic products, vegetables, flavors, and restorative fragrant spices (which might be subbed with intensified cancer prevention agent), and phenolic qualities can usually expand the cell reinforcement action in ice-cream.[4,16]

The research study on the effect of grape wine storage on common form of antioxidant function and cell reinforcement properties of ice-cream indicated that grape wine storage accumulates the cancer prevention agent movement at expanding fixations and cancer prevention agent exacerbates stays steady. In a report[31] on red kiwis, high amount of water-dissolvable nutrients and inhibitor trademark were utilized in the manufacturing of frozen pastry.

The best absolute phenolic gum content (such as gallic corrosive), cancer prevention agents, and water-solvent nutrient substances were resolved inside the frozen dessert containing red kiwis or the green kiwis, which has potent antioxidant properties. Researchers have also examined cancer-preventing agent properties and organic cycle estimations of sugarless, fat-decreased milk-based ice-creams. In an investigation, absolute phenolic substance and inhibitor movement of the frozen treat made with *canella* and green tea extracts were valuable.[2,32]

5.5.3 PEPTIDES, BIOACTIVE PROTEINS, AND AMINO ACIDS

Proteins, peptides, and amino acids provide energy and basic food sources during nutrition, but also have a variety of natural abilities, such as development, antihypertensive, antimicrobial, and cancer prevention properties due to anticarcinogenic mechanism[5,33]; for example, milk proteins lactoferrin, lactoperoxidase, and immunoglobulin have antiviral, antibacterial, antifungal, and parasitic properties and have significant potential within a considerable safe composition of these individual proteins of the serum, because the fractions of the peptides are utilized due to the hydrolysis of these protein compounds to enhance the dietary benefits from different food sources and to improve their basic properties and provide health benefits from functional foods. Some researchers have used fish macromolecule powder (protein source) at various levels of 10%, 20%, and 30% to antibodies to frozen ice-creams.[26]

Nowadays, the altered strategies (such as inverted assimilation, microfiltration, ultrafiltration), whey powder, molecular trade, demineralized whey, whey macromolecular structures, whey protein concentrates, low lactose whey proteins, and hydrolyzed whey are regularly hydrolyzed before their use.[22,23] Whey proteins with good food quality among all dietary proteins have a good collection of precious food components. Whey and its derivatives are used in functional ice-creams to create dry matter generalization, to refine flavor and natural flavor, to increase protein contents, and to improve the structural properties.[2,11]

5.5.4 BIOACTIVE LIPIDS

In addition to the taste of making ice-cream from the functional food sources, fat provides an important part of the energy derived from the food phytochemicals that regulate the transport of saturated fatty foods and the endpoint of base oil (which is an essential component of ice-cream). Functional ingredients play an important role in many metabolic cycles to enhance the aesthetic properties of the ice-cream. Milk fat is a critical supplement regarding nourishment physiology because of its unsaturated fats, which has high nutritional values consisting of the vital fat-soluble vitamins (such as A, D, E, and K, which are equivalent PUFA (Polyunsaturated unsaturated fatty acids)), cholesterol, phospholipids, glycolipids, and gangliocytes[1,23] of fats (which are unsaturated) are integral parts of the milk-fats to help in building texture of sustenance physiology. The linoleic omega-6, linolenic omega-3, and arachidonic fats having omega-6 structure are fundamental unsaturated fats that cannot be mimicked by artificial procedures.[15]

5.6 QUALITY AND SAFETY OF FUNCTIONAL ICECREAMS

Ice-cream is a favorable media for microbial development because of high supplement esteem, practically unbiased pH esteem (pH ~6 to 7), and its long stockpiling span. Nonetheless, purification, freezing, and solidifying steps in the creation can dispose of most of the microbiological dangers.[33] Purification process is regularly applied treatment in the dairy business. This can annihilate practically most of the pathogenic microbes in milk. The ensuing cycle that subjects the combinations to frigid temperature can also hinder the development of any excess greenery. Solidifying process is an additional significant control point that further decreases the dangers. Moreover, as programmed machines are usually utilized for ice-cream making in dairy industry, the possibility of tainting through direct hand control can be decreased.[16,25]

All issues considered, there are few stages in the creation of ice-cream that can prompt the microbiological risks. Warmth treatment by purification can devastate the greater part of the microorganisms that posture danger to general health. Nonetheless, the potential microbiological perils found in the end results can in any case be presented after purification through adding polluted fixings and ill-advised interventions with methods. This is particularly significant in the planning of delicate ice-cream as its last phase of the creation is done at a retail location. For ice-cream items, *L. monocytogenes* is of huge food handling issue around the world.[28]

The manufacturing of ice-cream is characterized in Part II of Schedule-1 of the Food and Drug Regulations. As per legal mandates, the ice-cream should contain at least 5% fat, 10% sugar, and 7.5% milk solids other-than-fat; also, the ice-cream containing any organic product, natural product mash or natural product puree will either adjust to the aforementioned standard, or then again, the complete substance of fat, sugar and milk solids other-than-fat will not be under 25% of the ice-cream including the natural product, natural product mash or natural product puree; all things considered, and such all of the substance of fat, sugar, and milk solids other-than-fat will incorporate at least 7.5% fat, 10% sugar, and 2% milk solids other-than-fat. With the end goal of the standard identifying with ice-cream, sugar signifies sucrose, sugar, or solids of any improving material from starch, given that no ice-cream should contain fewer than 7.5% sucrose.[1,32]

For quality control of ice-cream with respect to all the parameters, premises fabricating ice-cream should be covered by legitimate Frozen Confection Factory Permit under the Frozen Confection Regulation. Frozen Confection grants are likewise needed for retail outlets selling ice-cream in mass or cone in unique coverings. Every one of these premises is likewise needed to go along and notice important permitting necessities and conditions. The Frozen Confection Regulation specifies the prerequisite for the assembling of frozen ice-cream and the microbiological guidelines. What is more, as far as possible for *Listeria monocytogenes* in the rules of ready-to-eat food is utilized for the observations. For imported ice-creams, the Frozen Confections Regulation requires that all frozen sugary treats imported from a wellspring of production should be approved by the Food and Environmental Hygiene Department.[5,29]

5.7 CURRENT TRENDS AND UPCOMING CHALLENGES

The global ice-cream industry is expanding year by year with the compounded annual growth rate (CAGR) of 4%. The manufacturers of the ice-cream are serious and exploring different avenues regarding arrangements and fixings. Arrangement patterns are communicated in both ways.[29] Ice-creams are presented as in serving with and as bites; for example, ice-cream sandwiches or softies are presented with the fried potato chips. Moving ice-cream variants has combination of new flavors and better suitable building ingredients including sugar-less and reduced sugar assortments and considerations, which include vanilla, chocolate, and other flavors. The tactile properties play an indispensable part because it decides the overall adequacy of the result.[24,31]

When examining connected investigations, the notion of usefulness was certainly known in the American and European eras, and the inclusion of this idea has expanded the scope of the studies on the functionality of ice-cream by incorporation of various active bioingredients. The functional now has not limited to the scholarly level and concept, it is generalizing, and the collection of active functional ice-cream is not restricted to the typical Euro-American markets but also to the local Asian businesses. It is important to increase the amount of useful ice-cream in the market with new practices.[3,18]

5.7.1 ICE-CREAM BUSINESS AND ITS MARKET

The functional ice-cream industry is relatively nascent in India since traditional market has remained much more conventional product-oriented due to helpless overall revenues here and there even under 10%. Due to conscious customers, Indian ice-cream makers are giving business sector strong organic ingredients for improving the overall quality of the ice-creams.[25]

Probiotic ice-cream, which overhauls absorption, is essentially a useful food source in nature that relies on savvy calorie counting points. Huge Indian ice-cream makers (such as Hindustan Unilever, Amul quality associations, etc.) and sponsors are involved in finding the right market to adopt the functional products of the frozen desserts. The offers from Baskin Robbins, Walls, Vanilla, Mother Dairy, Amul, Cream Bell, etc., are ice-creams made from plain lacteal secretion at a reasonably low cost, ice-cream from Walls is for joy and retention, and in addition to the endless compelling flavor premium classes, Vanilla, Cream Bell, Amul, and Mother Dairy offer a portfolio in a very vast range of domains.[2,17]

The Indian ice-cream market having more than 40% share has a place with the coordinated area developing at about 15% on yearly premise. Amul stands out with around 36–38% piece of the overall industry (5% of its absolute incomes) trailed by Quality Walls and impulse the functional ice-cream buying and rest by family utilization at home and about 15% in parlor deals. In the Indian functional ice-cream industry, where cost is the prime factor and its sharing is far negligible, the functional ice-creams cannot be marketed solely based on a particular brand alone, but this should also coordinate to the customer cost strength and friendly relationships between the negotiators. It is not about just the ice-cream itself, rather the delicious taste, the compelling marketing channel of distribution, as well as the persistent outreach of the group that extensively separates one manufacturing organization from another and it drives the whole organization with a way forward.[8,32]

5.7.2 CHALLENGES

Stronghold is a decent strategy to apply on the off chance. Numerous times the stronghold is lost because of elements like temperature, warmth, and bundling material or if appropriate cleanliness is not kept up. Unit tasks utilized in handling of ice-cream do not debase nutrients, which are included in the structure. In any event, whipping and air circulation of the blend in frigid temperature does not cause a lot of oxidative misfortunes; however, the best handling misfortunes are because of sanitization of the blend. The timeframe of realistic usability of sustained ice-creams may diminish with a total change in its tactile properties. It is along these lines testing to sustain ice-creams remembering of both the nourishing elements and tactile properties of the Ice-cream.[24,29]

There are certain unavoidable issues associated with the functional ice-cream businesses, and it is very true that functional ice-cream businesses and the income derived out of them have been potentially disrupted in past years because of the enduring interest and potential market for ice-creams in populous tropical countries, such as India. In addition to these difficulties, the functional ice-cream industry needs to take a quick forward step for greater globalization of the functional products, influential changes in the public perception toward functional ice-cream and dairy products, constant increases in per capita financial aspects, and lively causes of the Indian population, under the spontaneous stimulus of special and innovative barriers, in nonindustrial countries and customs around the globe.[7,13]

5.7.3 DIGESTIBILITY OF THE FUNCTIONAL ICE-CREAM

Ice-cream is the most favored dessert usually consumed by all age groups after meal and enjoyed as a delicious treat. Its sweet, wonderful flavor, smooth surface, and trademark coolness make it an exceptionally satisfactory food. Its high attractiveness invigorates the progression of stomach-related juices, which upgrades processing.[21,33] Besides, homogenization utilized in its production encourages assimilation as it occurs with homogenized milk, as the stomach-related squeezes promptly follow up on formation of the tiny globules because of the breaking down of large fat globules. Similarly, it remains relatively constant for the milk-fat mixture in the functional ice-cream when confined through the homogenizer, such as the current business functional ice-cream plant training.[6,17]

5.8 SUMMARY

Ice-cream business is one of the quickest growing sectors of food processing and dairy farm industry because ice-cream is popular among each youngster and adult. The seasoned ice-creams are quite common within the current situation because of its low price of production. A functional ice-cream is rising at an ever-increasing pace in new foodstuff development across the planet because of its organic process edges additionally to instant satisfaction to the consumer. The most important aim of functional ice-cream is to introduce useful biocompounds into the body through daily dietary intake. It is a smart carrier of all useful ingredients because of its low storage temperature, stabilization of ingredients, and recognition among consumers. This idea flips the thought of ice-cream being associate degree unhealthy and turns Ice-cream into a positive and essential healthy treat. The market of ice-cream is turning into a lot of health aware day by day and rigorous not just for the supply of energy, however conjointly for the wonderful delivering of some nutrients and bioactive compounds. It includes all supplements of functional nutrients like vitamins, minerals, antioxidants, fiber, probiotics, omega-3 fatty acid fatty acids, and prebiotic. The nutritionally sound and nutraceutical enriched ice-cream is superimposing the present commerce to underline their demand at national and international markets.

KEYWORDS

- **antioxidants**
- **functional ingredients**
- **health**
- **ice-cream**
- **nutraceuticals**
- **probiotics**

REFERENCES

1. Akalın, A. S.; Kesenkaş, H.; Dinkci, N.; Ünal, G.; Özer, E.; Kınık, O. Enrichment of Probiotic Ice-Cream With Different Dietary Fibers: Structural Characteristics and Culture Viability. *J. Dairy Sci.* **2018,** *101*(1), 37–46.
2. Al-Sherajia, S. H.; Ismail, A.; Manap, M. Y.; Mustafa, S.; Yusof, R. M.; Hassan, F. A. Prebiotics as Functional Foods: A Review. *J. Funct. Foods* **2013,** *5*, 1542–1553.
3. Anonymous. World Ice-cream consumption. [Online] 2018. https://www.dunya.com/dunya-gida/dondurmatuketimi-son-10-yilda-4-kat-artti-haberi-376170 (accessed Aug 24, 2021).
4. Arslaner, A.; Salık, M. A.. Determination of Some Quality Properties of Low-Calorie Ice-Cream Produced With Walnut Paste and Dried Mulberry Powder. *Atatürk Univ. J. Agric. Fac.* **2017,** *48*(1), 57–64.
5. Arslaner, A.; Salık, M. A. Functional Bioactive Components of Milk. *Erzincan Univ. J. Sci. Technol.* **2019,** *12*(1), 124–135.
6. Arslaner, A.; Salık, M. A. *Functional Ice-cream Technology and Nutraceutical Components Used in Production*, Unpublished Paper at Third International Conference on Advanced Engineering Technologies, Bayburt, Turkey, Sept 19–21, 2019; p 8.
7. Corradini, S. A. S.; Madrona, G. S.; Visentainer, J. V.; Bonafe, E. G.; Carvalho, C. B.; Roche, P. M.; Prado, I. N. Sensorial and Fatty Acid Profile of Ice-Cream Manufactured With Milk of Crossbred Cows Fed Palm Oil and Coconut Fat. *J. Dairy Sci.* **2014,** *97*(11), 6745–6753.
8. Dayısoylu, K. S.; Gezginç, Y.; Cingöz, A. Functional Food or Functional Component? Functionality in Foods. *Foods* **2014,** *39*(1), 57–62.
9. Douglas, L. C.; Sanders, M. E. Probiotics and Prebiotics in Dietetics Practice. *J. Am. Diet. Assoc.* **2008,** *108*, 510–521.
10. FAO/WHO. *Joint FAO/WHO (Food and Agriculture Organization/World Health Organization) Working Group Report on Drafting Guidelines for the Evaluation of Probiotics in Foods*, FAO/WHO: Ontario, Canada, 2002, p 110.
11. Gabbi, D. K.; Bajwa, U.; Goraya, R. K. Physicochemical, Melting, and Sensory Properties of Ice-Cream Incorporating Processed Ginger (*Zingiber officinale*). *Int. J. Dairy Technol.* **2018,** *71*(1), 190–197.
12. Gibson, G. R. Fiber and Effects on Probiotics (the Prebiotic Concept). *Clin. Nutr. Suppl.* **2004,** *1*(2), 25–31.
13. Gibson, G. R.; Roberfroid, M. B. Dietary Modulation of the Human Colonic Microbiota: Introducing the Concept of Prebiotics. *J. Nutr.* **1995,** *125*, 1401–1412.
14. Goff, H. D., Hartel, R. W. *Ice-cream*, 7th ed.; Springer: Heidelberg, Dordrecht, 2013; p 476.
15. Granato, D.; Branco, G. F.; Cruz, A. G.; Faria, J. A. F.; Shah, N. P. Probiotic Dairy Products as Functional Foods. *Compr. Rev. Food Sci. Food Saf.* **2010,** *95*, 455–470.
16. Granato, D.; Branco, G. F.; Nazzaro, F.; Cruz, A. G.; Faria, J. A. F. Functional Foods and Non-dairy Probiotic Food Development: Trends, Concepts, and Products. *Compr. Rev. Food Sci. Food Saf.* **2010,** *9*(3), 292–302.
17. Kailasapathy, K.; Harmstorf, I.; Phillips, M. Survival of *Lactobacillus acidophilus* and *Bifidobacterium animalis* ssp. Lactis in Stirred Fruit Yogurts. *LWT-Food Sci. Technol.* **2008,** *41*, 1317–1322.

18. Mahmoudi, R.; Fakhri, O.; Farhoodi, A.; Kaboudari, A.; Rahimi, S. F.; Tahapour, K.; Khayyati, M.; Chegini, R. Review on Probiotic Dairy Products as Functional Foods Reported From Iran. *Int. J. Food Nutr. Saf.* **2015,** *6*(1), 1.
19. Nadeem, M.; Situ, C.; Abdullah, M. Effect of Olein Fractions of Milk Fat on Oxidative Stability of Ice-cream. *Int. J. Food Prop.* **2015,** *18*(4), 735–745.
20. Oliveira, R. P. Z.; Perego, P.; Oliveira, M. N.; Converti, A. Effect of Inulin As Prebiotic and Synbiotic Interactions Between Probiotics to Improve Fermented Milk Firmness. *J. Food Eng.* **2011**, *107*(1), 36–40.
21. Pandiyan, C.; Annal-Villi, R.; Kumaresan, G.; Murugan, B.; Gopalakrishnamurthy, T. R. Development of Symbiotic Ice-Cream Incorporating *Lactobacillus Acidophilus* and *Saccharomyces Boulardii. Int. Food Res. J.* **2012,** *19*(3), 1233–1239.
22. Ratnam, D. V.; Ankola, D. D.; Bhardwaj, V., Sahana, D. K., Ravi Kumar, M. N. V. Role of Antioxidants in Prophylaxis and Therapy: A Pharmaceutical Perspective. *J. Control. Release* **2006,** *113*(3), 189–207.
23. Roberfroid, M. B. A European Consensus of Scientific Concepts of Functional Foods. *Nutrition* **2000,** *16*(7/8), 689–691.
24. Salem, S. A., Hamad, E. M., Ashoush, I. S. Effect of Partial Fat Replacement by Whey Protein, Oat, Wheat Germ and Modified Starch on Sensory Properties, Viscosity and Antioxidant Activity of Reduced Fat Ice-cream. *Food Nutr. Sci.* **2016,** *7*, 397–404.
25. Schrezenmeir, J., Vrese, M. Probiotics, Pebiotics and Synbiotics-Approaching a Definition. *Am. J. Clin. Nutr.* **2001,** *73*(2), 361–364.
26. Sezen, A.G. Effects of Prebiotics, Probiotics and Synbiotics Upon Human and Animal Health. *Atatürk Univ. J. Vet. Sci.* **2013,** *8*(3), 248–258.
27. Shahidi, F. Functional Foods: Their Role in Health Promotion and Disease Prevention. *J. Food Sci.* **2004,** *69*(5), 146–149.
28. Tsevdou, M., Aprea, E., Betta, E., Khomenko, I., Molitor, D., Biasioli, F., Gaiani, C., Gasperi, F., Taoukis, P., Soukoulis, C. Rheological, Textural, Physicochemical and Sensory Profiling of a Novel Functional Ice-cream Enriched With Muscat de Hamburg (vitis vinifera l.) Grape Pulp and Skins. *Food Bioprocess Technol.* **2019,** *12*(4), 665–680.
29. Tsuchiya, A. C., Silva, A. G. M., Brandt, D., Kalschne, D. L., Drunkler, D. A., Colla, E. Lactose-reduced Ice-cream enriched with whey powder. *Semina: Ciências Agrárias.* **2017,** *38*(2), 749–758.
30. Tur, J. A., Bibiloni, M. M. Functional Foods. In *Encyclopedia of Food and Health,* 2016; pp 157–161.
31. Turgut, T., Çakmakçı, S. Investigation of the Possible use of Probiotics in Ice-cream Manufacture. *Int. J. Dairy Technol.* **2009,** *62*(3), 444–451.
32. Ullah, R., Nadeem, M., Imran, M. Omega-3 Fatty Acids and Oxidative Stability of Ice-Cream Supplemented with Olein Fraction of Chia (*Salvia hispanica* L.) Oil. *Lipids Health Dis.* **2017,** *16*(34), 1–8.
33. Xavier, A. A. O.; Mercadante, A. Z. The Bioaccessibility of Carotenoids Impacts the Design of Functional Foods. *Curr. Opin. Food Sci.* **2019,** *26*, 1–8.

PART II

PHYSICOCHEMICAL CHARACTERIZATION OF MILK AND MILK PRODUCTS

CHAPTER 6

PHYSICOCHEMICAL CHARACTERISTICS OF MILK

SUNIL MEENA, PARTHA PRATIM DEBNATH,
SUVARTAN G. RANVIR, and DINESH CHANDRA RAI

ABSTRACT

Milk contains almost all nutrients required by our body. The high nutritive value of milk is primarily because of presence of fats, proteins, lactose, vitamins, and minerals. Although milk contains almost 85–87% water, yet its physicochemical properties can vary as compared to water. These properties of milk are directly affected by processing parameters, animal species, and health of animals. This chapter presents review on density, specific gravity, freezing and boiling point, acidity, pH, surface tension, optical, thermal properties, etc. This chapter has significant importance for budding dairy professionals for easy handling and processing of milk in the dairy industry.

6.1 INTRODUCTION

Milk is a complex fluid, which consists of several types of macro- and micronutrients, such as fat, protein, milk sugar (lactose), vitamins, minerals, and several types of enzymes. In milk, fat is present in oil-in-water emulsion, milk protein, and some minerals are in colloidal state and lactose, whey protein and some other milk minerals are in true solution. Milk has whitish color due to dispersed milk constitutes, such as milk fat, protein, and some minerals, while cow milk is of slightly yellowish color because of presence of β-carotene. Milk is a rich source of various micronutrients, such as calcium, phosphate, and different vitamins.[28]

The Chemistry of Milk and Milk Products: Physicochemical Properties, Therapeutic Characteristics, and Processing Methods. Megh R. Goyal, Suvartan Ranvir, & Junaid Ahmad Malik (Eds.)

Physicochemical properties of milk differ slightly from water as it contains numerous solutes (fat, protein, lactose, and minerals). These properties are influenced by milk composition and structure, handling, processing, and storage conditions, etc. The assessment of physicochemical properties of milk is important for examining the milk and milk components, studying microstructure, designing processes and equipment, and understanding complex chemical changes. Many researchers have reviewed physicochemical properties of milk and have reported that processing treatments (such as heat transfer processes, fluid flow, churning, homogenization, and emulsification) could affect the physicochemical properties of milk and milk products.[36,38]

This chapter focuses on various physicochemical properties of milk. In addition, the effect of different milk components on physicochemical properties and their impact on different milk processing operations have also been elaborated.

6.2 DENSITY AND SPECIFIC GRAVITY OF MILK

The density (ρ is a symbol for density) is a mass per unit volume (kg/m^3 = 1000 g/m^3). Reciprocal of density is called specific volume of a milk product. Ratio of density of a milk product to the density of water at 4°C is called a specific gravity of a product (dimensionless). The density of milk and other dairy products are mainly useful in mass and volume conversions, solid content estimation, and in the determination of milk and milk product's physical properties (such as kinematic viscosity). Density is dependent on temperature of product at the time of testing, temperature during processing, product composition (majorly affected by fat content), and entrapped air in the product (mainly observed in higher viscous products).[32]

Addition of milk solids (such as lactose, protein, and mineral) will increase the density of final mix, while addition of milk fat or other oils decreases the density of mixture. Temperature significantly affects the density; thus, temperature measurement is important during density determination. Temperature profile (handling temperature) also influences density of milk and should be warmed to 40 to 45°C temperature (liquefying milk fat) to minimize such an effect on milk followed by cooling to measure the temperature (generally at 20°C). The density of milk ranges from 1.027 to 1.033 g/m^3 at 20°C.[14,15]

Specific gravity is a unit-less property and the effect of temperature on it is quite lower as compared to the density. Specific gravity is also known as relative density.[5] The relationship between specific gravity, density of water, and any substance is given in the following equation:

$$Specific\,Gravity = \frac{Density\,of\,substance(\rho s)}{Density\,of\,water(\rho w)} \tag{6.1}$$

The specific gravity of a milk is generally measured by a lactometer.[34] Lactometer scale graduation marking is directly termed specific gravity. Equation (6.2) can be applied for rapidly analyzing the total solids (TS) using specific gravity of a milk from lactometer readings and fat content of milk.

$$TS = \frac{(SG_{SNF} - SG_{fat})}{SG_{fat}(SG_{SNF} - 1)} + \frac{SG_{SNF}(100SG_{milk} - 100)}{SG_{milk}(SG_{SNF} - 1)} \tag{6.2}$$

Milk processing steps (e.g., pasteurization, homogenization, and sterilization) have negligible effect on density.[1,33] Nowadays, in-line measurement of density by using automatic measuring instrument for process and quality control purpose is available, for example, in-line measurement of TS of concentrated milk in evaporation process for TS control purpose.[39] Values of specific gravity of milk and milk constituents are given in Table 6.1.

TABLE 6.1 Specific Gravity of Various Milk and Milk Constituents.

Different milk and milk constituents	Specific gravity
Buffalo milk	1.030–1.032
Cow milk	1.028–1.030
Milk fat	0.918
Milk protein	1.346
Milk salts	4.120
Milk sugar (lactose)	1.780
Skimmed milk	1.036
Solids-not-fat (SNF)	1.616

6.3 FREEZING POINT (FP) AND BOILING POINT (BP) OF MILK

The freezing point (FP) and boiling point (BP) are colligative properties of milk that are determined by the use of molarity of solutes. The FP of normal milk is at −0.522°C, and BP of normal milk is at 100.15°C. The FP of milk is affected by concentration of water-soluble constituents of milk (such as lactose and salts (calcium, magnesium, chlorides, potassium, phosphate, lactate, citrate, etc.)), while FP is not affected by constituents, such as fat globules, casein micelles, and whey proteins.[34] Among all the salts in milk, major contribution is by chloride ions, which is around 25% of total freezing point depression. Lactose contributes around 55% and salts other than chloride contribute around 20% in freezing point depression (FPD).[5,6] Cole et al. reported that lactose and chloride are responsible for 75% of total freezing point depression (FPD).[3]

Some of the variations in FP may be due to season, environment, feed and fodder, stage of lactation, breed, milking time, water intake, and mastitis disease (clinical). Vacuum treatment of milk slightly increases FP; it may be due to degassing of milk. Cooling and heating (pasteurization) of milk causes changes in state of salts (colloidal to true state vice-versa), but there is no significant effect on FP, perhaps maybe these changes are slow and reversible over the time. UHT treatment (direct type; steam addition in milk) causes little increase in FP; it may be due to removal of gasses such as CO_2 during additional water. In the UHT treatment, precipitation of some phosphates occurs resulting in increase in freezing point. Fermentation of milk also effects freezing point; during fermentation of lactose, 4 mol of lactic acid gets formed on the breakdown of one mol lactose. The citrate fermentation also effects FP.[6,15,39]

The BP of milk is slightly higher as compared to water as milk contains various types of dissolved solids. The BP of milk ranges from 100.2 to 101.0°C. The BP of milk is influenced by addition of solids, salts, sugar, acids, etc. The boiling point of milk also varies with the composition of milk and pressure.[34] The milk secretion process from udder is controlled by osmotic pressure of milk and equilibrium is maintained between milk and blood. Constant osmolality in milk is sustained by transfer of blood components into mammary glands. If any changes occur in lactose content of milk, then osmolality changes proportionally as it is adjusted by concentration of chloride and sodium. The constant osmolality of milk is responsible for relative constancy of FPD.

Osmotic pressure is the difference between the freezing point of a solution and the solvent (water). The FPD is measured based on osmotic pressure of milk. Initially, the Hortvet cryoscope was most widely used for the estimation of FP. Later thermistors, which measure FDP based on changes in osmotic pressure, became popular. According to Raoult's law, equation (6.3) can be used for analyzing the effect of concentration of milk constituents on freezing point depression (FPD).[23,34,39]

$$T_f = K_f M \tag{6.3}$$

where T_f = Freezing point depression (FPD), K_f = Constant for molar depression (Value for ideal aqueous solution is 1.868°C), and M = Molarity of aqueous solution.

The FP determination is commonly used to check adulteration of water in milk. Milk is considered adulterated with water when FP of milk is below −0.525°C. Milk is a complex solution that comprises numerous solutes. The amount of added water (%) can be calculated with the help of the following equation (6.4).

$$Added\,water(\%) = \frac{0.550 - \Delta T}{0.550} \times (100 - TS) \tag{6.4}$$

where ΔT = freezing point depression (FPD) observed in the milk sample and TS = total solids of milk (in %).

Estimation of added water in milk depends upon constancy of FP of milk, but adulteration of milk with isotonic solutions, such as ultrafiltration (UF) permeates (which is being considered for standardization of the protein content of milk) will not be detected by this technique.[4]

6.4 ACID–BASE EQUILIBRIA OF MILK

The soluble calcium phosphate, citrate, bicarbonates, and casein are principal constituents in milk responsible for its buffering ability. Buffering capacity of milk is increased for pH <6.6 and is maximum at pH of 5.1, due to the dissolution of the colloidal calcium phosphate (CCP), especially at pH of <5.6.[23,34]

Processing of milk influences the acid–base equilibria of milk. Pasteurization of milk causes little changes in pH value because of slight precipitation of colloidal calcium phosphate ($Ca_3(PO_4)_2$) and loss of CO_2. Heating of milk above 100°C results in reduction of pH value because of Maillard reaction (which produces formic acid by degradation of lactose). Freezing also has an impact on the pH value of milk. The pH value of milk declines with slow freezing due to formation of large ice crystals; however, quick freezing does not cause any affect. Evaporation results in the depletion of pH as CCP gets broken down or degraded (under extensive heat treatment) and this degraded CCP mixes up with the serum phase, resulting in increased quantity of soluble calcium phosphate in milk. Dilution of milk also results in transformation of CCP into soluble forms, with a parallel reduction in [H^+].

Titratable acidity of milk is often used for the determination of buffering capacity. The effect of milk salts, casein, and whey protein on buffering capacity of milk has been reported as 58.6%, 36.0%, and 5.4%, respectively.[6,23,37] The soluble salts, casein including CCP, and whey proteins that contributed to milk buffering capability are approximately 47%, 32%, and 21%, respectively, for a pH range of 6.7–4.0. Ultrafiltration (UF) process increases the buffering capacity of milk, because the retentate contains high amounts of colloidal salts, casein, and whey proteins.[15,23,39]

6.4.1 TITRATABLE ACIDITY AND PH OF MILK

Acidity and pH are generally used for monitoring the milk freshness. Acidity and pH have inverse relationship. Acidity of fresh milk ranges from 0.14 to 0.16% of lactic acid. Natural acidity has been described as fresh milk's acidity, which is contributed by casein, citrate, phosphate, carbon dioxide, and to some extent albumin and globulin. On the contrary developed or real acidity arises because of bacterial hydrolysis of lactose into lactic acid.[23]

The pH and acidity of milk are affected by species, breed, individuality and stage of lactation, and health of animal. Lipolysis of milk fat in high fat products affects estimation of acidity, as lipolysis liberates free fatty acids in milk. Speed of titration is another prominent factor, which must be considered during the estimation of acidity. Various units (such as °SH (Soxhlet Henkel degrees), °Th (Thörner degrees), °D (Dornic degrees), percent (%) of L.A. (Lactic acid)) can be used for expressing the titratable acidity, depending on sodium hydroxide's strength used for titration. Among all the units, % of lactic acid is widely used for indicating titratable acidity in milk.[6,39] Cow milk and buffalo milk's natural acidity ranges from 0.13 to 0.14% and 0.14 to 0.15% of lactic acid, respectively.[34]

The pH is the indication of hydronium ion concentration of a milk solution, and it is defined as the negative logarithm of the hydronium ion $[H^+]$ concentration in a solution as shown in equation (6.5).

$$pH = -\,log[H^+] \tag{6.5}$$

The pH value of milk represents true acidity and its range in normal milk lies between 6.5 and 6.7. The pH of milk is generally measured at 25°C by using pH meter. The pH value of fresh cow and buffalo milk ranges between 6.4 and 6.6 and 6.7 and 6.8, respectively. The pH value on higher side of milk represents abnormal milk because of udder infection, while the pH values at lower side represents development of acidity due to bacterial action.

The pH value of milk depends on temperature because solubility of calcium phosphates is temperature dependent.[23] Colostrum milk exhibits a lower pH (as low as 6.0) as compared with normal milk because of the presence of high amounts of soluble salts and protein.

Mastitis leads to the elevation in pH value of milk, as the membrane permeability of mammary gland increases, leading to release of more blood components in milk. The pH of cow's blood is around 7.4; this variation in the pH value of milk and blood results in active transport of ions from blood to milk, followed by precipitation of CCP during synthesis of casein micelles, increased quantification of acidic groups in milk, and lower buffering capacity between 6 and 8 pH range.[23,25]

6.5 OXIDATION–REDUCTION EQUILIBRIA OF MILK

Oxidation is a loss of electrons by any substance, while reduction is a gain of electrons by a substance. A component causes oxidation of other components known as oxidizing agents, while a component, which causes reduction of the other, is known as a reducing agent. This tendency (loss or gain of electron) of a substance can be measured by using galvanic or voltaic cells.[24,34] Nernst equation (6.6) is used for evaluating the relationship between concentration of oxidized and reduced forms of a substance and redox potential, as follows:

$$E_h = E_0 + \frac{2.3RT}{nF} log\left(\frac{[Ox]}{[Red]}\right) \tag{6.6}$$

where E_h = Redox potential of substance (in V); E_0: = Standard redox potential (the potential when the reactant and product are at unit activity: in V); T = Absolute temperature (°K); R = Universal gas constant (= 8.31 J $K^{-1}mol^{-1}$); F = Faraday constant (96500 $C.mol^{-1}$); n = Number of electrons transferred per molecule; and [Ox] and [Red] = Molar concentrations of the oxidized and reduced forms, respectively.

At normal pH value, the E_h of milk sample (if sample in equilibrium with air) ranges from +0.25 to +0.35V at 25°C.[13,23,39] The lactose-pyruvate, ascorbate, and riboflavin are involved in the redox system of milk, whereas constituents, such as fat, lactose, and protein, do not contribute to redox reactions of milk. The ascorbate in milk is present in the reduced form when milk is drawn from udder, but later it gets oxidized to dehydroascorbate, and this reaction is entirely dependent on quantities of copper (Cu), ferrous (Fe), and oxygen in milk. Being present in a very minimal amount, the effect of riboflavin on the E_h value of milk is negligible.

The bacterial activity, heat treatment, oxygen concentration, extent of contamination of milk with ferrous and copper, and light exposure also play crucial role in the redox reactions of milk.[13,23] Heat treatment results in the drop of E_h values due to the liberation of sulfhydryl (-SH) groups from β-lactoglobulin, which gets further oxidized by atmospheric oxygen. Maillard reaction also affects the redox potential of milk, and it is mainly observed in condensed and dried milks.[11] In addition, pasteurization by the HTST method also lowers the E_h value of milk due to less oxidation of ascorbic acid and higher retention of disulfide-reducing substances in milk.

Removal of oxygen prior to heat treatment decreases the E_h value, as considerable retention of reducing substance occurs when heating is done in the presence of air or in equilibrium with oxygen.[24,34] The lowering of E_h value by bacterial activity can be due to the consumption of dissolved oxygen in medium during the bacterial metabolism process. Addition of any dye or color in milk also plays a part in lowering the redox potential. On the other hand, contamination of milk with copper and iron results in the rise of E_h value of milk.[23,24,39]

6.6 OPTICAL PROPERTIES OF MILK

6.6.1 *ABSORPTION AND SCATTERING OF LIGHT THROUGH MILK*

When electromagnetic radiation passes through the chemical components (atoms, ions, or molecules), it can either get absorbed at specific frequency or get scattered or it can emit fluorescent radiation.[24,34] Milk is a complex colloidal system, in which fat globules, casein micelles, and whey proteins are dispersed in aqueous phase of lactose, salts, and other minor compounds. Milk as a complex colloidal system not only absorbs the light at various wavelengths but also scatters the light as milk contains fat globules and casein micelles.

In the visible region, the riboflavin in milk absorbs radiation at 470 nm and emits maximum fluorescent radiation at 530 nm, while the β-carotene in milk absorbs the radiation close to 460 nm. In ultraviolet (UV) region, aromatic amino acid residues of milk proteins, such as tyrosine and tryptophan, absorb radiation close to 280 nm and emit it as fluorescent radiation at 340 nm.[15]

Milk protein content can be quantified by measuring the UV fluorescent intensity. In infrared (IR) region, absorption of radiation occurs by amide (II) groups (CONH) of proteins (at 6465 nm), hydroxyl groups (OH) of milk sugar (at 9610 nm), and ester carbonyl groups (C=O) of milk fat (at 5723 nm).

There are numerous commercial IR analyzers for measurement of fat, milk sugar, and protein in the milk. The milk samples should be properly homogenized prior to analysis to reduce the size of fat globules. Linear Beer-Lambert plot is drawn for each milk component (0–6% range) and the sample thickness should be around 37 μm. Co-relation of IR method and chemical method (Standard method) should be done by repeated sample preparation, dilution, and instrument calibration. The casein micelles and fat globules will scatter light and by the measurement of the transmitted and scattered light, the fat content of milk, size distribution of casein micelles, and fat globules can be analyzed.[24,33,34] Similar methods have been utilized for evaluating the homogenization efficiency of milk and milk products.[7,12] Milko-tester follows the same principle and is widely used in dairy industry for the determination fat and protein content on the basis of turbidity.[23]

Appearance of milk is also affected by light scattering and absorption properties.[39] The slight yellowish color of cow milk is due to the presence of β-carotene. Skim milk has a bluish tinge due to scattering of blue light at short wavelengths. Color of milk sample is determined by transmission and reflectance techniques. Tri-stimulus colorimeter has been used for color determination and results are expressed in terms of L^*(dark-light axis), a^*(green-red axis), and b^*(blue-yellow axis).[16]

6.6.2 REFRACTIVE INDEX (RI) OF MILK

The RI is a ratio of the velocity of light in air to the velocity of light in the medium (sample). It is normally determined for transparent (in which light passes) sample. Refraction properties of a ray of light depend on the density of medium. When rays pass from lower to higher density, then it bends toward the normal, while in case of high to less density rays of light it bends away from the normal.[23] The RI (n) can be expressed by equation (6.7)

$$n = \frac{\sin i}{\cos R} \tag{6.7}$$

where i = Angle between the ray in the first medium and the normal to the dividing surface (incidence); and R = Exit angle (refraction).

The RI value of milk is checked by light of constant wavelength (589.3 nm) and at constant temperature (20°C). The RI values of water and milk at 20°C are 1.330 and 1.344 to 1.35, respectively.[15,27] The RI value of milk fat varies from 1.4537 to 1.4552 at 40°C. This difference in refractive index values of milk and water can be explained because of the presence of dissolved constituents in milk, such as lactose, whey proteins, casein micelles, and minerals.[34,39] RI can be used for rapid detection of SNF content. The purity of milk fat can also be quickly judged from RI values. The measurement of refractive index (RI) is difficult due to scattering of light by milk constituents, such as casein micelles and fat globules. The RI is also used to check adulteration of milk with water, but it is less accuracy.[23]

6.7 HEAT STABILITY (HS) OF MILK

The HS of milk is checked by the addition of alcohol (68% v/v) or by heating of milk. The HS is measured through heat coagulation temperature (HCT), heat coagulation time, viscosity measurement, heat exchanger's fouling, sediment formation, or casein micelle size. Whole milk has 20 to 30 min HCT at 140°C temperature and natural pH. Low pH and higher concentration of milk reduce the HS of milk. The heat-induced changes depend on intensity of heat treatment. On coagulation, milk proteins get aggregated due to destabilization of κ-casein micelles. Addition and removal of calcium and phosphate in milk affects HS of milk. The HS of milk and milk products can be improved by the addition of stabilizer and controlling the pre-denaturation of whey protein.[35]

6.8 VISCOSITY OF MILK

Viscosity is an important physical property of milk and milk products, as it gives an idea about mouth feel, sensory acceptability, and processing characteristics of the product. The viscosity of any substance is defined as resistance to flow. Viscosity of milk can be described as the internal friction or its ability of resistance to flow.[24,34] The viscosity of milk is generally measured in centipoise (= 10^{-2} poise) and its SI unit is Pa.s (Pascal-second). The kinematic viscosity is commonly determined for fluid flow. Whole milk and skim milk shows Newtonian behavior (except cooled raw milk), for which viscosity of milk is independent of the shear rate. Cooled raw milk and cream shows non-Newtonian behavior, for which viscosity changes according to the shear rate. Rheological behavior of milk can be defined by coefficient of viscosity, which is temperature-dependent.[15,39] The Newton's law of viscosity is given by equation (6.8):

$$\tau = \gamma\,(\eta) \tag{6.8}$$

where τ = shear stress (Pa), η = Newtonian coefficient of viscosity (Pa-s), γ = shear rate (s^{-1}).

Viscosity of milk depends on milk composition, concentration, temperature, temperature history, and milk processing. Milk sugar (lactose) and whey proteins have negligible impact on viscosity, while milk fat and protein (casein) have relatively significant impact on viscosity.[15,34] Effects of milk composition and concentration can be determined by using Eilers's semiempirical equation. Viscosity of milk escalates with increase of total solids in milk. Viscosity depends on temperature of milk, as the state (solid to liquid or *vice versa*) of milk fat changes according to the temperature of milk and composition of fat globule membrane deviates with temperature. Lowering the temperature of milk results in the increase in viscosity because of increases in voluminosity of casein micelles and dissociation of casein (especially β-casein) occurs at low temperature. On the other hand, viscosity again increases above 65°C because of denaturation of whey proteins.[34]

The pH value of milk also affects the viscosity of milk as voluminosity of casein micelles changes according to increase or decrease of pH. Ageing of milk increases the viscosity due to slight changes in ionic equilibrium.[15] Homogenized milk is more viscous as compared with un-homogenized milk, because of uniform size distribution of fat globules in homogenized milk. Pasteurization of milk causes nonsignificant changes in viscosity.[30]

The viscosity of milk and milk product can be determined by using Coaxial cylinder (Brookfield), Falling sphere (Hoeppler), and Capillary tubes (Ostwald's) viscometers. In addition, more advanced viscosity measurement tools are available, such as rheometer and consistometer.[15] Viscosity of different milk and milk products is given in Table 6.2.

6.9 SURFACE TENSION (ST) OF MILK

The forces, which lead to reduction in interfacial or surface area, can be defined as interfacial or surface tension, respectively. The ST of any liquid varies with the degree of purity of liquid or the temperature of liquid. The ST of milk is affected by the addition of surfactants, because surfactants have hydrophilic and hydrophobic groups.[10] Gibbs equation for relating surfactant activity and interfacial tension is as follows:

$$dT = -RKC_s dln(a) \tag{6.9}$$

where C_s = excess concentration of solute at membrane in bulk solution, a: = activity of solute in the continuous phase, R = universal gas constant, and K = temperature (in Kelvin).

TABLE 6.2 Viscosity of Milk and Milk Products.

Milk product	Viscosity (mPa.s)
Evaporated milk	17
Lactose suspension (5%)	1.150
Skimmed milk	1.790
Stirred yoghurt	100
Whey (Rennet)	1.250
Whole milk	2.127

The ST of milk can be measured by the use of static, dynamic, and semi-dynamic methods. ST is commonly measured by the static method in which capillary rise, the DuNuoy ring, the Wilhelmy plate, or bubble pressure methods are used. Ring or plate method is generally used for the determination of ST of milk. The falling drop method (in which known volume of milk drops and number of drops are counted) has been described as a dynamic method. The ST of whole milk ranges from 46 to 47.5 N m^{-1} at 20°C, while values for skimmed milk, cream (25% fat), and sweet-cream buttermilk range from 52 to 52.5, 42 to 45, and 39 to 40 Nm^{-1}, respectively.[15,24]

Milk fat and milk serum are continuous phases of milk and air-serum, and fat-serum are common interfaces. If milk has a dissolved air, ice, or lactose crystal, then it will be also considered milk phase. In milk, many constituents act as surfactants, such as some milk proteins, phospholipids, and free fatty acids. Lactose and minerals are not affected by the interfacial or ST of milk. Buttermilk has lower surface tension because it contains high amounts of surface-active proteins, phospholipids, and fat globules membrane. The ST is majorly affected by the fat globular membrane material as it has very low interfacial tension. The ST value of whole milk is slightly lower than the skimmed milk because it contains higher amounts of fat globular membrane material and free fat as compared to the skimmed milk. Thus, it can be inferred that higher is the fat content of milk, lower will be the surface tension. Lipolysis of milk fat also reduces surface tension of milk.[34] Bacterial growth also affects the ST of milk. By determining the surface tension values, it is possible to check the adulteration of milk with water or starch.[2]

Apart from milk's composition, the method of processing also has a considerable impact on surface tension values of milk. ST of milk decreases with increase in temperature. Homogenization of milk influences the ST value, due to free fatty acid formation and lipolysis of milk (in case lipase remains active). Milk pasteurization slightly increases ST of milk. Sterilization of milk also slightly increases ST of milk because of coagulation and denaturation of milk proteins, which show low surfactant activity.[18] The ST of different types of milk at different temperatures is shown in Table 6.3.

TABLE 6.3 Surface of Various Milk and Milk Products at Different Temperatures.

Type of milk	Surface tension (mN m^{-1}) at		
	10°C	25°C	40°C
Skimmed milk	48.55–51.19	40.84–42.94	40.55–42.57
Whole milk	48.86–50.58	46.08–48.5	39.35–45.09

6.10 THERMAL CHARACTERISTICS OF MILK

6.10.1 HEAT CAPACITY (HC) OF MILK

The HC (J kg^{-1} K^{-1}) of milk can be described as the required quantity of heat energy to increase the temperature of 1 kg of milk by 1 K. Sometimes, the word-specific heat is simultaneously used with HP. The ratio between the HC of milk and water at 15°C can be defined as specific heat. Both HC and specific heat have similar numerical values.[23] Equation (6.10) can be used for determining the HC:

$$HC\ (J\ kg^{-1}\ K^{-1}) = \frac{Q}{M \times \Delta T} \tag{6.10}$$

where Q = quantity of heat (J) = M C_p (ΔT), M = quantity of milk (kg), and ΔT = change in temperature (K).

Calorimeter incorporated with electric heater is one of the best setups for the measurement of HC. Although differential scanning calorimetry (DSC) is more user-friendly, yet it is commonly used for the HC determination.

The specific heat values of milk and milk products can also be calculated by using two equations (6.11) and (6.12):

$$c = \left(0.5\,MF + 0.3MSNF + MW\right)x4.18, \left(\text{KJ kg}^{-1}\text{K}^{-1}\right) \quad (6.11)$$

$$c = 4.18\ \text{MW} + 1.4\ \text{MC} + 1.6\ \text{MP} + 1.7\ \text{MF} + 0.8\ \text{MA} \quad (6.12)$$

where MW = mass fraction of water, MF= mass fraction of fat, MSNF = mass fraction of SNF, MC = mass fraction of carbohydrate, MP = mass fraction of protein, and MA = mass fraction of ash.

As temperature rises from 1 to 50°C, the skim milk's specific heat rises from 3.906 to 3.993 KJ kg^{-1} K^{-1}. Many research studies have stated the skim and whole milk's specific heat values are about 4.052 and 3.931 KJ kg^{-1} K^{-1}, respectively, at 80°C.[34] Above the freezing point, the specific heat values of low-fat cheese, cheese containing 37 to 38% moisture content, cream (40% fat), ice cream (containing 34–36% total solid), milk and butter are 2.68, 2.09, 3.56, 3.27, 3.85, and 2.05 KJ kg^{-1} K^{-1}, respectively.[20] An inverse relationship has been observed between specific heat of milk and its total solids content; however, at 70 to 80°C irregularities were reported. The specific heat value of skim milk powder lies between 1.172 and 1.340 KJ kg^{-1} K^{-1} at 18 to 30°C, whereas milk fat has about 2.177 KJ kg^{-1} K^{-1}. Therefore, it can be interpreted that as compared to solids-not-fat content, the fat concentration in milk is the principal factor to influence the specific heat of milk. This phenomenon is more evident in case of fat-rich dairy products especially cream.[24,34]

Specific heat in simple terms is sensible heat, where with energy input, only change in temperature is observed without any phase alteration. However, latent heat follows the reverse mechanism, where shift in phase occurs at persistent temperature. Milk fat is a mixture of fatty acids, where each type of fatty acid has different melting points. Therefore, the fatty acid profile of milk is a crucial factor as it absorbs the latent heat. The thermal history of fat samples is also another factor. As a result, for measuring specific heat of milk fat, both sensible and latent heat should be included. In this aspect, the terminology apparent specific heat is widely used, which is the summation of actual specific heat and energy absorbed in melting of fat (in the form of latent heat). Milk fat's apparent specific heat value is temperature dependent and milk fat's melting and crystallization temperatures give maximum variations. Therefore, it can be inferred that the term apparent specific heat should be used in case of milk fat and in fat-rich dairy products, whereas the term specific heat is suitable in skim milk, as it involves no latent heat.[17]

6.10.2 THERMAL CONDUCTIVITY OF MILK

Thermal conductivity (K) is defined as the heat transmission rate by conduction across a substance. It can also be described as the heat transfer rate across one-meter thickness of one-meter square area of a substance for one kelvin temperature change. This can be represented by equation (6.13):

$$K = \frac{Q \times D}{\Delta T \times A} \tag{6.13}$$

where K = value of thermal conductivity (W m^1 K^{-1}), Q = heat transfer rate (J s^{-1} or W), D = material's thickness (m), A = area of the material (m^2), and ΔT = change in temperature (K).

Solid products are most ideal materials for measuring thermal conductivity; however, it can also be applied in liquid products if the convection factor is not considered. Temperature has a wide significance in the thermal conductivity determination of milk. A linear correlation has been observed between temperature and thermal conductivity factor. However, as the content of total solids-not-fat content increases, accordingly the thermal conductivity of milk decreases. In addition, as the level of concentration of milk product increases, its thermal conductivity declines. The thermal conductivity values (at 20°C temperature) of water, skimmed milk having 0.1% fat content, whole milk having 2.9% fat content, and cream containing 30.7% fat are 0.603, 0.568, 0.559, and 0.384 W m^1 K^{-1}, respectively.[15,23,24] Equation (6.14) can be utilized for determining the thermal conductivity of heat-concentrated whole milk[26]:

$$K = (0.59 + 0.0012\ T)(1 - 0.0078\ TS) \tag{6.14}$$

where K = thermal conductivity, T = temperature at which thermal conductivity is determined, and TS = total solids.

However, equation (6.14) is only applicable for milk having total solid content range of 37–72% at 40–90°C. Among all the constituents of milk or food products, water is the governing factor, which influences the product's thermal conductivity. The higher thermal conductivity of water as compared with other milk constituents is the underlying reason behind this phenomenon. Air content in milk powder is another parameter influencing thermal conductivity. Therefore, it can be inferred that bulk density also plays a pivotal role apart from milk powder's constituents, while estimating its thermal conductivity. Skim milk powders having bulk densities in the range of 292–724 kg m^{-3} exhibited thermal conductivities of about 0.036 to 0.109 W m^{-1} K^{-1} at 12 to 50°C.[22,23]

6.10.3 THERMAL DIFFUSIVITY OF MILK

Thermal diffusivity is mainly applicable in unsteady state of heat transfer as it helps in the measurement of how rapidly a change in temperature occurs with time during the heat treatment and chilling processes. In simple terms, it can be inferred that thermal diffusivity measures the temperature circulation rate in a substance during its unsteady state, cooling, and freezing cycles. Thermal diffusivity (μ, m^2s^{-1}) can be more specifically described as fraction of thermal conductivity (K) and volumetric specific heat (density multiplied with the material-specific heat). This can be represented by equation (6.15)[15,23]:

$$\mu = \frac{K}{DC} \tag{6.15}$$

where μ = thermal diffusivity (m^2s^{-1}), K = value of thermal conductivity (W m^{-1} K^{-1}), D = material density (kg m^{-3}), and C = value of specific heat of the material (J kg^{-1} K^{-1}).

Thermal diffusivity of milk is 1.25×10^{-2} m^2 s^{-1} at 15–20°C.[14] However, another study reported milk and cream's thermal diffusivity value of about 9.7×10^{-8} m^2 s^{-1} at 0–22°C.[24]

6.11 ELECTRIC AND DIELECTRIC CHARACTERISTICS OF MILK

6.11.1 ELECTRICAL CONDUCTIVITY OF MILK

Milk contains a variety of soluble elements which get dissociated into ions and allow the passage of electric current through these. Electrical resistivity and conductivity can be described as the capability of a substance to transmit electric current through it. In solid foods, the term resistance is widely used; however, in liquid products, the term conductance is mostly preferred. Equation (6.16) can be used for the calculation of specific resistivity (S_r, Ohm-m) of a material.[15,23,39] Similarly, the inverse of specific resistivity (S_r) is specific conductance (G), which is represented by equation (6.17).

$$S_r = \frac{A \times R}{L} \tag{6.16}$$

$$G = \frac{1}{S_r} = \frac{L}{R \times A} \tag{6.17}$$

where A = material's area (m^2), L = length of the material (m), and R = resistance (Ohms).

The unit of specific conductance (G) is mho/m^{-1} or S/m^{-1}, where S stands for Siemens. Measurement of electrical conductivity can be carried out by passing electric current (AC voltage should be applied for avoiding electrolysis) through two platinum electrodes inserted in the milk product, followed by the resistance or conductance measurement. Milk products have poor electrical conductivity value. The electrical conductivity of bovine milk ranges from 0.0040 to 0.0055 mho/cm^{-1} at 25°C; however, it depends on the animal breed.[14] Conductivity and temperature have linear relationship: as the milk temperature increases by 1°C, its conductivity also increases by 0.0001 mho/cm^{-1}.[29] This is mainly because with rise in temperature, milk's viscosity reduces, which results in higher breakdown of electrolytes and faster movement of ions. The viscosity of the milk also has a considerable impact on electrical conductivity. The higher the viscosity, the lower the electrical conductivity as it impedes the ion's mobility.

Dilution of milk diminishes the conductivity value because amounts of ions present in a fixed volume of milk deplete, even though because ionic mobility increases with dilution. Thus, measuring electrical conductivity value of milk can be a method of detecting added water (adulterant) in milk. However, this has not been approved as a satisfactory method of detection. In contrast, detection of neutralizers by the electrical conductivity method is a suitable mode of analysis. Concentration of milk instead increases conductivity value as compared with dilution, due to increase in total solid (mainly mineral) content. This rise in conductivity with total solid level is only up to 28%, because of highly complicated salt balance among colloidal and soluble parts.[15,23,39]

The mineral profile of milk is the main governing factor to influence the conductivity values of milk. Milk ions, such as sodium, potassium, and chloride, are greatest contributors. This has formed the basis for the detection of mastitis milk, as it has high chloride content, which will ultimately lead to higher conductivity value. Both lactose and casein play a negligible part in the electrical conductivity. However, milk proteins sometimes tend to decrease the conductivity values as it obstructs the ion movement due to its role in increasing viscosity by water binding.[9] In contrast, if the colloidal calcium phosphate bridges of casein micelles get dissociated under acidic conditions, it leads to release of calcium ions in the surrounding medium, which ultimately increases the electrical conductivity. This is the underlying reason behind the higher conductivity values of fermented milk and milk having low pH (about 5) due to microbial activity.

The fat content shows a prominent effect on the electrical conductivity properties of milk. The milk fat globules (poor electrical conductor) restrict the ion movement by filling up the volume, thus resulting in decreased conductivity values. Skim milk's conductivity is approximately 10% higher as compared with whole milk and in cream, the amount of fat will determine the conductivity value.[30] Pasteurization process in full cream milk leads to higher conductivity values. In contrast, homogenization does not increase the conductivity value, specifically in the milk fat range of 1–35% (w/w).[31]

6.11.2 DIELECTRIC CHARACTERISTICS OF MILK

The ready-to-cook convenience foods are more preferred by consumers these days, which has resulted in significant usage of microwaves. The dielectric characteristic of food plays a pivotal role in the microwave food processing and radio-frequency heating. The three major parameters for dielectric characteristics are dielectric loss tangent (tan δ), permittivity (ε'), and dielectric loss factor (ε").

Permittivity or dielectric constant (ε') is the component's capability in storing electromagnetic energy, whereas the capability of a component to dissipate electromagnetic energy in heat form is described as dielectric loss factor (ε"). The dielectric tangent loss ($\tan \delta = \varepsilon'' / \varepsilon'$) is the component's capability to absorb microwaves. All these parameters (ε', ε", tan δ) are dependent on frequency, that is, 915 or 2450 MHz for microwave processing or 13.56, 27.12, and 40.68 MHz for radio-frequency heating. In addition, permittivity and dielectric loss factor values decline with increasing temperature at the above frequencies.[15,23,39] Table 6.4 presents the dielectric characteristics of various milk products at different frequencies and temperature.[8,19,21,38]

Dielectric properties also vary with milk composition. The moisture content and mineral (ionic) profile of milk are the governing factors in dielectric heating. In the presence of electromagnetic field, the polar molecules (water) of milk rotate and this rotation generates heat.[38] In addition, the ions or minerals present in milk get accelerated under electromagnetic forces and undergo collision with other molecules leading to heat generation. Therefore, it can be inferred that the higher the water and mineral content of milk product, the better will be its dielectric properties. In contrast, the nonionic constituents of milk (such as fat, protein, and lactose) have a negative impact on its dielectric characteristics. It has been observed that measurement of dielectric properties of milk is extremely useful to evaluate the approximate composition of milk in vast quantities in less time.[29] However, in case of milk powder, it might not be practical as the particle density of milk powder also has a considerable impact on its dielectric characteristics.[15]

TABLE 6.4 Dielectric Characteristics of Various Milk Products.

Milk product	Temperature (°C)	Frequency (MHz)	Permittivity (ε')	Dielectric loss factor(ε")
Human milk	5.0	2450	58.87	15.32
Human milk	30.0	2450	51.37	8.68
Milk having 1% fat	20.0	2450	70.6	17.6
Milk having 2% fat	20.0	2450	69.4	17.8
Milk having 3.25% fat	20.0	2450	67.9	17.6
Milk powder	20.0	20	1.9	0.5
Pre-mixed yoghurt	22.0	915	71.0	21.0
Skimmed milk	25.0	3000	68.0	18.0
Skimmed milk	55.0	3000	59.0	16.0
Unsalted butter	50.0	3000	22.5	4.6
Water	20.0	2450	80.2	13.4
Whole cow milk	5.0	2450	69.78	19.11

6.12 ACOUSTIC PROPERTIES OF MILK

Sounds occurring in the frequency range of $16 \times 10^{-3} - 20$ kHz can be easily detected by human ear, while infrasound (frequencies less than 16×10^{-3} kHz) and ultrasound (frequencies higher than 20 kHz) cannot be perceived by human ears. Ultrasound has extensive applications in the dairy industry. Ultrasound has been applied to milk and food products either as low intensity (power levels below one W cm^{-2}) or as high intensity (power levels in range of 10–1000 W cm^{-2}). Low-intensity ultrasound has a nondestructive effect and is mainly used in food characterization and in-line flow rate measurement. High-intensity ultrasound is used in processing of dairy and food products.[24]

Ultrasound is also used for analyzing milk's composition (especially milk fat and solids-not-fat), fat globules size, rate of milk coagulation by enzymes (rennet), and for estimating microbial spoilage in packed products (nondestructive method). In addition, it is utilized to estimate the rate of cream separation in liquid milk products. Milk adulterated with hydrogen peroxide and sodium bicarbonate can be detected by ultrasound. As ultrasound-based analysis methods are of nondestructive type, these can be used for concentrated and nontransparent samples as well. Quick and accurate results can be achieved by ultrasound-based analysis and the process or analysis is also economical.[29]

Pulse-echo techniques are used to determine the ultrasound properties of any substance, in which an ultrasound pulse of specific frequency is applied to one side of sample and successive echoes are detected and recorded on opposite side of the sample by using sample-holding cell. The velocity and attenuation coefficient are calculated by measuring echoes and amplitude time intervals. The acoustic properties change with frequency of pulses and with the sample's temperature.[24]

6.13 COLOR OF MILK

Cow milk exhibits yellowish creamy white color, while buffalo milk is creamy white in color. The colloidal casein particles and fat globules in milk are the two primary factors influencing the white color of milk as both are capable of scattering the visible light. The yellowish creamy white color of cow milk is contributed by the presence of β-carotene and up to some extent xanthophylls. The yellowness of cow milk depends on animal breed, feed type, amount of milk fat, and fat globules size. Whey has a greenish color due to the presence of riboflavin.[15,23]

Homogenization increases the whiteness of whole milk because of generation of smaller size fat globules, which increases the scattering of light in homogenized milk. Mild heating of milk gives more whitish color as compared with raw milk. This is because on heat treatment, casein micelle's particle size increases, and whey protein denatures, and both affect the reflectance of light in milk. Intense heat treatment of milk tends to develop brownish color because of Maillard reaction. Numerous factors affect the browning of milk, such as heating time and temperature, pH of sample, and water activity.[29]

The yellow color of butter and cheese is because of presence of carotenoid pigment, whose major source in milk is cattle feed. Sometimes colorants are also added to dairy products. Annatto (E160b) is mostly used in butter, cheese, and many others. Annatto is yellow-orange in color. It contains apo-carotenoid pigment, which is obtained from the pericarp of the seeds of *Bixa Orellana* (tropical shrub). Annatto has pigments like cis-bixin and nor-bixin, and these pigments mainly contribute to the yellow-orange color in milk products.[15,23]

6.14 SUMMARY

The individual constituents in milk affect its physicochemical properties differently, lactose content elevates milk's density, whereas fat content depletes it. Similarly, electrical conductivity of milk increases with its mineral content; however, a declining trend in conductivity values is observed with the rise in fat content. Temperature plays a prominent role in the determination of the physicochemical parameters. Various processes involved in milk handling (such as homogenization, pasteurization, sterilization, and cooling) also have a significant impact on the physicochemical parameters. Thus, it can be inferred that all the above-mentioned parameters must be considered for the smooth operation of dairy industry and for designing a process equipment and during milk processing and handling.

KEYWORDS

- **acidity**
- **acoustic properties**
- **density**
- **milk**
- **optical properties**
- **pH**
- **refractive index**
- **specific gravity**
- **surface tension**
- **thermal properties**
- **viscosity**

REFERENCES

1. Bertsch, A. J. Specific Heat Capacity of Whole and Skimmed Milk Between 50 and 140°C. *Lait* **1982,** *63,* 265–275.
2. Calandron, A.; Grillet, L. Measurement of the Surface Tension of Certain Milks With a Nouy Tensiometer. *Lait* **1964,** *44*, 505–509.

3. Cole, E. R.; Mead, M. 599. The Lactose-Chloride Contribution to the Freezing-Point Depression of Milk. *J. Dairy Res.* **1955,** *22*, 340–344.
4. Elisses, J.; Zee, B. The Freezing Point of Authentic Cow's Milk and Farm Tank Milk in Netherlands. *Neth. Milk Dairy J.* **1980,** *34*, 162–180.
5. Figura, L.; Teixeira, A. A. (Eds.) *Food Physics: Physical Properties-Measurement and Applications*. Springer Science & Business Media: Boston, MA, 2007; p 513.
6. Fox, P. F.; McSweeney, P. L.; Paul, L. H. (Eds.) *Dairy Chemistry and Biochemistry*; Springer: Cham, 1998; p 584.
7. Goulden, J. D. S.; Sherman, P. Simple Spectro Turbidimetric Method for the Determination of the Fat Content of Homogenized Ice-Cream Mixes. *J. Dairy Res.* **1962,** *29*, 47–53.
8. Guo, W.; Zhu, X.; Liu, H.; Yue, R.; Wang, S. Effects of Milk Concentration and Freshness on Microwave Dielectric Properties. *J. Food Eng.* **2010,** *99*, 344–350.
9. Hamann, J.; Zecconi, A. (Eds.) Evaluation of the Electrical Conductivity of Milk as a Mastitis Indicator. Bulletin 334; International Dairy Federation: Brussels, Belgium, 1998; p 22.
10. Harkins, W. D.; Thomson, G. Determination of Surface and Interfacial Tension. *Phys. Methods Organic Chem.* **1949,** *1*, 355–426.
11. Harland, H. A.; Coulter, S. T.; Jenness, R. The Interrelationships of Processing Treatments and Oxidation-Reduction Systems as Factors Affecting the Keeping Quality of Dry Whole Milk. *J. Dairy Sci.* **1952,** *35*, 643–654.
12. Haugaard, G.; Pettinati, J. D. Photometric Milk Fat Determination. *J. Dairy Sci.* **1959,** *42*, 1255–1275.
13. Higginbottom, C.; Taylor, M. M. The Oxidation-Reduction Potential of Sterilized Milk. *J. Dairy Res.* **1960,** *27*, 245–257.
14. Jenness, R. Biosynthesis and Composition of Milk. *J. Invest. Dermatol.* **1974,** *63*, 109–118.
15. Jenness, R.; Patton, S. (Eds.) *Principles of Dairy Chemistry*. John Wiley & Sons Inc.: New York, 1959; p 446.
16. Kneifel, W.; Ulberth, F.; Schaffer, E. Tristimulus Color Reflectance Measurement of Milk and Dairy Products. *Le Lait* **1992,** *72*, 383–391.
17. Konrad, H.; Rambke, K. Physical Properties of Liquid Milk Products, Part IV: Heat Conductivity of Milk, Cream and Milk Concentrates. *Die Nahrung* **1971,** *15*, 269–277.
18. Kristensen, D.; Jensen, P. Y.; Madsen, F.; Birdi, K. S. Rheology and Surface Tension of Selected Processed Dairy Fluids: Influence of Temperature. *J. Dairy Sci.* **1997,** *80*, 2282–2290.
19. Leite, J. A.; Quintal, V. S.; Tadini, C. C. Dielectric Properties of Infant Formulae, Human Milk and Whole and Low-Fat Cow Milk Relevant for Microwave Heating. *Int. J. Food Eng.* **2019,** *15*(3), 101–110.
20. Lewis, M. J. Physical Properties of Foods and Food Processing Systems. Ellis Horwood Series in Food Sciences & Technology 13, VCH Verlag: Chichester, England, 1987; pp 103–107.
21. Lide, D.R. (Ed.) *CRC Handbook of Chemistry and Physics*; CRC Press: Boca Raton, 2004; vol 85, p 2712.
22. Maccarthy, D. The Effective Thermal Conductivity of Skim Milk Powder. *Eng. Food* **1984,** *1*, 527–538.

23. Mathur, M. P.; Mathur, M. P.; Roy, D. D.; Dinakar, P. (Eds.) *Textbook of Dairy Chemistry*. Directorate of Information and Publication of Agriculture, Government of India: New Delhi, 2005; p 456.
24. McCarthy, O. J. Physical and Physicochemical Properties of Milk. In *Encyclopedia of Dairy Sciences*, Elsevier: London, 2002; pp 467–477.
25. McIntyre, R. T.; Parrish, D. B.; Fountaine, F. C. Properties of the Colostrum of the Dairy Cow. VII. PH, Buffer Capacity and Osmotic Pressure. *J. Dairy Sci.* **1952,** *35*, 356–362.
26. More, G. R.; Prasad, S. Thermal Conductivity of Concentrated Whole Milk. *J. Food Process Eng.* **1988,** *10*, 105–112.
27. O'brien, J. *Bulletin*: *Heat-Induced Changes in Lactose-Isomerization, Degradation, Maillard Browning*; International Dairy Federation: Bruxelles Belgium, 1995; pp 134–170.
28. Pereira, P. C. Milk Nutritional Composition and Its Role in Human Health. *Nutrition* **2014,** *30*, 619–627.
29. Pinkerton, F.; Peters, I. I. Conductivity of Lactose, and Freezing Point of Milk. *J. Dairy Sci.* **1958,** *41*, 392–397.
30. Prentice, J. H. Dairy Rheology: a Concise Guide; Google Scholar [Online] 1992. https://scholar.google.com/scholar_lookup?title=Dairy%20Rheology&author=J.H.%20Prentice&publication_year=1992 (accessed Apr 25, 2021).
31. Robin, O.; Britten, M.; Paquin, P. Influence of the Dispersed Phase Distribution on the Electrical Conductivity of Liquid O/W Model and Dairy Emulsions. *J. Colloid Interf. Sci.* **1994,** *167*, 401–413.
32. Rowland, S. J.; Wagstaff, A. W. The Estimation of the Total Solids and Solids-Not-Fat of Milk from the Density and Fat Content. *J. Dairy Res.* **1959,** *26*, 83–87.
33. Rutz, W. D.; Whitnah, C. H.; Baetz, G. D. Some Physical Properties of Milk, Part I: Density. *J. Dairy Sci.* **1955,** *38*, 1312–1318.
34. Sherbon, J. W. Physical Properties of Milk. In *Fundamentals of Dairy Chemistry*; Wong, N. P., Ed.; Springer Nature: Switzerland, 1988; pp 409–460.
35. Singh, H. Heat Stability of Milk. *Int. J. Dairy Technol.* **2004,** *57*, 111–119.
36. Singh, H.; McCarthy, O.; Lucey, J. Physicochemical Properties of Milk. In *Advanced Dairy Chemistry*; Fox, P. F., Ed.; Chapman & Hall: New York, USA, 1997; vol 3, pp 469–518.
37. Srilaorkul, S.; Ozimek, L.; Wolfe, F.; Dziuba, J. The Effect of Ultrafiltration on Physicochemical Properties of Retentate. *Can. Inst. Food Sci. Technol. J.* **1989,** *22*, 56–62.
38. Tıraş, B.; Dede, S.; Altay, F. Dielectric Properties of Foods. *Turk. J. Agric. Food Sci. Technol.* **2019,** 7(11), 1805–1816.
39. Walstra, P.; Jenness, R. (Eds.) *Dairy Chemistry & Physics*; John Wiley & Sons: New York, 1984; p 467.

CHAPTER 7

PHYSICOCHEMICAL CHARACTERISTICS OF CONCENTRATED MILK

SONU K. SHIVANNA, LAXMANA NAIK, and PRIYANKA SINGH RAO

ABSTRACT

Milk contains almost all essential nutrients (such as carbohydrates, lipids, proteins, minerals, and vitamins), which are needed for nourishment, growth, and maintenance of our body. Due to its high perishable nature, it is not possible to preserve milk for longer duration. To preserve the milk nutrients for longer life, it has been converted into various forms of products since ancient times. Concentration of milk is one such approach to preserve milk constituents by partial removal of water. Reduction of water activity along with the preservation method involved in concentrated milk has a detrimental impact on chemicals, microbiological, and biochemical changes. Thus, shelf stability of concentrated milk can be increased. Broadly, concentrated milk is categorized as sweetened condensed milk (SCM), in which added sugar can act as a preservative; and evaporated milk, which requires heat and its sterilization for the preservation purpose.

7.1 INTRODUCTION

Milk is regarded as near complete food since it contains all necessary dietary requirements of an individual. Milk contains several nutrients, such as fat, protein, lactose, vitamins, and minerals. The high-water content with rich nutritional profile of the milk makes it an excellent medium for the microbial growth and hence it has shorter shelf-life. To preserve the milk for longer

The Chemistry of Milk and Milk Products: Physicochemical Properties, Therapeutic Characteristics, and Processing Methods. Megh R. Goyal, Suvartan Ranvir, & Junaid Ahmad Malik (Eds.)

time, various approaches have been utilized in the dairy industry. One such approach is the reduction of water content. The lower-water content has detrimental impact on both microbial growth and chemical changes; hence, there is an increase in the shelf-life. Concentrated milk are milk products prepared via partial removal of water and sustained either by the addition of sucrose in case of sweetened condensed milk or by the sterilization process in case of evaporated milk (EM). Due to its high nutritional value, concentrated products are used either as an ingredient in different food formulations, consumed as such or used as intermediate product in the preparation of dried powders.

Preparation of concentrated milk (to reduce the bulkiness and to preserve milk solids) dates backs to Napoleon times. For the first time in 1810, Nicolas Appert introduced the process that involved the heating of milk in an open pan to reduce its volume by one third, followed by packing in an airtight sealed vessel; and further vessel was cooked in water bath to extend the storage stability of milk. Eventually, Louis Pasteur described the reason behind the extended shelf-life of milk, that is, by inactivation or killing of microorganisms via heat treatment. However, the product developed by the Nicolas method was often to coagulate during heat sterilization. Nicolas Appert is named as father of canning and name appertization was given for in-container or in-can sterilization. In 1856, Gail Borden succeeded in the preparation of concentrated milk under vacuum condition, and it was sold under the brand name Eagle, and he was known as the father of process of milk condensing. Borden used sucrose to prepare sweetened condensed milk. Afterward, John Meyenberg successfully prepared sterilized evaporated milk in 1884. In 1909, concentrated milk was homogenized to overcome the fat separation during the storage of the product. The Nestle company in 1911 was the world's biggest condensed milk factory in Dennington, Victoria, Australia. In India, Anand milk producer's Union Ltd. bagged credit for first ever commercial production of SCM in 1961.

The various physicochemical properties are exhibited by the milk system because of the different equilibria existing in milk, such as equilibria among soluble and colloidal mineral system, interactions of proteins with milk minerals, and association of milk proteins.[26] In milk, equilibrium exists among alkaline and acid equivalents. The sum of the alkaline equivalents (such as CaO, MgO, K_2O, and Na_2O) is approximately equal to the sum of the acid equivalents (such as P_2O_5, Cl, SO_3, CO_2, citrates, casein, albumin, and globulin). The acid equivalents show a marked stabilizing effect on protein system in milk, whereas the alkaline equivalents (especially Ca^{2+} and Mg^{2+})

perform in an opposite manner, leading to aggregation, destabilization, and precipitation of the casein micelles.

The existing native milk equilibria may get disturbed by various processing parameters (such as preheat treatment, concentration, and sterilization process involved in the preparation of concentrated milk, etc.), and hence may cause changes in physicochemical properties (increase in total solids content, decrease in water activity, denaturation of whey protein, change in minerals phases from soluble to colloidal, dephosphorylation of casein, lactose degradation, decrease in pH, Maillard browning, age thickening, etc.) of the milk. Multiple factors, such as change in pH, protein rearrangement, and change in mineral equilibria, need to be addressed during the manufacturing of a concentrated milk. The understanding of chemistry involved in preparation and further thermal processing of concentrated milk is important to get desirable product with shelf stability.

Changes in physicochemical properties (increase in total solids content, decrease in water activity, denaturation of whey protein, change in minerals phases from soluble to colloidal, dephosphorylation of casein, lactose degradation, decrease in pH, Maillard browning, and age thickening, etc.) of the milk are discussed in this chapter along with the composition and manufacturing methods of concentrated milks.

7.2 DEFINITION, CLASSIFICATION, COMPOSITION, AND STANDARDS OF CONCENTRATED MILK

Concentrated milk is a dairy product prepared from either whole milk, skim milk, full fat milk, or partially skimmed milk via partial removal of water. Based on the method of preservation, concentrated milk is broadly classified as evaporated milk and sweetened condensed milk (SCM). Typically, EM (evaporated milk) is a heat-sterilized unsweetened milk, but SCM is unsterilized sweetened milk in which added sugar acts as the preservative agent. Inactivation or destruction of microorganisms and enzymes by sterilization process makes EM shelf-stable, whereas the decrease in water activity and higher osmotic pressure of SCM make it shelf-stable.

The approximate composition of concentrated milk[30] has been depicted in Table 7.1. The SCM contains >71% total solids, in which sucrose alone constitutes around 40 to 45%. The added sugar forms saturated aqueous sugar solution (63–64%) resulting in decrease of a_w to 0.83, and osmotic pressure is increased to about 23 MPa. Hence, growth of the most microorganisms may be inhibited, except for certain bacteria and few osmophilic yeasts.

TABLE 7.1 Approximate Composition (%) of Concentrated Milk.[30]

Product	Water	Carbohydrates	Fat	Proteins	Ash
Evaporated skimmed milk	79.5	11	0.3	7.6	1.6
Evaporated whole milk (American standard)	72.7–74.7	9–10	7.5–8.0	6.5–7.1	1.3–1.6
Evaporated whole milk (British standard)	67–69	11.0–12.5	9–10	8–9	1.9–2.1
Skimmed milk	90	4.9	<0.1	3.4	0.75
Sweetened condensed milk	27	55	9	8	1.8
Sweetened condensed skimmed milk	28	59	0.3	10	2.3
Whole milk	87	4.8	3.7	3.3	0.7

7.2.1 THE CODEX DEFINITION OF THE EM[12] AND SCM[13]

Evaporated milk (EM) is a milk product where water is eliminated partially via heat treatment or any other equivalent process, which results in the characteristic product of similar composition. The product should meet the standards prescribed by the *codex regulations* (Table 7.2) in Section 3 of this standard. Whey protein to casein ratio should not be altered during the standardization process. The permitted additives by *Codex* are presented in Table 7.3. Sodium orthophosphate, calcium chloride, calcium carbonate, or soy lecithin are used to regulate the heat stability, while polyphosphates are used to prevent gelation in UHT-treated EM. Carrageenan may be added to retard creaming.

If milk fat is replaced by vegetable fat in the evaporated milk, then this type of milk is called filled evaporated milk. Usually, coconut oil or palm oil is most frequently used in filled evaporated milk. According to *Codex*, filled evaporated milk is a blend of evaporated skimmed milk and vegetable fat.

SCM is a sweetened dairy product prepared by incomplete removal of water with sugar addition or prepared by any other equivalent process, which results in a characteristic product. Adjustment of fat and/or protein content may be done to meet the specific composition (Table 7.2). Any addition or withdrawal of milk constituent should not change the whey protein to casein ratio of the milk being adjusted.[49] Physicochemical properties of concentrated milk[49] are shown in Table 7.4.

TABLE 7.2 *Codex* Standards for Concentrated Milk.

Product	Milk fats	Milk solids	MSNF	Protein in SNF
	% (m/m)			
Evaporated high fat milk (min)	15	–	11.5	34
Evaporated milk (min)	7.5	25	–	34
Evaporated partly skimmed milk	1–7.5	20 (min)	–	34 (min)
Evaporated semi skimmed milk	4–4.5	20 (min)	–	–
Evaporated skimmed milk	1 (max)	20 (min)	–	34 (min)
Sweetened condensed high fat milk (min)	16	–	14	34
Sweetened condensed milk	8	28	–	34
Sweetened condensed partly skimmed milk	1–8	24 (min)	20 (min)	34 (min)
Sweetened condensed semi skimmed milk	4–4.5	28 (min)		
Sweetened condensed skimmed milk	1 (max)	24 (min)	–	34 (min)

Note: For all types of SCM, the amount of sugar to be added is monitored by GMP to a minimum value that ensures the keeping quality of the product and a maximum value above which sugar crystallization may occur.

TABLE 7.3 Additives Permitted by *Codex* Standards for Concentrated Milk.[12,13]

INS no.	Name of additive	Maximum level
	Firming agents	
508	Potassium chloride	Singly 2000 mg/kg or in combination 3000 mg/kg, expressed as anhydrous substances
509	Calcium chloride	
Stabilizers		
331	"Sodium citrate"	Singly 2000 mg/kg or in combination 3000 mg/kg, expressed as anhydrous substances
332	"Potassium citrate"	
333	"Calcium citrate"	
	Acidity regulators	
170	Calcium carbonates	Singly 2000 mg/kg or in combination 3000 mg/kg, expressed as anhydrous substances
339	Sodium phosphates	
340	Potassium phosphates	
341	Calcium phosphates	
450	Diphosphates	
451	Triphosphates	
452	Polyphosphates	
500	Sodium carbonates	
501	Potassium carbonates	
	Thickeners	
407	Carrageenan	150 mg/kg
Emulsifier	Lecithin	GMP

TABLE 7.4 Physicochemical Properties of Concentrated Milk.[49]

Parameter	Units	Sweetened condensed milk (SCM)	Evaporated milk (EM)
Specific gravity	(at 16°C)	1.303	1.083
Freezing point	°C	−14.9	−1.3
Water activity	–	0.83	0.986
Viscosity	MPa.-sec at 20°C	2000	6.3
Specific heat	kJ/kg/k	2.35	3.2
Latent heat of fusion	kJ/kg	91	247
Coefficient of thermal conductivity	W/m/k	–	0.45

7.3 MANUFACTURING PROCESS FOR EVAPORATED MILK (EM)

The sequential steps involved in the preparation of EM are depicted in Figure 7.1. The process begins with the receiving of good quality raw milk, which must pass the alcohol stability test. The milk is standardized with respect to fat and protein content, if necessary. Then, it is subjected to preheat treatment to increase the heat stability, inactivate native enzymes, and curtail the microbial growth. The adaptation of suitable preheating temperature–time combination based on the several factors (such as nature of milk and its heat stability, microbial and enzymatic activity, initial heat stability), and further heat processing conditions to be involved.[42] Although different preheating temperatures are being used for different durations, industrial applicable temperature may range from 80 to 95°C/10 to 30 min or above 100°C/few seconds to a min.[17]

UHT treatment (135–140°C/2–4 s) is getting more popular, because it kills spores along with vegetative cells and causes minimum heat damage to milk constituents particularly to milk proteins. Subsequently, preheated milk is subjected to the evaporation to remove water.

Usually, evaporation is carried out under vacuum at 45 to 75°C. Falling film evaporators are frequently used to evaporate bulk water and then are switched to circulating evaporator to remove remaining water.[49] Alternatively, membrane filtration and freeze concentration are also employed to concentrate the milk. In membrane filtration (reverse osmosis), milk is pumped through a semipermeable membrane under pressure of 3–4 MPa. The maximum achievable concentration is restricted to about 26% TS in whole milk and to 30% TS in skim milk; beyond these limits, fouling of membrane may take place. Although membrane process is energy efficient (approximately three times) than the evaporation, yet it is commercially unsuccessful because of higher membrane cost, frequent membrane fouling, and requirement of high-pressure pump to achieve higher concentration.[48] Another feasible, but more expensive, concentration method is freeze concentration, where milk is allowed to freeze to form ice crystals, which are further removed mechanically. High-cost requirement of this process limits its utilization at commercial level.

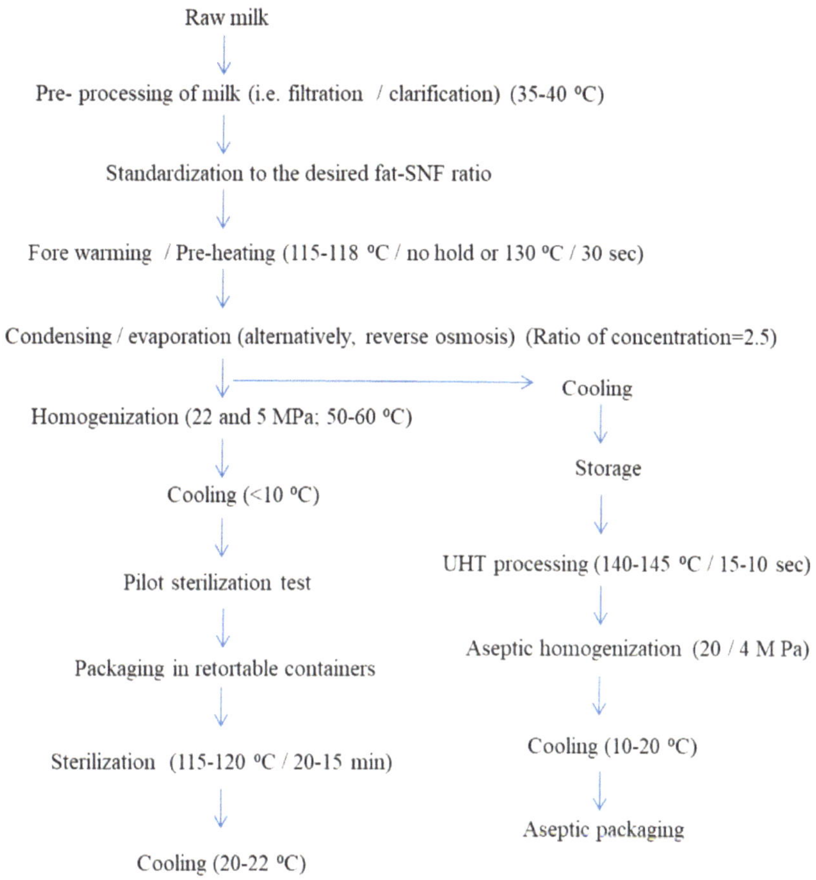

FIGURE 7.1 Flow diagram for the manufacturing of evaporated milk.[42]

Refractometer or density meters are frequently used to monitor TS content during the evaporation process. Once the desired TS has been reached, EM is further homogenized and stabilized with the additives.[17] Two stage homogenizers (20 MPa followed by 5 MPa) are being used for reducing fat globule size to < 2 μm to prevent coalescence and cream separation in the product during storage. The EM is stabilized by monitoring salt balance and pH to maintain the colloidal stability and to achieve desired viscosity. Stabilizers are often used to serve the purpose. Piolet sterilization test is employed for each batch to determine the quantity of stabilizers that are to be added. Trisodium citrate and disodium hydrogen phosphate are frequently used stabilizers.

Finally, the milk is filled in presterilized cans (care has to take to provide sufficient headspace to accommodate thermal expansion of the product), vacuum sealed, and sterilized at 121°C for 15 min or any other suitable time–temperature combination to curb the microbial growth and enzyme activities to preserve the quality of the product. Since sterilization is a static process, usually cans are agitated to reduce coagulation and gelation.[17] In these days, the UHT sterilization process (135–140°C/10–20 s) is gaining more attention compared with in-container sterilization. In fact, UHT sterilization is more advantageous than the in-container sterilization, that is, it minimizes the undesirable physical, chemical, and organoleptic changes.

Adequate mixing during sterilization ensures the prevention of aggregated proteins attachment on the cans surface, which otherwise will reduce the heat transfer efficiency. Sterilized product is cooled and stored at room temperature. Some of the issues normally encountered during the preparation of EM are fouling of heat exchangers and evaporators that requires high maintenance cost, instability of casein micelles during sterilization, which results in their aggregation, and subsequent age-gelation during the storage of the product.

7.4 MANUFACTURING OF SWEETENED CONDENSED MILK (SCM)

The superior quality milk with negative alcohol test is used for the preparation of the product. The steps involved in this process are shown in Figure 7.2. The standardization of milk or milk concentrate is carried out in three stages: (1) initially milk is standardized to achieve required fat to SNF ratio; (2) then standardization is done with respect to sugar to milk solids to maintain proper sugar ratio; and (3) final standardization of finished product to meet the regulatory aspects. Preheating is carried out by suitable time–temperature combinations to have a control over microbial growth (killing pathogens, coliforms, osmophilic yeasts, and molds) to inactivate native milk enzymes and to control viscosity and age-thickening defect during storage.

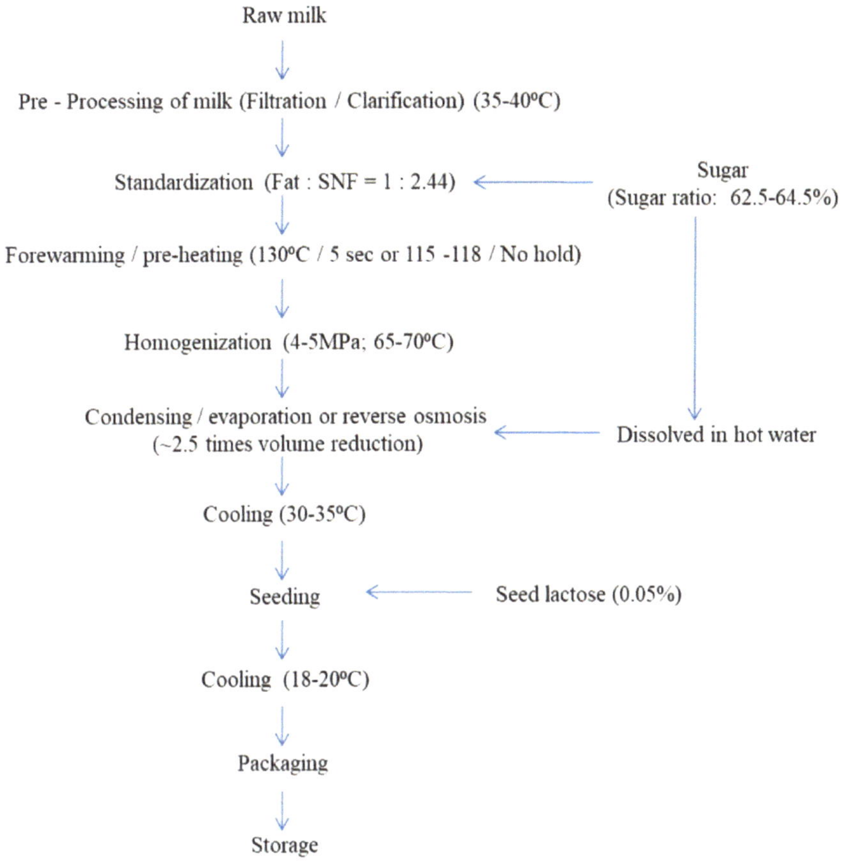

FIGURE 7.2 Flow diagram for the manufacturing of sweetened condensed milk (SCM).[42]

Evaporation is achieved using multiple-effect evaporators. Sugar may be added at different stages (before or after preheating or at the end of condensing process). The refined sugar is preferred, and it must be devoid of invert sugar, which will hasten the browning reaction. If sugar is added to raw milk before evaporation, it may lead to extensive Maillard reaction and may increase the viscosity, which makes the product prone to age-gelation during storage; hence, it is not preferred to add sugar at this stage. Heat-treated concentrated sugar syrup with sugar ratio 62.5 to 64.5 g/100 g is usually added at the final stage of concentration. The sugar increases the osmotic pressure and lowers the a_w (0.83–0.84) of SCM. The sugar index or sugar ratio may not be less than 0.61 and not more than 0.64 (beyond this limit, crystallization of sugar takes place).[34] Hydrometer is used to measure TS during the process. At the outlet of evaporator, the product will have a density of 1.35 kg dm^{-2} for skimmed milk and 1.30 kg dm^{-2} for the whole milk. Due to high viscosity, fat separation and coalescence are not an issue during storage; hence, homogenization is not practiced. However, homogenization at low pressure (2–6 MPa at 70°C) is often practiced in the preparation of low viscous product to prevent fat and sugar separation.[43]

Cooling and crystallization of lactose are concern parameters in the preparation of SCM. Immediate cooling to 30 to 32°C is preferred to monitor uncontrolled way of lactose crystallization, which otherwise forms thicker lactose crystals responsible for sandiness defect. The lactose content in SCM is 38 to 45 g/100 g water. The solubility of lactose is 18 g/100 g water at room temperature, which again reduced to half in SCM because of sucrose content. Thus, lactose present in supersaturated state by the end of evaporation is likely to be crystallized (25% lactose is present in soluble form and remaining 75% tends to crystallize at ambient temperature). Consequently, high viscosity of the product retards the nucleation that results in the formation of few localized, large-sized crystals responsible for sandiness defect. Hence, forced crystallization is initiated by adding seeding material at 30–32°C to induce rapid nucleation to get small-sized crystals in more number. The size of the lactose crystal must be <10 um, which is not normally detectable by tongue. The formation of nuclei and crystal growth can be altered by the factors like rate of cooling, amount of seeding material and temperature at which it is adding, viscosity, agitation involved, and storage condition.

The different seeding materials like skim milk powder (0.5%), whey powder (0.2–0.3%), and formerly centrifuged SCM (1%) of previous batch are used to achieve the target. More often, α-lactose monohydrate after converting into anhydride form (vacuum heat at 93°C and then grind to get powder of size <1 um which ultimately sterilized at 130°C for minimum one hour) is used as seeding material. The amount of seeding material used is 0.4% or 0.004 times of the lactose to be crystallized (approximately, used at level 0.04–0.05% on the basis of milk or 0.1–0.2% of condensed milk). Lastly, the product is filled in suitable containers considering minimum head space to control molds and is stored.

7.5 HEAT STABILITY OF CONCENTRATED MILK

The capacity of milk to resist for occurrence of visual coagulation and gelation under the employed processing conditions is known as heat stability.[44]

Milk proteins and minerals are key milk constituents, which influences the heat stability. Heat coagulation time is frequently to describe heat stability.

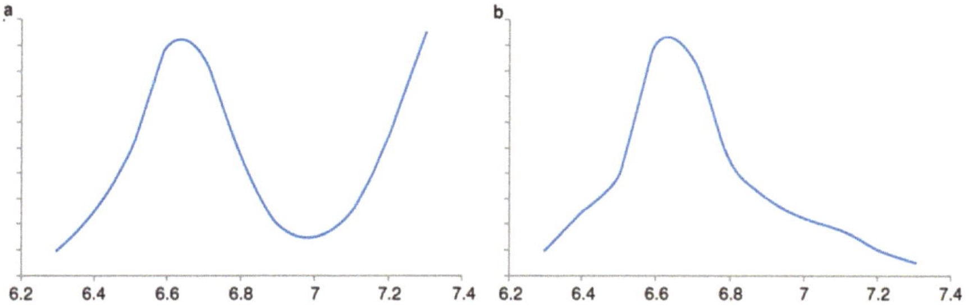

FIGURE 7.3 HCT versus pH profile of normal milk at 140°C (Fig. 7.3a) and concentrated milk at 120°C (Fig. 7.3b).[26]

HCT is the time taken between the start of heating process (temperature-controlled oil bath is used as a heating medium) and the appearance of visible protein-coagulated particles.[17] The experimental temperature is adopted to check heat stability of normal milk at 140°C, whereas for concentrated milk, it is 110–120°C. HCT of the sample is measured at the indicated temperature as a function of pH. In case of normal milk, HCT increases with pH and attains a maximum value at pH of 6.6–6.7, then it reaches to a minimum value at pH 6.9–7, and further HCT is increased at the alkaline side of pH. But, in case of concentrated milk, HCT increases initially with pH and shows maximum value at pH of 6.5–6.6, then further HCT decreases with increase in pH (on the alkaline side). Concentrated milk shows no HCT minimum value at alkaline pH of milk, which is found in normal milk.

The HCT versus pH curves of milk and concentrated milk are depicted in Figure 7.3. The dissociation of κ-casein from micellar casein under heat treatment may be ascribed to maxima and minima in the HCT-pH profile.[26] The occurrence of maximum in HCT-pH graph of normal milk is attributed to β-lactoglobulin, which interacts with micellar surface protein (κ-casein) and alters electrostatic and steric stabilization force, and hence prevents κ-casein dissociation from micelles.

The minima in the HCT-pH curve is linked to dissociation of complex (β-lactoglobulin and κ-casein) from micellar surface to serum portion so that prone κ-casein depleted micelles to calcium-induced coagulation. Several factors (such as milk composition, total solids level, preheat treatment, addition of additives, and homogenization pressure) will influence the heat stability of concentrated milk. Higher homogenization pressure decreases the heat stability, whereas pressure up to 5 MPa is observed to increase heat stability. With increasing level of TS, heat stability decreases.[19] The strategies being used in the improvement of heat stability of milk[21,50] are

- Addition of phospholipids.
- Adequate preheat treatment before concentration of milk.
- Adjusting the pH of milk by the addition of stabilizers.
- Chelating the excess soluble calcium using chelating agents like phosphates, citrates, and EDTA.
- Conjugation of milk proteins with reducing sugar.
- Modifying the charge of amino acids residue with the use of oxidizing agents like $H_2O_{2,}$ $KBrO_{4,}$ KIO.
- Preventing the dissociation of individual caseins from micelles by thermal, chemical (using formaldehyde and other aldehydes), or enzymatic (using transglutaminase) crosslinking reactions.
- Preventing the disulphide-thiol exchange reactions using reducing agents like dithiothreitol, β-mercapto-ethanol, or cysteine.

7.6 PHYSICOCHEMICAL CHANGES DURING PREPARATION OF CONCENTRATED MILK

7.6.1 *PREHEATING*

Preheating of milk is the most significant factor to be considered in the preparation of concentrated and dried milk. The different time–temperature combinations are employed for this purpose using either direct or indirect heating methods. The primary importance of preheating is to safeguard heat stability in concentrated milk system. Preheating causes unfolding of whey proteins, which further interact with other proteins. Serum proteins like immunoglobulins and serum albumin start to denature at 65°C, whereas ß-lactoglobulin and α-lactalbumin exhibit significant denaturation at > 70–75°C.[14,37] The complete denaturation of whey proteins can be achieved at 90°C for 10 min. The reported sensitivity order of whey proteins to heat denaturation is:

immunoglobulins > serum albumin > ß-lactoglobulin > α-lactalbumin.

The denatured proteins make interaction either among themselves to form polymeric compounds or with the caseins. Usually, ß-lactoglobulin make complex with κ-casein via disulfide linkages. The location of the complex, which depends on pH of the system, influences the heat stability. Heat stability increases if the complex is positioned on the micellar surface, whereas heat stability decreases if the complex gets dissociated from micelles to serum phase. Preheating also causes precipitation of soluble calcium phosphate and hence may alter ionic configuration of milk.[40,41]

At >100°C, dephosphorylation of casein will take place[7] along with evolution of non-protein nitrogen.[15] Skipping of preheat treatment may result in destabilization of milk protein system during subsequent heat treatments of milk concentrates. The probable reasons are presence of more amounts of caseins, ionic calcium, and undenatured whey proteins, which can hasten rapid aggregation of whey proteins and their copolymerization reaction with casein micelles.[32]

Preheating also causes the glycation reaction in milk.[52] Preheating at 98°C for 10 min induces the glycation of ß-lactoglobulin and bovine serum albumin (denaturation exposes the glycation sites), whereas prolonged preheat treatment (98°C for 40 min) retards the glycation process due to aggregation of denatured proteins.[52] Contrarily, preheating exhibits limited influence on glycation of ß-casein.

7.6.2 PROCESS FOR CONCENTRATION OF MILK

Concentration of milk establishes new equilibrium in the milk by reducing the water content, altering the salt balance among soluble and colloidal phases, and reducing the distance between the colloidal particles.[49] The established equilibrium is distinguished by the decreased pH, increased total solids, increased ionic strength, and lowered electrostatic repulsion.[44] Changes occur during concentration by evaporation, such as

- Because of higher total solids content, heat stability of milk concentrates decreases.
- Closer packing of casein micelles.
- Decrease in pH due to shift in ionic equilibria.
- Increase in casein micelles size due to whey proteins association with micelles and aggregation of some micelles.
- No change in denaturation of whey proteins is noticed during evaporation unless temperature used in preheat treatment is very low, which causes further denaturation of whey proteins.
- The concentration of whey proteins, minerals, and lactose are increased.

Membrane concentration of milk by ultrafiltration can cause the depletion of CCP (colloidal calcium phosphate), which leads to loosening of micellar structures and further swelling of micelles will also take place. With increase in concentration, native intact micelles are converted into diffusive swollen micelles, which may be further disintegrated into smaller micellar fragments.[43]

7.6.2.1 CHANGES IN WATER ACTIVITY AND pH

During concentration process, free water of milk is removed extensively compared to the intra-micellar water[28] that results in decreased soluble calcium and increased colloidal calcium. The removal of water may also lead to compaction of micellar particles.[10] Water activity decreases with concentration, and it will result in increased hydrophobicity of particles. The water activities of milk, EM, and SCM are 0.993. 0.986, and 0.83, respectively. The pH of milk is decreased during concentration, but change in pH is less pronounced (since lower temperature was involved in evaporation) compared to preheat treatment and sterilization process. The pH is decreased by 0.3 and 0.5 units for 2 and 3 concentration factors, respectively. The probable reason for the acidification of concentrated milk includes the formation of organic acids (because of lactose degradation), increase of micellar mineral content (shift of soluble form of minerals to colloidal form), and release of hydrogen ions.[25] Since low temperature is involved in concentration process, lactose degradation is less pronounced, but concentration of lactose increases that would initiate its interaction with other molecules, for example, lysine of protein. The effect of mineral on milk pH during concentration is more prominent, since incorporation of colloidal calcium phosphate into micelles leads to release of hydrogen ions, which eventually cause acidification of concentrated milk.[49]

7.6.2.2 CHANGES IN VISCOSITY AND TURBIDITY

Viscosity of concentrated milk is an important property that will lead to the maximum concentration during evaporation and ultrafiltration process. Viscosity increases with increase in TS. The liquid shows Newtonian flow behavior upto 20% TS, but it shows pseudoplastic (shear thinning) flow behavior at >30% TS. Thus, concentrated milk exhibits shear thinning or pseudoplastic flow characteristics.[31] The most important contributor for change in viscosity is heat treatment of proteins.[31] Increase in the volume of casein micelles, which appear as hydrated spherical-shaped gel particles, leads to change in viscosity.[39] Turbidity of EM increases due to the increase in lactose and mineral contents associated with casein micelles, and product becomes denser with increased refractive index.[29]

7.6.2.3 PARTICLE SIZE

Specific volume of casein micelles increases because of: (1) the attachment of whey proteins with micellar surface,[10] (2) shifting of soluble calcium and casein to colloidal form,[28] and (3) aggregation of some micelles.[43]

Particle diameter may decrease with increase in TS due to micellar shrinkage, which may occur because of micellar saturation with CCP and dehydration of micelles.[28,9] However, decrease in size of micelles is a reversible process even at higher TS and hence mean particle diameter increases. Therefore, increase in TS leads to evolution of bigger particles, which will be linked to noncovalent interaction due to reduced net negative surface charge of micelles and decreased electrostatic repulsion.[8,36]

7.6.2.4 CHANGES WITH RESPECT TO PROTEIN COMPONENT OF MILK

During concentration, only slight changes have been noticed in native whey protein level compared to preheat treatment.[38] The whey protein resists to denaturation over the concentration process and that is ascribed to lower temperature of concentration process and higher TS of milk. The higher lactose concentration has a protective role in whey protein denaturation by enhancing its interaction with protein and an increased protein hydration.[6]

The casein micelle structures may get collapsed and form swollen diffused micelles that would be fragmented into smaller structures at higher TS level.[43] This structural change promotes new reaction between proteins and causes micellar reformation during the sterilization process. Cao et al.[10] showed the presence of aggregates during evaporation and they also observed the reduction in SH^- group of concentrated milk, which may not be found in the normal milk (reduced SH^- group in concentrated milk may be linked to aggregation of ß-lactoglobulin with α-lactalbumin through disulfide interchange reaction or evolution of ß-lactoglobulin-κ-casein complex).

Anema et al.[4] reported that there is a slight dissociation of individual caseins from micelles in milder heating condition, which again depends on pH of the system. ß-casein and α_s-casein may be dissociated in concentrated milk at respective pH of 6.5 and 6.3. This pH-dependent solubilization of caseins may be attributed to destabilization of calcium phosphate in saturated casein micelles.

7.6.2.5 CHANGES WITH RESPECT TO MINERAL CONSTITUENTS OF MILK

Minerals of milk are found in both soluble and colloidal forms. The evaporation of milk shifts significant proportion of calcium and phosphate from soluble to colloidal states during concentration that may cause decrease in soluble casein and calcium in the serum phase. The possible mechanism for the shift of soluble calcium from serum phase to colloidal phase includes the lower solubility of calcium phosphate at higher temperature and removal of water during concentration that affects the ability of salt system to continue phosphate and calcium in soluble state.[28] The ionic strength increases with concentration process.[27] However, ionic calcium content slightly decreases with evaporation process, which may be linked to lower pH of condensed milk.[24,35] On the other hand, soluble sodium level may increase and soluble magnesium is decreased with concentration.[35]

7.6.3 EFFECT OF HOMOGENIZATION

The total surface protein of concentrated milk fat globule will increase after homogenization, which would be influenced by temperature of preheat treatment and homogenization pressure. Preheating of milk induces the association of fat globule surface protein with serum protein or with serum protein-casein complex, which ultimately increases the surface protein of fat globules. Higher preheating temperature may induce whey protein association with micellar casein and with MFGM native proteins via disulfide bonds that may be adsorbed on the surface of fat globule during homogenization. Thus, level of whey protein increases with total protein content on the surface of fat globules. Roughly, whey protein level is 1.2 mg/g of fat associated with fat globular membrane during preheat treatment,[51] whereas the amount may increase to 12 mg/g of fat after the homogenization (total surface protein constitutes 6–8 mg/m^2 or 60 mg/g of fat).

7.6.4 EFFECT OF STERILIZATION ON MILK CONSTITUENTS

Various physicochemical changes have been noticed during the sterilization process that includes the casein micelles disintegration, whey protein denaturation, variation in protein content among serum and colloidal phase, dephosphorylation, Maillard reaction, and scaling of heat exchangers (due to protein coagulation and gelation).

The pH of the concentrated milk decreases throughout the manufacturing process. Decrease in pH is more intense during sterilization temperature because of generation of organic acids (principally formic acid) from lactose degradation, precipitation of calcium phosphate, which may lead to liberation of hydrogen ions, and casein dephosphorylation.

Whey proteins are denatured at sterilization process and further involve in aggregation and formation of gel in the concentrated milk, whereas gelation is not normally found in normal milk due to lower serum protein content.[26] The instability of serum proteins is influenced by sterilization temperature. For example, 100% denaturation will occur in in-container sterilization, whereas denaturation is less severe in UHT treatment.[17] Low temperature (<70°C) denaturation is reversible, but denaturation at higher temperature (>70°C) is irreversible at which native whey protein molecules start to unfold to expose both hydrophobic residues (involved in some hydrophobic reaction) and highly active SH^- group at $cysteine_{121}$ (responsible for covalent interactions).[17]

Denatured serum proteins may either interact with other proteins or involve in the formation of soluble aggregates. This reaction is largely impacted by the pH and TS levels compared with sterilization. Thus, heating at pH > 6.6 results in serum protein-κ-casein aggregates of soluble in nature, while heat treatment at pH <6.6 leads to attachment of serum proteins on micellar surface. Micellar casein surrounded by denatured whey protein networks are stable to coagulation depending on the pH of the system.[26]

Denaturation of whey protein at 95°C is retarded at higher solid concentration due to interaction between lactose with free thiol group of whey protein that inhibits protein aggregation. Heat- induced whey protein aggregation in concentrated milk is linked to the presence of four different forms of whey protein, such as native whey proteins, soluble denatured complexes, whey protein-casein complex, and co-sediment of whey protein casein aggregates.[20]

At the industrial level, preheating is employed to reduce the whey protein aggregation, that is, preheat treatment at the acidic pH side of HCT-pH maximum (pH < 6.7) prior to concentration results in the formation of gel-like structure due to denaturation of whey protein, which further prevents heat-induced gelation of casein micellar during sterilization process and thus increases heat stability of concentrated milk.[26] Preheating at the alkaline pH side (> 6.9) has less effect on heat stability.

Micellar caseins are stabilized by hairy brush configuration of κ-casein, that is, c-terminal of κ-casein is involved in stabilization, which protrudes toward the serum phase and carries net negative charge because of more glutamine residues.[26,16] The dissociated carboxyl and ester phosphate group also contributes for the net negative charge of micellar surface (−13 mV at 20°C). Thus, the negative surface charge results in electrostatic repulsion of micelles and provides steric stabilization to micelles. Any disturbance, which alters the electrostatic repulsion and steric stabilization of casein micelles, results in the disintegration of micelles.

The change in pH during heat treatment is a major factor responsible for dissociation of caseins from micelles. However, the factors like whey protein interaction with casein micelles, mineral composition, and total solids content also contribute to dissociation of caseins.[1,2,21,45,46] As a result, any deviation in the optimum heat stability pH (6.5–6.6) of concentrated milk results in the dissociation of micellar caseins, which causes destabilization and coagulation of micelles.[1,2,26,45,46]

The hydrophobic and electrostatic bonds involved in the association of α-casein and ß-with the CCP play a prominent role in the micellar integration.[44] CCP of concentrated milk has got its new form due to precipitation of soluble minerals into colloidal state during the evaporation. This new form of CCP has shown a lower ability to crosslink the caseins inside the micelle and hence exhibits lesser stability during sterilization compared to stability of native micelles of normal milk.

Dephosphorylation of casein is another contributing factor for micellar instability. Loss of phosphates from casein results in the decreased binding ability of calcium phosphate, which favors caseins dissociation from micelles. The dissociation of caseins from micelles is dependent on the factors, such as pH of milk before sterilization, presence of whey proteins, composition of micelle and serum, and concentration of total solids.[26]

The native mineral equilibria between serum and colloidal phases depend on the pH, temperature, and total solids content. Shift in mineral equilibria is more pronounced during sterilization of concentrated milk in comparison with the evaporation process. The new form of CCP will form during evaporation of milk and it is affected by the precipitation of ionic calcium and phosphate into micelles at sterilization treatment. Thus, the transfer of soluble minerals to colloidal form may result in additional saturation of casein micelles with calcium and phosphate, which later behave differently than the native micelles, and are more prone to destabilization in sterilization condition.

Concentration of ca^{2+} (ionic calcium) will experience slight decrease during evaporation, while it is further decreased significantly during heat treatment of concentrated milk.[11,35,44] Thus, it promotes the association of ca^{2+} with carboxyl ester group of micellar surfaces and neutralizes negative charge of micellar surface, which finally cause destabilization of micelles.[22] Nevertheless, ca^{2+} interacts with phosphate group released during dephosphorylation reaction (occurs during heat treatment) to form a new stabilization system in concentrated milk.[11] The ca^{2+} modification reactions are reversible to some extent after cooling.[17]

Heating of concentrated milk will fluctuate the size of casein micelles, which again depend on the factors, such as, preheat treatment prior to concentration, total solids content, and pH. The voluminosity of particles may increase due to conformational changes and interaction among proteins. Heating of milk at pH 6.5 and 6.7 may increase the diameter of micelles by 25–30 nm and 5–10 nm, respectively.[5]

The Maillard reaction may take place at sterilization of EM. Lysine (ε-amino group) reacts with carbonyl group of lactose to induce the reaction. Usually, the reaction is not acceptable in dairy products, since it results in change in color, alteration in flavor profile by the formation of unacceptable burnt flavor, and decrease in nutritional value of proteins.

7.7 MANUFACTURING AND STORAGE DEFECTS IN EVAPORATED MILK

The fresh EM has thin cream like consistency with slight dark color and owns pleasant acceptable flavor with cooked or heated characteristic. UHT treated EM has less intense in color and flavor deviation than the fresh milk. Microbial growth is rarely found in EM since the product is heat sterilized, whereas heat stable residual enzymes, which survive the heat treatment, may cause physicochemical changes during the storage.

Bacillus stearothermophilus spores (heat resistant) may germinate under storage conditions (37°C) and cause cheesy flavor or acid coagulation of proteins.[42] Stale flavor, age gelation, browning, fat separation or creaming, and settling or deposit formation are generally occurred during storage. Cooked flavor of fresh EM may be gradually replaced with stale flavor upon storage. Color and flavor defects are attributed to Maillard reaction in the products. Compounds like methyl ketones and o-amino acetophenone impart staleness.

Fat separation during storage is a serious issue in EM. The problem is more pronounced at high temperature storage for longer periods. Fat molecules may rise to the top and form a viscous leathery structure that further limits the pourability of product. Protein and fat globules are covered by thick layer of protein in sterilized EM that will aggregate to some extent. Extent of aggregation will be strong just before the occurrence of coagulation. Small sized fat globules with higher protein content may aggregate to settle at the bottom, whereas larger fat globules with low protein load are aggregated to form creamy layer on the surface of EM. Thus, 6 months older product may contain three layers of which top and bottom layer will respectively be characterized by more fat and protein than middle layer. Generally, two stage homogenization has been practiced for overcoming this problem. However, inadequate homogenization process, high temperature storage for longer period, and improper handling of product may lead to fat separation.

Age thinning and age thickening are frequently noticed in storage defects in EM. The decrease in viscosity in earlier weeks of storage may be attributed to decreased volume of protein aggregates that are ascribed to change in irregular shaped micelles into spherical forms. The decrease in viscosity is subsequently followed by further increase in viscosity depending on shear rate and this phenomenon is referred as age thickening. Age thickening is an event that occurs before formation of gel, and it is characterized by higher viscosity. Upon further storage, age thickening may lead to gel formation, which is called as age gelation. It is caused by the formation of 3-D protein network, which results in the loss of fluidity of the product.[3]

Age gelation is not normally seen in the in-can sterilized EM, besides the EM is subject for cold storage for minimum of 2 days before sterilization (cold storage leads in the production of rennin-like enzymes that result in age gelation). However, it takes 4–6 months. On the other hand, UHT treated EM is more prone to age gelation, which often thickens and forms gels within 10 days of storage. Some unknown physicochemical changes of micelles like dissociation of protein may cause the gelation. Heat stable proteinases and lipases produced by the microorganisms may also be involved in the gel formation. Electron microscopic images confirm the thread like structure of protein particles in EM, which may induce gel formation by linking of particles.

7.8 MANUFACTURING AND STORAGE DEFECTS IN SWEETENED CONDENSED MILK (SCM)

The fresh product of SCM has smooth creamy texture, glossy appearance with light yellow in color and exceptionally white in color (shaded with greenish yellow) if buffalo milk is used. Lactose crystals are main contributors for the white color of SCM, whereas micellar casein and fat globules are responsible for whiteness of milk and other dairy products. The product should be free from lumps, gritty, and coarse particles.

Microbial spoilage, chemical changes, and lactose crystallization are possible reasons for storage defects (such as: mold button, age thickening, brown discoloration, rancidity, metallic flavor, and sandiness) in SCM. Even though product is not sterile, high sugar content and low water activity limits the microbial spoilage. However, the product may spoil if packed product contains or is contaminated with osmophilic yeast, molds spores, and micrococci. Growth of molds (e.g., *Aspergillus glaucus*, *Aspergillus repens*, Penicillium spp.) on the surface of the product may lead to flavor problems along with mold buttons defect, while growth of yeasts may be involved in the gas production that eventually cause protrusion of cans and off flavor (fruity flavor).

Growth of osmotolerant Micrococci may develop off flavor and increase the product viscosity. Maillard browning may also occur with storage and hence cause flavor defects. The conditions like low water activity, high intense heating, increase in total solids, and higher storage temperature will hasten the Mallard reaction. Prolonged storage of the product leads to auto-oxidation of the product.

The lactose related problem associated with SCM is crystallization of lactose and its insolubility. When the SCM is cooled, it forms lactose crystals of varying sizes, which may affect the product characteristics (such as: density, water activity, viscosity, melting point, freezing point, and browning reaction product).[23] Sandiness defect is characterized by the formation of large sized lactose crystals. Lower viscosity in conjunction with excessive sandiness leads to sugar separation and crust formation. Optimization of cooling method, addition of seeding materials, and forced crystallization may be the remedy for this problem.[42]

Age thickening is another important storage defect in SCM, which occur either due to the chemical changes or by the production of rennin-like enzyme from microorganisms. The two important factors influencing the age thickening are the irreversible change in the protein molecular structure and the formation of weak network structure among casein micelles.[47] The viscosity of the product increases with storage time, unlike the evaporated milk in which viscosity decreases initially and then increases with storage time. Age thickening with prolonged storage life causes age gelation. Preheating temperature of 80–100°C promotes the faster age gelation, whereas temperature of >100°C reduces the age gelation rate.

Additives and their stage of addition may also influence the age gelation. For example, addition of calcium after concentration increases the age gelation rate, but addition of calcium in between preheating and concentration delays the age gelation rate. Likewise, addition of citrate after concentration may delay the age gelation and it is less effective when added before concentration. Addition of EDTA and orthophosphates can hasten the age gelation irrespective of the stage of addition.[33,34] Storage of the product at higher temperature (>20°C) increases the age gelation.

7.9 SUMMARY

EM and SCM are concentrated dairy products which differ by their preservation principle, while the former one is preserved by heat sterilization and the later one is preserved by the addition of sugar content. Preparation of both concentrated milks includes preheating, evaporation, and preservation steps. Preheating is primarily done to improve the heat stability during further heat treatment of concentrated milk. Evaporation process removes the water content partially. Heat sterilization of EM preserves the product, while SCM is preserved by the addition of sugar. Various physicochemical changes will occur during the preparation and storage of the product. The major changes include the increase in total solids content, whey proteins denaturation and their interaction with other milk proteins, shift in mineral equilibrium from soluble to colloidal phase, dephosphorylation of caseins, heat degradation of lactose, decrease in pH, crystallization of sugar in SCM, Maillard browning and age thickening. The storage defects (e.g., cream separation, age thinning, age thickening, sandiness, etc.) can be prevented by optimizing the process parameters, addition of permitted additives or stabilizers, and proper storage conditions.

KEYWORDS

- **condensed milk**
- **evaporated milk**
- **heat stability**
- **physicochemical change**
- **storage defects**
- **sweetened condensed milk**

REFERENCES

1. Anema, S. G.; Klostermeyer, H. The Effect of pH and Heat Treatment on the Kappa-Casein Content and the Zeta-Potential of the Particles in Reconstituted Skim Milk. *Milchwissenschaft* **1997a,** *52*(4), 217–223.
2. Anema, S. G.; Klostermeyer, H. Heat-Induced, pH-Dependent Dissociation of Casein Micelles on Heating Reconstituted Skim Milk at Temperatures Below 100°C. *J. Agric. Food Chem.* **1997b,** *45*(4), 1108–1115.
3. Anema, S. G. Age Gelation, Sedimentation, and Creaming in UHT Milk: A Review. *Compr. Rev. Food Sci. Food Saf.* **2019,** *18*(1), 140–166.
4. Anema, S. G.; Li, Y. Association of Denatured Whey Proteins with Casein Micelles in Heated Reconstituted Skim Milk, and Its Effect on Casein Micelle Size. *J. Dairy Res.* **2003a,** *70*(1), 73–83.
5. Anema, S. G.; Li, Y Effect of pH on the Association of Denatured Whey Proteins With Casein Micelles in Heated Reconstituted Skim Milk. *J. Agric. Food Chem.* **2003b,** *51*(6), 1640–1646.
6. Arakawa, T.; Timasheff, S. N. Stabilization of Protein Structure by Sugars. *Biochemistry* **1982,** *21*(25), 6536–6544.
7. Belec, J.; Jenness, R. Dephosphorization of Casein by Heat Treatment, Part II: Skim Milks. *J. Dairy Sci.* **1962,** *45*(1), 20–26.
8. Bienvenue, A.; Jimenez-Flores, R.; Singh, H. Rheological Properties of Concentrated Skim Milk: Importance of Soluble Minerals in the Changes in Viscosity During Storage. *J. Dairy Sci.* **2003a,** *86*(12), 3813–3821.
9. Bienvenue, A.; Jimenez-Flores, R.; Singh, H. Rheological Properties of Concentrated Skim Milk: Influence of Heat Treatment and Genetic Variants on The Changes in Viscosity During Storage. *J. Agric. Food Chem.* **2003b,** *51*(22), 6488–6494.
10. Cao, J.; Zhang, W.; Wu, S.; Liu, C.; Li, Y.; Li, H.; Zhang, L. Effects of Nanofiltration and Evaporation on the Physiochemical Properties of Milk Protein During Processing of Milk Protein Concentrate. *J. Dairy Sci.* **2015,** *98*(1), 100–105.
11. Chandrapala, J.; McKinnon, I.; Augustin, M. A.; Udabage, P. The Influence of Milk Composition on pH and Calcium Activity Measured *in Situ* During Heat Treatment of Reconstituted Skim Milk. *J. Dairy Res.* **2010,** *77*(3), 257–264.

12. Codex. Codex Standard 281-1971 for Evaporated Milks. Codex Alimentarius Commission, FAO: Rome, 2010; p 4.
13. Codex. Codex Standard 282-1971 for Sweetened Condensed Milks. Codex Alimentarius Commission, FAO: Rome, 2010; p 5.
14. Dannenberg, F.; Kessler, H. G. Reaction Kinetics of the Denaturation of Whey Proteins in Milk. *J. Food Sci.* **1988,** *53*, 258–263.
15. Davies, D. T.; White, J. C. D. In *Determination of Heat-Induced Changes in the Protein Stability and Chemical Composition of Milk.* 15th International Dairy Congress Proceedings, 1959, vol 3, pp 1677–1684.
16. De Kruif, C. G.; Holt, C. Casein Micelle Structure, Functions, and Interactions; Chapter 5. In *Advanced Dairy Chemistry*; Fox, P. F., McSweeney, P. L. H., Eds.; Springer: Massachusetts, USA, 2003; vol 1A; pp 248–276.
17. Deeth, H.; Lewis, M. Protein Stability in Sterilized Milk and Milk Products; Chapter 10. In *Advanced Dairy Chemistry*; McSweeney, P. L. H., O' Mahony. J. A., Eds.; Springer: New York, USA, 2016; vol 1B; pp 233–247.
18. Dumpler, J. (Eds.) *Heat Stability of Concentrated Milk Systems: Kinetics of the Dissociation and Aggregation in High Heated Concentrated Milk Systems.* Springer: London, 2017; p 201.
19. Dumpler, J.; Kulozik, U. Heat Stability of Concentrated Skim Milk as a Function of Heating Time and Temperature on a Laboratory Scale-Improved Methodology and Kinetic Relationship. *Int. Dairy J.* **2015,** *49*, 111–117.
20. Dumpler, J.; Kulozik, U. Heat-Induced Coagulation of Concentrated Skim Milk Heated by Direct Steam Injection. *Int. Dairy J.* **2016,** *59*, 62–71.
21. Dumpler, J.; Wohlschläger, H.; Kulozik, U. Dissociation and Coagulation of Caseins and Whey Proteins in Concentrated Skim Milk Heated by Direct Steam Injection. *Dairy Sci. Technol.* **2017,** *96*(6), 807–826.
22. Fox, P. F.; Morrissey, P. A. The heat Stability of Milk. *J. Dairy Res.* **1977,** *44*(3), 627–646.
23. Fox, P. F.; Uniacke-Lowe, T.; McSweeney, P. L. H.; O'Mahony, J. A. (Eds.). Lactose. In *Dairy Chemistry and Biochemistry*; Springer: Cham, 2015; pages 21–68.
24. Hardy, E. E.; Muir, D. D.; Sweetsur, A. M.; West, I. G. Changes of Calcium Phosphate Partition and Heat Stability During Manufacture of Sterilized Concentrated Milk. *J. Dairy Sci.* **1984,** *67*(8), 1666–1673.
25. Huppertz, T. Chemistry of the Caseins; Chapter 4. In *Advanced Dairy Chemistry*; McSweeney, P. L. H., Fox, P. F., Eds.; Springer: Massachusetts, USA, 2013; vol 1A, pp 135–160.
26. Huppertz, T. Heat Stability of Milk, Chapter 7. In *Advanced Dairy Chemistry*; McSweeney, P. L. H., O'Mahony. J. A., Eds.; Springer: New York, USA, 2016; vol 1B, pp 179–196.
27. Lewis, M. J. The Measurement and Significance of Ionic Calcium in Milk - A Review. *Int. J. Dairy Technol.* **2011,** *64*(1), 1–13.
28. Liu, D. Z.; Dunstan, D. E.; Martin, G. J. Evaporative Concentration of Skimmed Milk: Effect on Casein Micelle Hydration, Composition, and Size. *Food Chem.* **2012,** *134*(3), 1446–1452.
29. Liu, Z. Fundamental Study of Physical and Biochemical Alterations to Casein Micelles During Milk Evaporation and Ultrafiltration. Doctoral Degree, Thesis, University of Melbourne, Victoria, Australia, 2013; p 126.

30. Mehta, B. M. Chemical Composition of Milk and Milk Products, Chapter 17. In *Handbook of Food Chemistry*; Cheung, P. C. K., Mehta, B. M. Eds.; Springer: Berlin, Germany, 2015; pp 511–553.
31. Morison, K. R.; Phelan, J. P.; Bloore, C. G. Viscosity and Non-Newtonian Behavior of Concentrated Milk and Cream. *Int. J. Food Prop.* **2013,** *16*(4), 882–894.
32. Morr, C. V. Chemistry of Milk Proteins in Food Processing. *J. Dairy Sci.* **1975,** *58*(7), 977–984.
33. Nieuwenhuijse. J. A. Concentrated Dairy Products: Evaporated milk. In *Encyclopedia of Dairy Sciences*; Fuquay, J. W., Fox. P. F., McSweeney, P. L. H., Eds.; Academic Press: San Diego, USA, 2011; vol 1, pp 862–868.
34. Nieuwenhuijse. J. A. Concentrated Dairy Products: Sweetened Condensed Milk. In *Encyclopedia of Dairy Sciences*; Fuquay, J. W., Fox. P. F., McSweeney, P. L. H., Eds.; Academic Press: San Diego, USA, 2011; vol 1, pp 869–873.
35. Nieuwenhuijse, J. A.; Timmermans, W.; Walstra, P. Calcium and Phosphate Partitions During the Manufacture of Sterilized Concentrated Milk and Their Relations to the Heat Stability. *Nethe. Milk Dairy J.* **1988,** *42*(4), 387–421.
36. Nieuwenhuijse, J. A.; van Boekel, M. A. J. S.; Walstra, P. On the Heat-Induced Association and Dissociation of Proteins in Concentrated Skim Milk. *Nethe. Milk Dairy J.* **1991,** *45*(1), 3–22.
37. Oldfield, D. J.; Singh, H.; Taylor, M. W. Association of β-Lactoglobulin with the Casein Micelles in Skim Milk Heated in an Ultra-High Temperature Plant. *Int. Dairy J.* **1998,** *8*(9), 765–770.
38. Oldfield, D. J.; Taylor, M. W.; Singh, H. Effect of Preheating and Other Process Parameters on Whey Protein Reactions During Skim Milk Powder Manufacture. *Int. Dairy J.* **2005,** *15*(5), 501–511.
39. Olivares, M. L.; Achkar, N. P.; Zorrilla, S. E. Rheological Behavior of Concentrated Skim Milk Dispersions as Affected by Physicochemical Conditions: Change in pH and $CaCl_2$ Addition. *Dairy Sci. Technol.* **2016,** *96*(4), 525–538.
40. Pyne, G. T. The Heat Coagulation of Milk, Part II. Variations in Sensitivity of Casein to Calcium Ions. *J. Dairy Res.* **1958,** *25*(3), 467–474.
41. Rose, D.; Tessier, H. Effect of Various Salts on the Coagulation of Casein. *J. Dairy Sci.* **1959,** *42*(6), 989–997.
42. Sharma, P.; Patel, H.; Patel, A. Evaporated and Sweetened Condensed Milks; Chapter 13. In *Dairy Processing and Quality Assurance,* 2nd ed.; Chandan. R. C., Kilara, A., Shah, N. P., Eds.; John Wiley & Sons, Ltd.: New Jersey, USA, 2016; pp 310–332.
43. Singh, H. Interactions of Milk Proteins During the Manufacture of Milk Powders. *Le Lait* **2007,** *87*(4–5), 413–423.
44. Singh, H. Heat Stability of Milk. *Int. J. Dairy Technol.* **2004,** *57*, 111–119.
45. Singh, H.; Creamer, L. K. Aggregation and Dissociation of Milk Protein Complexes in Heated Reconstituted Concentrated Skim Milks. *J. Food Sci.* **1991a,** *56*(1), 238–246.
46. Singh, H.; Creamer, L. K. Influence of Concentration of Milk Solids on the Dissociation of Micellar K-Casein on Heating Reconstituted Milk at 120°C. *J. Dairy Res.* **1991b,** *58*(1), 99–105.
47. Sone, T. Abnormal Flow Properties of Foodstuffs, Chapter 4. In *Consistency of Foodstuffs*; Sone, T., Ed.; Springer: Dordrecht, Netherlands, 1972; pp 149–158.
48. Stabile, R. L. Economics of Reverse Osmosis and Multistage Evaporation for Concentrating Skim Milk from 8.8 to 45% Solids. *J. Dairy Sci.* **1983,** *66*, 1765–1772.

49. Walstra, P.; Wouters, J. T.; Geurts, T. J. (Eds.). *Dairy Science and Technology*, 2nd ed.; CRC Press: Boca Raton, FL, 2005; p 768.
50. Wu, J.; Li, H.; Ayun, Q.; Doost, A. S.; De Meulenaer, B.; Vander Meeren, P. Conjugation of Milk Proteins and Reducing Sugars and its Potential Application in the Improvement of the Heat Stability of (Recombined) Evaporated Milk. *Trends Food Sci. Technol.* **2021,** *108*, 287–296.
51. Ye, A; Singh, H; Oldfield, D. J.; Anema, S. Kinetics of Heat-Induced Association of Β-Lactoglobulin and Α-Lactalbumin with Milk Fat Globule Membrane in Whole Milk. *Int. Dairy J.* **2004,** *14*, 389–398.
52. Zhao, D.; Xu, D.; Sheng, B.; Zhu, Z.; Li, H.; Nian, Y.; Wang, C.; Li, C.; Xu, X.; Zhou, G. Application of Preheating Treatment in Up- and Down-Regulating the Glycation Process of Dietary Proteins. *Food Hydrocoll.* **2020,** *98*, E-article 105264.

CHAPTER 8

BUTTER OIL (*GHEE*): COMPOSITION, PROCESSING, AND PHYSICOCHEMICAL CHANGES DURING STORAGE

MITUL R. BUMBADIYA, SOMA MAJI, KHUSBHU SAO, and SUVARTAN G. RANVIR

ABSTRACT

Fats and oils have been key component of human nutrition and possess numerous therapeutic properties. They supply vital elements to the body and provide the highest energy content among all components of food. Ghee (heat clarified anhydrous butterfat or butteroil or clarified butterfat) is quite popular and widely consumed throughout the world. Ghee is a rich source of energy-filled fat-soluble vitamins, vital fatty acids, and other health-promoting compounds, making it ideal for a variety of therapeutic uses. Ghee holds important position among all the dairy products because of characteristic rich, nutty flavor, and pleasant aroma, which imparts superior mouthfeel and textural properties to dairy products. Ghee flavor components are lactones, carbonyls, and free fatty acids (FFA), which are affected by temperature of clarification, ripening of cream, and storage period. Different methods have been used for manufacturing of ghee and these methods differ on the basis of usage of raw ingredients (milk or butter), intermediary processing of raw ingredients, and treatment of fully formed or partially finished ghee. Ghee is known as complex lipid of free fatty acids, tri-glycerides, sterols, hydrocarbon, phospholipids, and carotenoids. Fatty acid composition of ghee includes saturated or unsaturated carbon chain of an even number of carbon atoms from 4 to 20 which greatly affects the physicochemical constants of ghee. The keeping quality of ghee is a major quality criterion

The Chemistry of Milk and Milk Products: Physicochemical Properties, Therapeutic Characteristics, and Processing Methods. Megh R. Goyal, Suvartan Ranvir, & Junaid Ahmad Malik (Eds.)

for its acceptance by the consumers. Ghee can deteriorate during storage by the development of oxidized or rancid flavors, which are mainly dependent on the method of manufacture, high moisture content, ripening of cream, and temperature of clarification. Several physicochemical constants (such as FFA content, peroxide value, acid value, conjugated diene, etc.) are used to determine the quality of ghee upon storage.

8.1 INTRODUCTION

Lipids are a vital component in our diet and possess numerous therapeutic and nutritional properties. It is rich in principal energy components and essential elements required for nourishment of body. Dietary lipids are responsible for absorption of fat-soluble vitamins and good source of essential fatty acids.[16] Fats and oils in our daily diet are from different sources, such as milk and dairy products, meat, some vegetables, plant foods, poultry, and seafoods. Milk fat has been harvested since thousands of years before and it is considered the third major source of lipids in human nutrition.[4,21] Milk fats and milk fat-rich products have played a remarkable role in the dietary preferences in cultures throughout the world. The shelf-life of butter or cream is limited due to lipolysis and bacterial deterioration. It has been proposed that such types of spoilage exist at the interface or at water-fat-phase interfaces.[27,41]

According to ancient literature, butter was used as a foodstuff in India during 2000 B.C. and 1400 B.C. Documented history reveals manufacturing of butter from direct churning of milk, which is followed in many areas of Asia. Before the advent of modern techniques, especially refrigeration for enhancing shelf-life of milk, it had to be quickly consumed or some means had to be found out for its conversion into more stable edible product. Ghee satisfied these requirements and maximum surplus milk was converted into ghee. In tropical countries like India, milk quickly gets sour. The traditional method involves the souring of the milk by the use of a proper starter culture to prepare *dahi* (a fermented milk product) and further churned to butter, which too had a short storage life. As development occurred, practice grew to boil the butter to get ghee, which had greatly enhanced keeping quality.

Ghee, a clarified butterfat prepared from cow, buffalo, or mixed milk, is the most popular dairy product in the Indian subcontinent. It has characteristic aroma and flavor, which makes it highly prized dairy product among all the dairy products. The acceptance of ghee or butter by the consumers depends on the method of manufacturing and keeping quality. Upon storage, oxidized or rancid flavors may be generated as a result deterioration of ghee.

This chapter elaborates the chemistry of butter oil and ghee, classification and composition, and various manufacturing and processing methods and physicochemical changes during the storage of ghee or butter oil.

8.2 GHEE

Ghee is a traditional widely produced milk product in Asian, African, and Middle East countries.[66] Ghee is commonly known by other names, such as butteroil, Indian butteroil, anhydrous milk fat, butterfat and clarified butter,[33,46,59,66] and Indian ghee.[33]

8.2.1 DEFINITION OF GHEE

The IDF (International Dairy Federation),[26] FSSAI (Food Safety and Standards Authority of India),[18] *Codex Alimentarius*,[12] AGMARK[3] have defined the ghee and its requirements. Ghee is described as "*pure clarified fat derived solely from milk or curd, desi (cooking) butter or cream to which no coloring matter or preservative has been added*" according to FSSAI.[18] "*Ghee is a product exclusively obtained from milk, cream or butter from various animal species by means of processes which result in the almost total removal of moisture and solids-not-fat and which gives the product a particular physical structure.*"[26] IDF standard[26] also determined values for "*ghee to have 96% minimum milk fat, 0.3% maximum moisture, 0.3% maximum free fatty acids (FFA; expressed as oleic acid) and a peroxide value (PV) less than one.*" Moreover, to ghee, neutralizing substances may be added in trace amounts and it should not have an objectionable odor or taste. The standard definition of ghee as per IDF and *Codex Alimentarius* is the same, except the IDF standard provides extra information for maximum levels of FFA and PV.

Butter oil is defined as "*a concentrated source of fat obtained exclusively from butter or cream by the removal of all the nonfat solids and water portion.*" Other synonyms of butter oil are "milk fat, anhydrous milk fat, dry butter fat or dehydrated butter fat." It can be prepared by various methods, such as Alfa-Laval process or Westfalia process.[13,28] The FSSAI standards[19] of ghee, Anhydrous Butter Oil, Butter Oil are shown in Table 8.1.

TABLE 8.1 FSSAI Standards for Ghee, Butteroil, and Anhydrous Butter Oil.

Parameter	Ghee	Anhydrous butteroil	Butter oil
Butyro-Refractometer value (at 40°C)	40.0–45.0 (different for all states in India)	40.0–44.0	40.0–44.0
FFA (% Oleic acid)	NMT 3.0	NMT 0.3	NMT 0.4
Milk fat, % (m/m)	NLT 99.5	NLT 99.8	NLT 99.6
Moisture % (m/m)	NMT 0.5	NMT 0.1	NMT 0.4
Peroxide value (maximum)	–	NMT 0.3	NMT 0.6
Polenske value	–	1.0–2.0	1.0–2.0
RM value	Ghee below 28.0	NLT 28.0	NLT 28.0

8.3 PROCESSING METHODS OF GHEE

Various methods have been used for manufacturing of ghee which may differ with the use of raw material (milk, cream, butter), raw materials intermediary treatment, and handling of partially produced or completely formed ghee. There are five methods widely used for the preparation of ghee, such as traditional method, continuous method, creamery-butter method, direct cream method, and pre-stratification method.[43,64] Ghee lacks in characteristic flavor when it is produced from fresh cream by using method of direct cream, creamery-butter, or pre-stratification methods. The methods of ghee manufacturing are described in this section.

8.3.1 TRADITIONAL OR DESI OR INDIGENOUS METHOD

Traditionally, manufacturing of ghee is quite popular and represents a major contribution of ghee production. This method is well recognized for superior organoleptic quality, which involves simple technology, low-cost equipment, and small-scale operation. In India, this method is still employed in manufacturing of domestic ghee and in small-scale dairy industries. The principle of ghee manufacturing involves three steps.

- The first step involves fermentation of primary raw material (i.e., milk) by using lactic acid bacteria.
- The second one involves the concentration of milk fat in a concentrated form by mechanical process.

- The third step involves heating of the concentrated fat at a particular temperature for the removal of moisture and followed development of flavor by reaction of fermented residues of non-fat-milk with solids milk fat.

The development of taste of ghee, characteristic caramelized flavor, and color is based on the first and third steps.[43]

8.3.2 CREAMERY-BUTTER METHOD

This is the most basic and conventional method for ghee making and is widely used in dairy industry. The cream is separated from the milk using a centrifugal cream separator after it has been heated to nearly 40°C. After that, the cream is pasteurized, chilled, and ripened before being transformed into butter. The cream is generally ripened with lactic acid fermented culture to enhance the flavor of finished product. Buttermilk and creamery butter are made by churning ripened cream with water. Ghee residues are removed by filtering or using a ghee clarifier after the creamery butter has been clarified by boiling at 110–140°C.[10]

8.3.3 DIRECT CREAM METHOD

In this method, the ghee is made by clarifying the cream with the direct heating process. The entire procedure is divided into four steps, such as: (1) Reseparation of cream after diluting it with water to original volume of milk; (2) Using acidified or ordinary water to wash the cream; (3) Utilizing high fat percentage cream; and (4) Ripening cream by using starter cultural. All of these efforts are primarily aimed at the fat recovery in ghee and improvement in the flavor of ghee.

8.3.4 PRE-STRATIFICATION METHOD

Pre-stratification is a clarification step employed extensively during ghee manufacturing in industries. This method involves heating butter to 80–85°C for 15–30 min and then leaving it undisturbed overnight to separate three distinct layers. The lowest portion is made up of the serum component of the butter, which has the highest specific gravity. An intermediate stratum is formed by the top layer, which consists of fat and curd particles. This method makes it easier to separate fat from moisture and nonfat solids before clarification at high temperatures, which greatly increases ghee yield with low energy consumption.[10]

8.3.5 CONTINUOUS METHOD

The batch methods are widely used in the production of ghee for small- and medium-scale production. With rapid rise in demand of commercial production and wide scope for ghee export, there is a need for dairy industries to adopt a continuous ghee process. The first method employs three "Scraped Surface Heat Exchangers (SSHE)" in succession, accompanied by flashing into vertical vapor separators. The successive stages involve pumping of molten butter. Ghee from the previous phase is purified in a centrifugal clarifier to separate ghee residue, and the clarified ghee is kept in a tank for packaging. The continued ghee production approach was developed by Abichandani et al.[1] and it consists of a continuous butter-melting unit and the horizontal SSHE thin film for molten butter conversion into ghee.

8.4 PROCESSING METHODS OF BUTTER OIL

8.4.1 ALFA-LAVAL PROCESSING METHOD

Cream with 30 to 40% fat is initially heated to 55–58°C and is followed by transfer through pre-concentrator. As a pre-concentrator, hermetic separator is used to concentrate the cream to 70–75% fat content. The high-fat cream is a result of the use of a centrifixator, which is a particularly designed phase inversion chamber for changing "the oil-in-water emulsion into water-in-oil emulsion." The milk fat is separated in the next concentrator, which is cleaned, reseparated, and passed through a vacuum chamber set at 80 to 90°C to eliminate any leftover moisture. Afterward, it is cooled to 20–26°C, before being packaged.

If ghee is prepared from white butter, it is melted to about 50°C before being put through a plate heat exchanger to attain 70 to 80°C. The butter is moved to a separator to remove butter serum after being stored in a sealed tank and the butterfat is washed one more time with hot water. To make butteroil, washed butterfat is passed through a vacuum chamber to remove any remaining moisture.[36]

8.4.2 WESTFALIA PROCESSING METHOD

A standardized 40% fat cream is introduced into a cream concentrator, which separates the non-fat-milk solids component of the cream and the amount of fat is concentrated to 75% and it is followed by homogenization to break down the membrane around fat globules. In the centrifuge, major part of fat is separated from the solids and there exists as an aqueous phase. After being rinsed with water, the oil is centrifuged and vacuum-dried after it has been semidried. To make numerous fine crystals, butter oil should be cooled and crystallized under continual supervision. The desired outcome may be produced in the form of a smooth homogeneous mass that does not split into solid and liquid layers when left to stand by fast supercoiling at 13–18°C and agitating the mass during forced crystallization (by adding 5–15% of finely crystalline fat from the previous lot).[36]

8.5 BUTTER OIL VERSUS GHEE

The composition, flavor, and color of butter oil and ghee vary significantly as processing methods for both products are different. Ganguli and Jain[20] explain major differences between ghee and butter oil.

- Ghee has a nutty, pleasant, and mildly cooked flavor, whereas butter oil has a bland flavor.
- As compared to butter oil, ghee has more protein particles, less moisture, and a wider range of phospholipid and fatty acid concentration.
- Ghee is prepared at 100–140°C, whereas butter oil is produced by melting butter at temperatures below at 80°C.
- Ghee cannot be reconstituted because the flavor of cooked ghee is integrated into the final product, whereas butter oil may be reconstituted with skim milk powder.

8.6 CHEMICAL COMPOSITION OF GHEE

Among all Indian dairy products, ghee is known as a rich source of milk lipids. The composition of ghee varies and it depends on the type of methods used for manufacture. Ghee is an almost anhydrous fat developed exclusively in India.[20] Ghee chemically may be defined as "*complex lipids of triacylglycerol, together with small quantity of free fatty acids, phospholipids, hydrocarbons, carbonyl compounds, fat soluble vitamins (A, D, E, and K), carotenoid pigments, moisture, and traces of elements like copper and iron.*" According to Aneja et al.[6] and Battula et al., [9], general chemical composition of ghee is indicated in Table 8.2.

TABLE 8.2 Chemical Composition of Ghee.

Composition	Buffalo ghee	Cow ghee
Carotene (mg/g)	–	3.2–7.4
Cholesterol (mg/100 g)	209–312	302–362
Free fatty acids (%maximum)	2.8%	2.8%
Milk fat (%)	99–99.5%	99–99.5%
Moisture (%)	Less than 0.5%	Less than 0.5%
Tocopherol (mg/g)	18–31	26–48
Vitamin A (IU/g)	17–38	19–34

8.6.1 TRIGLYCERIDE COMPOSITION OF GHEE

As per chemical composition, ghee consists of 97–98% triglycerides, 0.37% di- and mono-glycerides, 0.31% cholesterol, 0.027% free fatty acid, and 0.6% phospholipids.[11] In milk fat, more than 200 triglyceride species have been found.[22] The synthesis of milk lipid glycerol takes place partially from glucose, partially from hydrolyzed blood lipids (monoglycerides and free glycerol) and trace amounts from free blood glycerol. These triglycerides are formed in endoplasmic reticulum, which is located near cell's basal membrane.

The melting behavior of triglycerides fractions is usually used for their differentiation. The melting point of a triglyceride is determined by its fatty acid composition and orientation of fatty acids in triglyceride. As a result, triglycerides can be classified into three groups, such as high melting, medium melting, and low melting fractions (HMF, MMF, and LMF, respectively). Milk Fat Globule Membrane (MFGM), a protective layer of fat, also has been reported with various glycerides in the lipid fraction. According to studies, MFGM has a lipid profile of triglyceride (83–88%), diglyceride (5–14%), and free fatty acids (1–5%).

8.6.2 FATTY ACID COMPOSITION OF GHEE

In bovine milk fat, about 400 different fatty acids have been identified. However, at concentrations >1%, only a limited number (about 12) have been identified.[29,30] Saturated, monounsaturated (MUFA), and polyunsaturated fatty acids (PUFA) are categorized as short chain (2–8), medium chain (8–12), and long chain (13–24) fatty acids based on the presence or absence of double bonds.[32] In the mammary glands, these acids are produced, but in the rumen, bacteria hydrogenate polyunsaturated fatty acids, resulting in a range of geometric and positional unsaturated acid isomers.[23]

Milk fat has a very high proportion of saturated fat (about 65%) and unsaturated fatty acids (about 35%) of total fatty acids. Among unsaturated fatty acids, major contribution is governed by oleic acid, which is about 30%.[71] Factors such as species, breed, diet, nutrition, region, season, and lactation stage have an impact on fatty acid concentration of milk fat.[24] Using GLC (Gas Liquid Chromatography) or Gas Chromatography-Mass Spectrometry,[25,31,45,48,63,68] several researchers have defined and measured fatty acid content of bovine milk fat.

Table 8.3 illustrates level and amounts of essential fatty acids in milk fat as determined by several investigators. According to some studies, the MUFA and PUFA percentage of ghee were found to be 32.8% and 2.9%, respectively.[35,66] The fatty acids like palmitic, stearic, and myristic were found in cow and buffalo ghee in amounts ranging from 24 to 28%, 9 to 14%, and 8 to 10%, respectively.[7] The fatty acid profile of ghee indicated presence of oleic acid and linoleic acid as the primary MUFA and PUFA. Butyric acid levels in buffalo ghee are much higher than cow ghee. Furthermore, cow ghee has much higher short chain fatty acids than buffalo ghee, whereas buffalo ghee has higher palmitic and stearic fatty acids.[15] Table 8.3 shows the fatty acid composition of buffalo and cow milk fat.[9,62]

8.6.3 CHEMISTRY OF GHEE FLAVOR

The chemistry of ghee flavor involves a wide range of components. Using GLC, more than 100 taste components responsible for ghee flavor have been documented.[70] Ghee flavor is composed of different flavor compounds, like carbonyls, reducing substances, FFAs, lactones, etc. Carbonyls and lactones are two of the most important components in ghee flavor. During fermentation, lactic, butyric, caproic, and caprylic acids, as well as carbonyls and other flavor constituents, are partially transferred to ghee and impart flavor.

The metabolic activity of starter bacteria on lactose, citrates, and glucose in milk/cream/butter can contribute to an increase in the production of ghee-flavor compounds (especially carbonyls and FFAs). Acidification of ripened cream or butter aids in the incorporation of flavor compounds. Furthermore, starter microorganism lowers the pH, which aids in the acceleration of the chemical process of caramelization during heat clarifying. Cream that has been allowed to ripen before being churned into butter then clarified into ghee with superior flavor.[43]

TABLE 8.3 Level of Fatty Acids in Buffalo and Cow Milk Fat.

Fatty acid Name (carbon no.)	Cow milk fat (%wt)	Buffalo milk fat (%wt)
Butyric acid (C4:0)	3.2	4.4
Caproic acid (C6:0)	2.1	1.5
Caprylic acid (C8:0)	1.2	0.8
Capric acid (C10:0)	2.6	1.3
Lauric acid (C12:0)	2.8	1.8
Linoleic acid (C18:2)	1.5	1.5
Linolenic acid (C18:3)	0.6	0.5
Myristic acid (C14:0)	11.9	10.8
Myristoleic acid (C14:1)	2.1	1.3
Palmitic acid (C16:0)	29.9	33.1
Palmitoleic acid (C16:1)	2.2	2.0
Stearic acid (C18:0)	10.1	12.0
Oleic acid (C18:1)	27.4	27.2

Browning constituents known as ghee residue are formed in ghee during heat clarification. It predominantly comprises volatile compounds like denatured milk proteins, caramelized lactose, entrapped fat, as well as phospholipids and minerals, which together considerably enhance the flavor profile of ghee. It contains numerous flavoring compounds found in ghee, such as lactones, carbonyls, and FFAs.[54,58,65] Table 8.4 indicates flavor compounds of cow and buffalo ghee.

8.6.3.1 FREE FATTY ACIDS

The characteristic flavor of ghee is due to a combination of various fatty acids. During the fermentation and processing of ghee, flavor compounds are created from fatty acid glycerides formed by lipolysis of milk or cream. Fresh ghee contains approximately 6–12 mg/g of FFAs. Lower fatty acids (C6–C12) contribute significantly to ghee flavor while being present in small amounts (0.4–1 mg/g) and accounting for just 5–10% of total free fatty acids.[69]

TABLE 8.4 Flavor Compounds of Cow and Buffalo Ghee.

Flavor compound	Total carbonyl (μmoles/g)	Volatile carbonyl (μmoles/g)	Headspace carbonyls (μmoles/g)	FFA (mg/g)	Lactones (ppm)
Cow ghee	7.200	0.330	0.033	5.0–12.3	30.3
Buffalo ghee	8.640	0.260	0.027	5.8–7.6	35.4
References	[43]			[70]	

8.6.3.2 CARBONYLS

Carbonyls are significant in the development of flavors and off-flavors in fat-rich milk products like ghee. Monocarbonyls and dircarbonyls are two types of compounds that fall under this category; "Alkan-2-ones, alkarnals, alk-2-enals, and alka-2,4-dienals" are all forms of monocarbonyls. Several dairy products, particularly those that have been heat processed, were found to contain methyl ketones or alken-2-ones.

Alkan-2-ones are produced by hydrolysis of ketogenic glycerides followed by decarboxylation of β-ketocarboxylic acids in various processing processes used in the production of ghee. The lipolysis of triglycerides by *Penicillium* molds during milk or cream fermentation can also result in the production of alkan-2-ones. When unsaturated fatty acids are oxidized during milk or cream fermentation or heat clarification of cream or butter, aldehydes, such as n-alkanals, alk-2-enals, and alka-2,4-dienals, are produced in ghee.[73]

Polar carbonyls (such as dicarbonyls, α-ketoacids, glyoxals, and furfurals) are produced as a result of fermentation and heat-clarifying stages of ghee manufacturing. It has been shown that ripening of cream and heating butter at clarification temperature increases the concentration of polar carbonyls.[49]

8.6.3.3 LACTONES

Lactones are recognized as major flavor-contributing compounds in heat-treated fat-based milk products. The production of lactones begins with the hydrolysis of lactogenic glycerides (hydroxyl), which is followed by dehydration (lactonization) of hydroxy acids. After the lipolysis of delta-hydroxy acid glycerides, it undergoes ring closure to create lactones and as a result free delta-hydroxy acids are produced. Delta ketoacid glycerides, on the other hand, are lipolyzed to generate delta keto acids, which are then reduced to produce hydroxyl acids that are then converted into lactones.[5]

8.7 PHYSICOCHEMICAL CHANGES DURING STORAGE OF GHEE

The development of off-flavor during storage of ghee is a major issue. The shelf-life of ghee is about 6–8 months, depending on raw material quality and storage conditions.[2,66] It undergoes oxidative degradation processes likes oxidation, hydrolysis, and polymerization during storage, resulting in the formation of acceptable and unacceptable secondary compounds that can affect the quality of ghee.[67]

8.7.1 RANCIDITY

It is a pronounced defect of ghee occurring during prolonged storage. There are three types of rancidity, such as hydrolytic, oxidative, and ketonic. The rancidity in ghee occurs mainly during the storage, although it can also occur in freshly prepared ghee if rancid raw material is utilized in ghee production. The presence of volatile chemicals that cause unpleasant scents in trace levels characterizes this defect. Milk fat hydrolysis is faster in liquid state than in solid form. Buffalo milk has more solid fat than the cow milk because of which buffalo milk fat hydrolyzes more slowly. As a result, during storage, cow ghee is more likely to develop a rancid taste. Different types of rancidity are discussed in this section.

8.7.1.1 HYDROLYTIC RANCIDITY

Lipoprotein lipase is a fat-splitting enzyme found in membrane of milk fat globules. It is responsible for milk fat hydrolysis and the generation of lower molecular weight fatty acids including butyric, caproic, and caprylic. Among these fatty acids, rancid flavor in ghee is formed particularly by butyric acid. High heat treatment is used to inactivate lipase enzyme during production of ghee. As a result, hydrolytic rancidity in ghee is unusual, unless high-quality raw materials (free from rancidity) are utilized. In comparison to ghee, butter oil has more prone to rancid flavor defect.[14]

8.7.1.2 OXIDATIVE RANCIDITY

The oxidation of PUFA in the existence of oxygen causes oxidation of butterfat or ghee. Free radical activation, proliferation, and termination are all involved in the reaction of oxygen with PUFA. The chain reaction in ghee and butter oil is catalyzed primarily by light, heat, ionization reaction, and trace metals (copper and iron). It causes lipid auto-oxidation, which results in the production of ketones, aldehydes, alcohols, hydrocarbons, and acids, among other end products.[14]

8.7.1.3 KETONIC RANCIDITY

It occurs in lauric acid oils, butterfat, and ghee. Ketonic rancidity develops when some fungi convert short and medium chain carbon fatty acids into methyl ketones. Ketonic rancidity does not occur in fats and oils that do not include fatty acids with carbon chains ranging from C6 to C14.[40]

8.8 MECHANISM OF LIPID OXIDATION

Oxidative rancidity is caused by lipid degradation, which is a significant cause of degradation of fat-rich dairy products. Lipid oxidation is a three-stage free radical chain reaction in which free radicals are initiated, propagated, and terminated. Hydroperoxides are odorless and tasteless compounds that are formed when unsaturated fatty acids are oxidized. Hydroperoxides are highly unstable compounds that break down into flavorful carbonyls and other compounds.

In autoxidation of UFA, the production of free radicals is a crucial and important step. Metal complexes, irradiation, enzymes, or active oxygen species all contribute to the generation of free radicals that start the oxidation process. The reaction is normally started by removing a hydrogen from methylene group adjacent to double bond in monounsaturated and nonconjugated PUFA in milk lipids. A peroxide-free radical is formed when a peroxide-free radical combines with ground-state molecular oxygen. This reacts with another unsaturated molecule, resulting in the formation of a hydroperoxide.[42,17]

- The first step in lipid peroxidation is the extraction of hydrogen atoms from a fatty acid. A small number of radicals, for example, transition metal ions or by high-energy irradiation photolysis produced radical, can remove hydrogen from lipid molecules, producing unstable lipid free radicals or fatty acid radicals.

$$\text{Initiation: } \rightarrow \text{RH R}^\bullet \text{ (RH is UFA)}$$

- Fatty acid radicals interact swiftly with molecular oxygen due to their instability, resulting in the creation of a peroxyl-fatty acid radical. This radical is a very fragile entity that reacts with another free fatty acid to create additional fatty acid radicals, as well as cyclic peroxide or lipid peroxide. The cycle begins when a fresh fatty acid radical acts in the same way, resulting in propagation.

$$\text{Propagation: } R^\bullet + O_2 \rightarrow RO^\bullet_2$$
$$RO^\bullet_2 + RH \rightarrow ROOH + R^\bullet$$

- As a result, the events form foundation of a chain reaction. More radicals and noxious aldehydes are generated as lipid hydroperoxide decomposes. The termination reaction starts when the substrate is drained. The unpaired electrons are combined by two radicals to form a nonradical product.

$$\text{Termination: } 2RO^\bullet_2 \rightarrow O_2 + RO_2R$$
$$RO^\bullet_2 + R^\bullet \rightarrow RO_2R$$
$$2R^\bullet \rightarrow R — R$$

Moreover, antioxidants (A–H) donate hydrogen to abstracts peroxyl radicals by combining with lipid radicals to halt further propagation of chain reactions.

8.9 PHYSICOCHEMICAL CHARACTERISTICS OF GHEE

The intrinsic flavor, color, and appearance of ghee will determine its quality. The flavor of ghee should be sweet, nutty, and slightly cooked. The flavor of ghee is better characterized as sweet rather than acidic with a lack of blandness. Ghee is widely accepted because of its golden yellow to light yellow color. Moreover as an important physical characteristic, granular appearance of ghee is also a critical consistency parameter as it is favored by the consumers. Apart from the sensory characteristics listed here, the chemical and other physical parameters are often assessed to determine the consistency of ghee and to restrict adulteration.

8.9.1 REFRACTIVE INDEX (RI)

The RI is defined as the ratio of light velocity in vacuum to light velocity in the sample medium. At a given wave length, the ratio of sine angle of incidence to sine angle of refraction, or the angle at which a ray passes from air to fat, is measured (typically 589.3 m, mean of sodium D-lines). At 40°C, the Abbe refractometer is used to test milk fat, with results ranging from 1.4157 to 1.4566. As compared to other fats and oils, ghee has a low RI value. The saturation and molecular weight of the component fatty acids have an impact on the RI of ghee. It can also be used as a test to see if ghee has been adulterated with a foreign fat.[20]

8.9.2 IODINE NUMBER

Under specific prescribed conditions, it is known as the number of grams of iodine absorbed by 100 g of fat. As a result, this constant is a measure of a fat's unsaturated linkages. The iodine content ranges between 26 and 35 for milk fat, which is minimal as compared to other fats and oils. Wij's method is widely used to calculate the iodine value. Each unsaturated linkage absorbs one molecule of halogen compound, which is represented as the equal number of grams of iodine absorbed by 100 g of fat.[60]

8.9.3 REICHERT-MEISSL VALUE (RM VALUE)

It is defined as the amount of mL of n/10 sodium hydroxide necessary to neutralize the steam volatile water-soluble fatty acids distilled from 5 g of ghee under the specific conditions of the method. It is utilized to determine amount of butyric and caproic fatty acid in milk fat. Milk fat has a value ranging from 17 to 35, which is more than the value of all other fats and oils. As a result, milk fat has a higher concentration of these acids than any other fat.[66] According to Patel,[49] buffalo ghee has a higher RM value than the cow ghee.

8.9.4 POLENSKE VALUE (PV)

It is defined as the number of mL of N/10 sodium hydroxide necessary to neutralize the steam volatile water insoluble fatty acids distilled from 5 g of fat under the specific conditions of the method. Caprylic and capric acids are primarily indicated in PV because they are steam volatile but mostly insoluble in water. Milk fat has a PV ranging from 1.2 to 2.4.[61] According to Patel,[44] the overall PV range for both pure cow and buffalo ghee was 1.30–1.80. Rakesh[43] also observed the similar results so that the PV could not be used to detect up to 20% adulteration in ghee.

8.9.5 SAPONIFICATION VALUE

It is the quantity of potassium (mg) required to saponify 1 g of fat. This value for ghee ranges from 210 to 233, with the most typical range being 225–230. This constant represents the average molecular weight of the fatty acid present. Since minerals oils, such as liquid paraffin, are unaffected by alkali and do not form a homogeneous solution on saponification, the saponification value is more useful in detecting their adulteration in ghee.

8.9.6 MELTING POINT

According to the literature, the melting point of milk fat varies from 30 to 41°C.[20]

8.10 METHODS TO DETERMINE PHYSICOCHEMICAL CHANGES IN GHEE

8.10.1 DURING STORAGE

Oxidation occurs in two stages (primary and secondary), each of which results in a different chemical change.[55] Several techniques for monitoring the oxidative degradation of various oils and fats have been documented, which are all based on chemical changes that occur during two phases of oxidation.[38]

The primary stage of ghee oxidation has been monitored using a variety of analytical methods, such as conjugated dienes,[39] weight gain,[34,52] iodine value,[51] Kreis number,[72] free fatty acids,[39] and peroxide value.[38,57]

8.10.1.1 PRIMARY OXIDATION

Peroxides, particularly hydroperoxides, are first compounds formed during the primary oxidation stage and they can form secondary oxidation products, such as aldehydes, ketones, hydroxyl compounds, epoxides, and polymers.

8.10.1.1.1 Weight Gain Method

The weight-gain approach is based on oxygen absorption during the early phases of autoxidation, resulting in a change in fat or oil weight. Mehta et al.[39] employed this procedure and observed that after 10 days of storage at 80°C, the weight of sample was decreased from 22.61 to 22.55 g. The first weight loss might be due to moisture and other volatile chemicals evaporating partially.

8.10.1.1.2 Conjugated Dienes (CD, %) Content

The presence of CD on lipids can be used to detect autoxidation of fat-acid moieties. Because CD moiety is a strong UV absorbing chromophore, it may be measured spectrophotometrically. According to Mehta et al.,[39] the CD of fresh ghee samples were between 0.27 and 0.31% with no change after 10 days of examination.

8.10.1.1.3 Kreis Number

Kreis number is used to determine the presence of aldehydes and ketones in rancid fat. In general, a value of 3.02 was found in fresh ghee samples and it was decreased significantly to 2.43 on the second day storage study at 80°C.[39] The Kreis value of ghee in consecutive ghee samples, on the other hand, progressively increased on storage.

8.10.1.1.4 Iodine Value

The iodine value is used to determine the degree of unsaturation in oil or fat. The iodine value is an essential parameter in the examination of oxidative rancidity of oils because the higher the percentage of unsaturation in the oil, the higher the risk of rancidity. Ghee lipid oxidation can be detected by a reduction in iodine levels.[51] Fresh ghee samples were found with initial iodine value ranging from 30.69 to 34.83. Following the fourth day of accelerated storage at 80°C, the iodine value exhibited a trend of declining (31.4) at a very slow rate.

8.10.1.1.5 Free Fatty Acids Content

A typical marker for measuring the amount of rancidity in oils and fats is the measurement of free fatty acids. According to Ashok and Bector,[8] FFA content in fresh ghee ranged from 0.09 to 0.28% oleic acid, with an average of 0.16% oleic acid, and it increased from 0.28 to 0.64% after 4 months of storage at 37°C.

8.10.1.1.6 Peroxide Value

Peroxide value is an indicator for the measurement of oxidation of lipids, fats, and oils and is expressed in mg per kg of peroxide oxygen (meq/kg). Fresh ghee samples had an initial peroxide value ranging from 0.02 to 0.37 meq of O_2/kg fat. When Mehta et al.[38] analyzed the peroxide value after 10 days of storage at 80°C, they found that it was gradually increased and reached highest at 1.85 meq O_2/kg of fat.

8.10.2 SECONDARY OXIDATION

Unpleasant odors, lipid discoloration, and oxidation can also reduce nutritional quality and safety in addition to developing rancidity. The carbonyl compounds are the key obstacles to the flavor of several food products associated with their rancidity.[56]

8.10.2.1 THIOBARBITURIC ACID (TBA TEST)

Malonaldehyde (MDA) interacts with TBA to generate a pink MDA-TBA complex, which is spectrophotometrically measurable at 530 nm and is used to evaluate lipid peroxidation. The average TBA value of a fresh ghee sample is 0.03 Malondialdehyde (pmol/mg), ranging from 0.01 to 0.07 pmol/mg. The TBA value is gradually increased to a highest on the sixth day of storage and is dropped on the eighth day of storage. The value is increased slightly on the 10th day of storage at 37°C at the conclusion of 4-month of storage period.[8] The TBA method has one drawback that it may react with a wide range of chemicals including other carbohydrates, nucleic acids, amino acids, and aldehyde. This has an impact on the TBA testing, resulting in considerable overestimation and inaccuracy in the results.[53]

8.10.2.2 P -ANISIDINE VALUE (P-ANV)

The p-AnV is a valuable method for analyzing nonvolatile α-unsaturated aldehydes during the secondary stage of oxidation. The content of aldehydes generated during the decomposition of hydroperoxides (mainly 2-alkenal and 2,4-alkadienal) is measured by this value. The initial p-AnV of fresh ghee sample ranges from 0.27 to 0.80 mmol/kg. On the fourth day of storage, p-AnV was increased gradually and then was decreased to 2.41 mmol/kg on the 10th day of the accelerated 80°C storage.[38]

8.10.2.3 MEASUREMENT OF TOTOX VALUE (TV)

TV is an overall measure of the oxidation that includes both primary and secondary oxidation products. It is a combination of "PV and p-AnV, for example, TV = 2PV + p-AnV." This reflects a level of oxidation at the early and later stages of the oxidation reaction. This equation estimates both hydroperoxides and their decomposition products and provides a better outcome to control the progressive oxidative deterioration of fats and oils.[37]

8.10.2.4 MEASUREMENT OF CARBONYLS CONTENT

Secondary oxidation products (such as carbonyl compounds, ketone, and aldehydes) are formed by hydroperoxide degradation and are presumed to be the dominant contributor to off-flavors in dairy foods. The total carbonyl content of various foods is measured using a colorimetric technique. Carbonyl levels in the original fresh ghee samples varied from 0.18 to 0.30 mol/g of fat. Carbonyl value was increased steadily throughout the storage and reached a maximum of 0.39 mol/g at 80°C at the end of the 10th day of the accelerated storage.[38]

8.11 FACTORS AFFECTING SHELF-LIFE/RANCIDITY

Various factors causing fat oxidation are raw product selection, storage, refining, manufacturing, etc., due to its constant exposure to oxygen to ambient air. Manufacturing of ghee from collected butter, which may have undergone deterioration on storage, is subjected to neutralization, pre-stratification, and refining. It is preferable to churn the butter without delay so as to prevent losses of water-soluble and volatile fatty acids, which can affect quality of ghee. Ghee preparation using neutralized curdled milk generally does not affect the analytical constants. Heat clarification of ghee at high temperatures (150°C) can lower its shelf-life, as it leads to a higher peroxide value and loss of antioxidant activity imparted by free sulfhydryl groups.

The ghee residue formed at high temperature can result in higher peroxide development as compared to low temperature (110°C). Compared to direct creamery and desi-butter ghee residues, ghee residue of creamery butter method shows higher antioxidant activity. Manufacturing, packaging, and storage of ghee should be under controlled hygienic conditions and should be kept at 21°C to get best quality till 9 months. Oxidative changes in ghee occur when left at ambient temperature for a long period.

8.12 PREVENTION OF OXIDATION OF GHEE

8.12.1 DECREASING THE RATE OF OXIDATION

The storage of ghee in cool, dark conditions significantly decreases the rate of oxidation. Exposure to prooxidants (such as oxidized products), heavy metals, UV, and visible light should be avoided.

8.12.2 ELIMINATION OF OXYGEN

Ghee when packed under nitrogen gas under vacuum or packaging with an oxygen scavenger can eliminate oxygen, which in turn will prevent the oxidation of ghee.

8.12.3 ELIMINATION OF THE SUSCEPTIBLE SUBSTRATE

Replacement of polyunsaturated part of fatty acid in ghee with less unsaturated fatty acid through process of interesterification can enhance its stability and thus prevent oxidation.

8.12.4 HYDROGENATION

This process uses catalyst to lower number of double bonds in the presence of suitable catalyst and this lowers degree of unsaturation, which reduces the susceptibility of ghee to oxidation during storage.

8.12.5 EFFECTS ON ANTIOXIDANTS

Ghee residue of creamery butter method shows higher antioxidant activity as compared to direct creamery and desi butter ghee residues. The largest proportion of phospholipids are transferred to ghee when ghee residue is heated with ghee in a 1:4 ratio at 130°C. The lysine, cysteine hydrochloride, proline, and tryptophan in the solid-not-fat portion of ghee have the highest antioxidant effects. Synthetic free radical scavengers (such as gallates (ethyl, propyl, octyl), Butylated Hydroxytoluene, Tertiary Butyl Hydroquinone, and Butylated Hydroxy Anisole) are used as antioxidants.[50]

8.13 SUMMARY

Ghee is a traditional Indian dairy product with functional ingredients and has proven to provide health benefits. The compositional, physicochemical, and flavor profile of ghee may vary depending on the method of manufacturing and utilization of raw material. Free fatty acids, carbonyls, and lactones are major contributors to ghee flavor and are greatly influenced by the temperature of clarification, process of manufacturing, and storage period. This chapter also documented physicochemical changes during storage of ghee, mechanism of lipid oxidation, various physicochemical constants of ghee, and prevention of oxidation of ghee. As a fat-rich dairy product, ghee is prone to hydrolytic and oxidative processes that cause rancidity. Furthermore, detailed studies are required for rapid detection of primary and secondary oxidation products formed during storage of ghee. Ghee is well known for its medicinal and nutritional benefits. However, more research or numerous clinical studies are needed to support the health benefits of ghee and their amount of consumption.

KEYWORDS

- **antioxidants**
- **butter oil**
- **free radical scavengers**
- **ghee**
- **physicochemical parameters**
- **rancidity**

REFERENCES

1. Abichandani, H.; Bector, B.; Sarma, S. C. Continuous Ghee Making System-Design, Operation and Performance. *Ind. J. Dairy Sci.* **1995,** *48*, 646–650.
2. Achaya, K. T. Ghee, Vanaspati and Special Fats in India. In *Lipid Technologies and Applications;* Gunstone, F. D., Padley F. B., Eds.; Marcel Dekker Inc.: New York, NY, 1997; chapter 14, pp 369–390.
3. AGMARK. *Ghee Grading and Marking Rules, 1938 (As amended in 1988)*. Office of Agricultural Marketing Advisor (AMA), Directorate of Marketing and Inspection (DMI), Government of India, Faridabad (Haryana), India, 1988; pp 1–6.
4. Aguedo, M.; Hanon, E.; Danthine, S.; Paquot, M.; Lognay, G.; Thomas, A.; Vandenbol, M.; Thonart, P.; Wathelet, J. P.; Blecker, C. Enrichment of Anhydrous Milk Fat in Polyunsaturated Fatty Acid Residues From Linseed and Rapeseed Oils Through Enzymatic Interesterification. *J. Agric. Food Chem.* **2008,** *56*(5), 1757–1765.
5. Alewijn, M.; Smit, B. A.; Sliwinski, E. L.; Wouters, J. T. M. The Formation Mechanism of Lactones in Gouda Cheese. *Int. Dairy J.* **2007,** *17*(1), 59–66.
6. Aneja, R. P.; Mathur, B. N.; Chandan, R. C.; Banerjee, A. K. Fat Rich Products. In *Technology of Indian Milk Products*; Dairy India Publication: New Delhi, India, 2002; pp 183–198.
7. Antony, B.; Sharma, S.; Mehta, B. M.; Ratnam, K.; Aparnathi, K. D. Study of Fourier Transform Near Infrared (Ft-Nir) Spectra of Ghee (Anhydrous Milk Fat). *Int. J. Dairy Technol.* **2018,** *71* (2), 484–490.
8. Ashok, K.; Bector, B. S. Comparative Study on the Determination of Oxidative Rancidity in Ghee by Different Methods. *Asian J. Dairy Res.* **1985,** *4*, 23–28.
9. Battula, S. N.; Naik, N. L.; Sharma, R.; Mann, B. Ghee, Anhydrous Milk Fat and Butteroil. In *Dairy Fat Products and Functionality*; Springer: Cham, 2020; pp 399–430.
10. Bhide, N. M. Effect of Modern and Traditional Methods of Preparation on the Composition and Flavor Profiles of Ghee. Master's Degree Thesis, Rutgers the State University of New Jersey, New Brunswick, USA, 2014; p 162.
11. Christie, W. W. Composition and Structure of Milk Lipids. In *Advanced Dairy Chemistry-2*; Fox, P. F., Ed.; Chapman and Hall: London, UK, 1995; pp 1–56.

12. *Codex Alimentarius* Draft Revised Standard for Milk Fat Products (A-2); CAC (FAO): Rome, 1997; 37–39.
13. De. S. Outlines of Dairy Technology; Oxford University Press: New Delhi, India, 1991; pp 143–173.
14. Deeth, H. C.; Fitz-Gerald, C. H. Lipolytic Enzymes and Hydrolytic Rancidity. In *Advanced Dairy Chemistry, Volume 2 - Lipids*; Springer: Boston, MA, 2006; pp 481–556.
15. Dorni, C.; Sharma, P.; Saikia, G.; Longvah, T. Fatty Acid Profile of Edible Oils and Fats Consumed in India. *Food Chem.* **2018,** *238*, 9–15.
16. FAO-WHO. Fat and Fatty Acid Requirements for Adults. In *FAO Food and Nutrition Paper - 91, Fats and Fatty Acids in Human Nutrition: Report of An Expert Consultation;* FAO: Rome, Italy, 2010; pp 55–62.
17. Fox, P. F.; McSweeney, P. L.; Paul, L. H. (Eds.) *Dairy Chemistry and Biochemistry*. Elsevier/Academic Press: Boston, MA, 1998; p 584.
18. FSSAI. Food Product Standard and Food Additives. Ministry of Health and Family Welfare, Government of India, New Delhi, [Online] 2017. https://fssai.gov.in/home/fss-legislation/fss-regulations.html (accessed Aug 31, 2021).
19. FSSAI. Food Safety and Standards (Food Products Standards and Food Additives) Regulations. Food Safety and Standard Authority of India (FSSAI), Ministry of Health and Family Welfare, Government of India, New Delhi, [Online] 2011. http://www.fssai.gov.in /Portals/0/Pdf/FSS_Gazete_Rules_2011.pdf (accessed Aug 31, 2021).
20. Ganguli, N. C.; Jain, M. K. Ghee: Its Chemistry, Processing and Technology. *J. Dairy Sci.* **1973,** *56*(1), 19–25.
21. Gnanasambandam, R.; Torres-Gonzalez, M.; Burrington, K. J.; Kapoor, R. *Milk Fat and Related Ingredients Serving Today's Marketplace. Unpublished Report by Department of Agriculture Economic Research Service (USDA), Washington, DC, 2017; vol 2017, pp* 1–16.
22. Gresti, J.; Bugaut, M.; Maniongui, C.; Bezard, J. Composition of Molecular Species of Triacylglycerols in Bovine Milk Fat. *J. Dairy Sci.* **1993,** *76*(7), 1850–1869.
23. Gunstone, F. D.; Harwood, J. L.; Padley, F. B. (Eds.). *The Lipid Hand Book*; Chapman and Hall: London, UK, 1986; pp 1–169.
24. Hanuš, O.; Samková, E.; Křížová, L.; Hasoňová, L.; Kala. R. Role of Fatty Acids in Milk Fat and the Influence of Selected Factors on Their Variability - A Review. *Molecules* **2018,** *23*(7), 1636–1641.
25. Hawke, J. C. The Fatty Acids of Butterfat and the Volatile Acids Formed on Oxidation. *J. Dairy Res.* **1957,** *24*(3), 366–371.
26. Illingworth, D.; Patil, G. R.; Tamime, A. Y. Anhydrous Milk Fat Manufacture and Fractionation. In *Dairy Fats and Related Products*; Wiley-Blackwell: Oxford, UK, 2009; pp 108–166.
27. International Dairy Federation (IDF). FIL/IDF Standard 68A: Anhydrous Milk Fat, Anhydrous Butteroil or Anhydrous Butterfat, Butteroil or Butterfat, Ghee: Standards of Identity. International Dairy Federation: Brussels, Belgium, 1977, p 43.
28. Jana, A. Quality Requirements of Manufacture of Fat Rich Dairy Products for Use in Recombined Dairy Products. In *Compendium of Lectures Delivered at Refresher Course on Technology of Fat Rich Dairy Product*; SMC College of Dairy Science, Gujarat Agricultural University, Gujarat, India, 1990; pp 188–199.
29. Jensen, R. G.; Ferris, A. M.; Lammi-Keefe, C. J. The Composition of Milk Fat. *J. Dairy Sci.* **1991,** *74*(9), 3228–3243.

30. Jensen, R. G.; Gander, G. W.; Sampugna, J. Fatty Acid Composition of the Lipids From Pooled, Raw Milk. *J. Dairy Sci.* **1962,** *45*(3), 329–331.
31. Jensen, R. G.; Newburg, D. S. Milk lipids. In Jensen, R. G., Eds.; *Handbook of Milk Composition*; Academic Press: New York, USA, 1995; pp 545–575.
32. Kostik, V.; Memeti, S.; Bauer, B. Fatty Acid Composition of Edible Oils and Fats. *J. Hyg. Eng. Des.* **2013,** *4*, 112–116.
33. Kumar, N.; Singhal, O. P. Effect of Processing Conditions on the Oxidation of Cholesterol in Ghee. *J. Sci. Food Agric.* **1992,** *58*(2), 267–273.
34. Lea, C. H., Rancidity in Edible Fats. Department of Scientific and Industrial Research, Food Investigation Special Report No. 46; Chemical Publishing Company: New York, 1939; p 230.
35. Manickavasagan, A.; Al-Sabahi, J. N. Reduction of Saturated Fat in Traditional Foods by Substitution of Ghee with Olive and Sunflower Oils - A Case Study with Halwa. *J. Assoc. Arab Univ. Basic Appl. Sci.* **2014,** *15*, 61–67.
36. Mehta, B. M. Butter, Butter Oil, and Ghee. *Gourmet and Health-Promoting Specialty Oils*; AOCS Press: Chicago, 2009; pp 527–559.
37. Mehta, B. M. Comparative Appraisal of Physical, Chemical, Instrumental and Sensory Evaluation Methods for Monitoring Oxidative Deterioration of Ghee. Doctoral Dissertation, AAU, Anand, India, 2014; p 148.
38. Mehta, B. M.; Darji, V. B.; Aparnathi, K. D. Comparison of Five Analytical Methods for the Determination of Peroxide Value in Oxidized Ghee. *Food Chem.* **2015,** *185*, 449–453.
39. Mehta, B. M.; Kumar Jain, A.; Darji, V. B.; Aparnathi, K. D. Evaluation of Different Methods to Monitor Primary Stage of Oxidation of Heat Clarified Milk Fat (Ghee). *J. Food Process. Preserv.* **2018,** *42*(8), E-article 13688.
40. Moran, D. P. (Ed.). *Fats in Food Products*; Springer Science & Business Media: New York, USA, 2012; p 415.
41. Mortensen, B. K. Butter and Other Milk Fat Products. In *Encyclopedia of Dairy Sciences;* Fuquay, J. W., Fox, P. F., McSweeney, P. L. H., Eds.; 2nd ed.; Elsevier/Academic Press: Boston, MA, 2011; p 502.
42. O'Connor, T. P.; O'Brien, N. M. Lipid Oxidation. In *Advanced Dairy Chemistry*; Springer Science & Business Media: New York, U.S.A. 1995; vol *2*, pp 309–347.
43. Pandya, A. J.; Sharma, R. S. In *Ghee-Its Chemistry, Technology and Nutrition-An Overview*, National Seminar on Role of Pure Ghee in Health and Nutrition Exploding Myths, Jointly Organized by Indian Dairy Association, Gujarat Chapter, Anand, GCMMF, Anand and J.S. Ayurveda College, Anand, June 13–14, 2002; pp 1–14.
44. Patel, A. M. Validation of Methods for Detection of Ghee Adulteration with Animal Body Fat. Master's Degree Thesis, National Dairy Research Institute (Deemed University), Karnal, India, 2011; p 102.
45. Patton, S.; Evans, L.; McCarthy, R. D. The Action of Pancreatic Lipase on Milk Fat. *J. Dairy Sci.* **1960,** *43*(1), 95–96.
46. Rajorhia, G. S. Ghee. In *Encyclopedia of Food Science, Food Technology and Nutrition;* Macrae, R., Robinson, R. K., Sadler, M. J., Eds.; Academic Press Ltd.: London, UK, 1993; vol 4; pp 2186–2192.
47. Rakesh, K. Evaluate the Quality of Milk Fat in Market Samples of Butter Ghee and Paneer. Master's Degree Thesis, National Dairy Research Institute (Deemed University), Karnal, India, 2016; p 129.

48. Ramamurthy, M. K.; Narayanan, K. M. Fatty Acid Compositions of Buffalo and Cow Milk Fats by Gas-Liquid Chromatography (GLC). *Milchwissenschaft* (Dairy Science), **1971,** *26*(11), 693–697.
49. Rao, D. V.; Ramamurthy, M. K. Polar Carbonyls in Cow and Buffalo Ghee. *Ind. J. Dairy Sci.* **1984,** *37*, 98–102.
50. Ray, P. R. Technological and Biochemical Aspects of Ghee (Butter Oil). In *Engineering Practices for Milk Products*; Apple Academic Press: New Jersey, USA, 2019; pp 83–109.
51. Reddy, K. J.; Jayathilakan, K.; Pandey, M. C.; Radhakrishna, K. Evaluation of the Physicochemical Stability of Rice Bran Oil and its Blends for the Development of Functional Meat Products. *Int. J. Food Nutr. Sci.* **2013,** *2*(2), 46–52.
52. Saad, B.; Wai, W. T.; Lim, B. P. Comparative Study on Oxidative Decomposition Behavior of Vegetable Oils and Its Correlation with Iodine Value Using Thermogravimetric Analysis. *J. Oleo Sci.* **2008,** *57*, 257–261.
53. Salih, A. M.; Smith, D. M.; Price, J. F.; Dawson, L. E. Modified Extraction 2-Thiobarbituric Acid Method for Measuring Lipid Oxidation in Poultry. *Poultry Sci.* **1987,** *66*(9), 1483–1488.
54. Santha, I. M.; Naryanan, K. M. Changes Taking Place in Proteins During the Conversion of Butter/Cream to Ghee. *Ind. J. Dairy Sci.* **1979,** *32*, 68–74.
55. Shahidi, F.; Wanasundara, U. N. Methods for Measuring Oxidative Rancidity in Fats and Oils. In *Food Lipids: Chemistry, Nutrition and Biotechnology*; Akoh, C. C., Min, D. B., Eds.; Marcel Dekker Inc.: New York, USA, 2002; pp 465–487.
56. Shahidi, F.; Zhong, Y. Lipid Oxidation: Measurement Methods. In *Bailey's Industrial Oil and Fat Products*; John Wiley & Sons, Inc.: New York, 2005; pp 357–385.
57. Shantha, N. C.; Decker, E. A. Rapid, Sensitive, Iron-Based Spectrophotometric Methods for Determination of Peroxide Values of Food Lipids. *J. AOAC Int.* **1994,** *77*(2), 421–424.
58. Sharma, R.S. Ghee Residue: Yield, Composition and Uses. *Dairy Guide* **1980,** *6*, 21–24.
59. Singh, S.; Ram, B. P. Effect of Ripening of Cream, Manufacturing Temperature and Packaging Materials on Flavor and Keeping Quality of Ghee. *J. Food Sci. Technol.* **1978,** *15*, 142–145.
60. Singhal, O.P. Adulterants and Methods for Detection. *Indian Dairyman* **1980,** *32*(10), 771–774.
61. Singhal, O.P. Studies on Ghee (Clarified Butterfat) and Animal Body Fats With a View to Detect Adulteration. PhD Thesis, Punjab University, Chandigarh, India, 1973; p 165.
62. Smink. W. Fatty Acid Digestion, Synthesis and Metabolism in Broiler Chickens and Pigs. PhD Thesis, Wageningen University, Wageningen, Netherlands, 2012; p 141.
63. Smith, L. M.; Ronning, M. Comparison of Fatty Acid Composition of Milk Fats Produced by Cows Fed Alfalfa, Oat, Or Ground, Pelleted Alfalfa Hay. *J. Dairy Sci.* **1961,** *44*, 1170–1174.
64. Srinivasan, M. R. Ghee Making in The Tropical Countries and Possibilities of its Industrial Production. *Indian Dairyman* **1976,** *28*, 279–283.
65. Srinivasan, M. R.; Anantakrishnan, C. P. *Milk Products of India*; Indian Council of Agricultural Research – Government of India, New Delhi, 1964; p 88.
66. Sserunjogi, M. L.; Abrahamsen, R. K.; Narvhus, J. A Review Paper: Current Knowledge of Ghee and Related Products. *Int. Dairy J.* **1998,** *8*(8), 677–688.
67. Tamime, A. Y. (Ed.). *Dairy Fats and Related Products*; Wiley-Blackwell: London, UK, 2009; p 344.

68. Thompson, M. P.; Brunner, J. R.; Stine, C. M. Characteristics of High-Melting Triglyceride Fractions from the Fat-Globule Membrane and Butter Oil of Bovine Milk. *J. Dairy Sci.* **1959,** *42*(10), 1651–1658.
69. Truong, T.; Lopez, C.; Bhandari, B.; Prakash, S. *Dairy Fat Products and Functionality*; Springer International Publishing: New York, USA, 2020; p 583.
70. Wadhwa, B. K.; Jain M. K. Chemistry of Ghee Flavor: A Review. *Ind. J. Dairy Sci.* **1990,** *43*, 601–607.
71. Walstra, P.; Geurts, T. J.; Noomen, A.; Jellema, A.; Van Boekel, M. A. J. S. (Eds.) *Dairy Technology: Principles of Milk Properties and Processes*; Marcel Dekker/CRC Press: New York, USA, 1999; vol 50–70, pp 501–510.
72. Watts, B. M.; Major, R. Comparison of Simplified, Quantitative Kreis Test With Peroxide Values of Oxidizing Fats. *Oil Soap* **1946,** *23*(7), 222–225.
73. Yadav, J. S.; Srinivasan, R. A. Advances in Ghee Flavor Research. *Ind. J. Dairy Sci.* **1992,** *45*, 338–348.

CHAPTER 9

BIOCHEMICAL CHARACTERIZATION OF CHEESE DURING RIPENING

RITA MEHLA, JYOTIKA DHANKHAR, SUVARTAN G. RANVIR, and SHAMIM HOSSAIN

ABSTRACT

Ripened or rennet coagulated cheeses undergo several biochemical changes during the period of ripening. The different microbiological and biochemical modifications contribute to the production of the texture in addition to distinctive taste in particular varieties of cheese during the maturation phase. Ripening is a complex biochemical pathway of three major events glycolysis, proteolysis, and lipolysis. Ripening is influenced by many variables, that is, the form and intensity of coagulant, lactic acid bacteria, indigenous and added enzyme. The composition and texture of the cheese are also influenced by factors, like cooking temperature, maturation temperature, and pH. Different cheese varieties are almost similar in chemical composition and texture in the initial stages of ripening. However, there are several variations occurred during the ripening process that affect the taste, texture, and odor of cheese. The primary reactions that significantly influence the biochemical reactions during ripening affect the changes in the texture and flavor of the cheese. Residual coagulant activity also affects the cheese quality. As cheese is considered to be a nutrient-dense medium and thought to be included in the diet by people of different regions. Therefore, studies were performed by different workers to understand the comprehensive process concerned with the biochemical and physicochemical reactions in cheese during ripening.

The Chemistry of Milk and Milk Products: Physicochemical Properties, Therapeutic Characteristics, and Processing Methods. Megh R. Goyal, Suvartan Ranvir, & Junaid Ahmad Malik (Eds.)

9.1 INTRODUCTION

Cheeses are a balanced diet, rich in energy and good quality protein. It is enriched with essential amino acids, vitamins, and minerals. The fermentation is generally carried out by microorganisms and is being used by dairy industries for thousands of years for an extended shelf life of milk.[27] Ripened cheeses undergo ripening process for a period of few months to 2 years and its examples are cheddar cheese and Parmigiano-Reggiano cheese.[24] Ripening is implicated with the several biochemical changes in the curd resulting formation in typical texture and taste. During ripening microbiological advances lead to lysis and death to starter bacteria, adventitious microflora (NSLAB, i.e., facultative *heterofermentative lactobacilli*).[38]

During the ripening process cheese, curd becomes softer as a result of casein micelle hydrolysis through proteolysis that leads to alter the water-binding capacity and pH in the curd. This change causes further calcium phosphate migration and precipitation.[37] During ripening, the biochemical changes are categorized into primary occurrences, such as glycolysis, lipolysis, and proteolysis. The secondary biochemical reactions are very essential for the production of various flavor-producing volatile compounds following these main events, including both fatty acids and amino acids metabolism. The biochemical characterization of cheese is a diverse research field and aspects of maturation have been extensively studied.[23]

The cheese ripening process and various technical approaches are used to improve the rate of ripening and development of flavor, along with measurement tools of several characteristics of cheese (Figure 9.1). The classification of different types of cheeses based on their ripening process is shown in Figure 9.2. The cheeses, such as, Brie, Roqueforti, and Camembert are examples of mold-ripened; and Limburger and Tilsit are examples of surface-ripened cheese.[29]

The internally ripened cheeses are classified into six categories (Figure 9.2) such as[38]:

- cheese with eyes (Gouda and Edam) and Swiss types (Emmental and Gruyere);
- extra-hard cheese (Parmesan and Asiago);
- hard cheese (Cheddar);
- high salt cheese (Feta);
- pasta filata (Mozzarella and Provolone);
- semihard cheese (Monterey Jack).

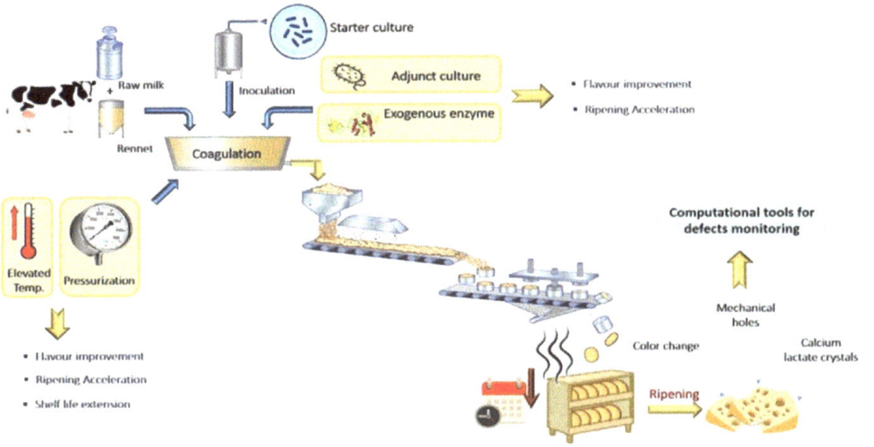

FIGURE 9.1 Graphical illustration of process of cheese ripening.[35]

Source: Reprinted with permission from Ref. [35]. © 2019 Elsevier.

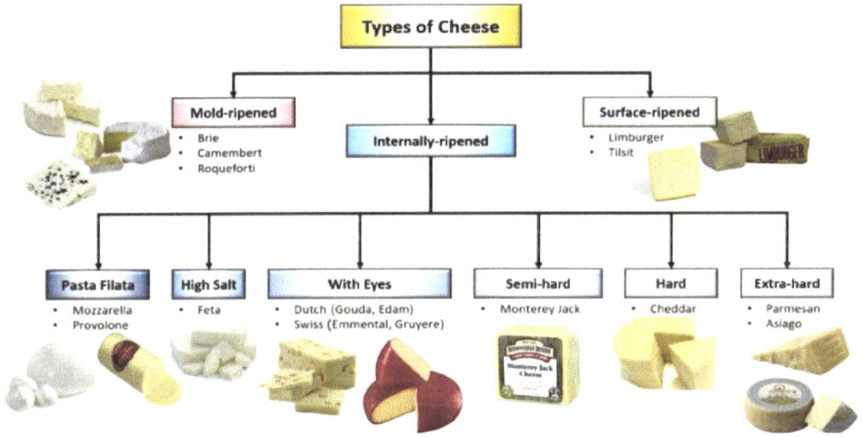

FIGURE 9.2 Classification of cheeses on the basis of ripening process.[35]

Source: Reprinted with permission from Ref. [35]. © 2019 Elsevier.

This chapter focuses on key aspects and biochemical characterization of cheese ripening, which are common to all varieties of ripened cheese.

9.2 CHANGES DURING RIPENING PROCESS

Throughout the ripening of cheese, there are numerous types of biochemical reactions that occur, mainly involving three main reactions such as "residual lactose metabolism and proteolysis, and lipolysis" that change the physical and chemical properties of cheese.[38] Enzymes involved in the maturation process come from many different sources. From milk "lipoprotein lipase survives pasteurization to participate in lipolysis." From coagulation, "some of the rennet enzyme (chymosin) is retained to participate in proteolysis."[47] The starter bacteria produce esterase and proteinase enzymes among other enzymes, apart from their major fermenting function. Due to high metabolic activity, NSLAB and secondary or adjunct cultures contribute to the ripening process.[6]

9.2.1 LACTOSE METABOLISM OR GLYCOLYSIS

Primarily during the fermentation process lactose to lactic acid is converted by lactic acid bacteria. Almost all lactose in the milk is lost with whey during the cheese manufacturing process. After cheese processing, the lactose content in cheese decreases, and a very lesser quantity remains in curd (0.8–1.0% particularly in cheddar cheese.[1] In cheese, lactose needs to be completely fermented to prevent unwanted secondary microflora growth. Lactose is metabolized in glycolytic pathways, depending on the type of starter bacteria. Lactose metabolism contributes to the formation of L or D Lactate or in combination with L or D lactate.

Several starter culture bacteria, like *Streptococcus thermophilus*, cannot metabolize the lactose specifically the galactose part and are therefore required to generate positive microorganisms with galactose (*Gal positive lactobacilli*).[4] Lactate adds flavor and most possibly adds flavor to ripened cheeses as well. D-lactate are mainly produced by nonstarter lactic acid bacteria (NSLB) by lactose fermentation or by racemization of L-lactate into D-lactate in a high population of NSLB.[51] Lactose metabolism in Swiss cheese is comparatively intricate.[22,50] The remaining lactose in cheese curd is metabolized rapidly by *Streptococcus thermophillus* with the development of L-lactate after the molding stage of cheese production. Initially and later, however, galactose accumulates along with lactose metabolized by the action of lactobacillili in D and L lactates.[22,38,51] During Swiss cheese maturation, owing to the development of propionic acid bacteria lactate has also been metabolized into acetate, propionate, CO_2, and H_2O.[46]

9.2.2 LIPOLYSIS IN CHEESE

Milk triglycerides are hydrolyzed into short and intermediate chain and free fatty acids by the action of bacterial and indigenous milk and this process in known as lipolysis. In cheese, oxidative degradations are less because of the low potential for oxidation/reduction (~250 mV).[8,18,19,39] In all cheese types, during maturation, native and/or bacterial produce lipases hydrolyses triglycerides which releases fatty acids. The milk fat of ruminant animals is rich in SCFAs which significantly contributes to the cheese flavor when it gets released during lipolysis. So many short-chain fatty acids like "hexanoic, octanoic, and decanoic acids" contribute to the aroma and flavor of the cheese. Sweaty, pungent, and rancid flavors are caused by hexanoic acid (Figure 9.3). The decanoic acid gives fatty citrus odors and the goaty waxy taste occurs due to octanoic acid.[6,25]

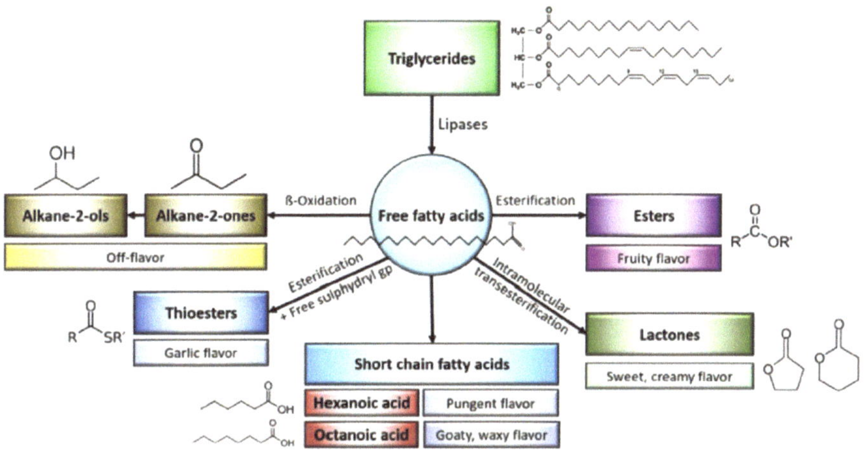

FIGURE 9.3 Lipolytic biochemical changes resulting from development of cheese flavor and off-flavor.

Source: Reprinted with permission from Ref. [35]. © 2019 Elsevier.

In blue cheese and some hard Italian varieties cheese, the extent of lipolysis is less. Lipolysis leads to the maturation of different types of cheeses, likes Swiss, Gouda, and Cheddar cheese, whereas enormous amounts of lipolysis can cause rancidity which is not desirable.[8,39] Fatty acids are essential for volatile flavor compounds production, also it directly contributes to the taste of cheese. Lipolysis in cheese was predominantly caused by the involvement of coagulant-associated lipolytic enzymes (rennet paste) and microorganisms (starter culture, nonstarter culture, and adjunct culture).[32] Milk includes indigenous lipases and lipoprotein lipases (55 kDa) that occur as a homodimer in milk. This enzyme plays a major role in triglyceride metabolism and is transferred through the blood in milk.[44]

Lipoprotein lipases can induce rancidity within 10s when the optimal conditions are reached.[6] The milk fat globule membrane helps to prevent milk fat from inducing lipoprotein lipase activity in normal milk. Casein micelle is connected to 90% of lipoprotein lipase. However, lipolysis occurs rapidly during homogenization, agitation, foaming due to mechanical action and contributes to the production of off-flavors.[16]

Lipoprotein lipase is position-specific and operates preferentially on MCFAs (C6–C12) at SN-3 and SN-1 positions.[44] Its operation is more important in raw cheeses than pasteurized cheeses as it is inactivated during pasteurization. For full inactivation of this enzyme, a temperature of 78°C for 10 s is needed.[10] PGE demonstrated greater activity against esterified SCFAs at SN-3 location comprising 0.5 M NaCl.[8]

In hard cheese types, this enzyme is responsible for a greater degree of lipolysis. Several ripened bacterial cheeses, that is, Cheddar, Gouda cheese, are prepared using heat-treated milk with no lipolytic agent and lipolysis in these cheeses takes place through the enzyme activity associated with starter and nonstarter microorganisms during ripening.[21]

Lactic acid bacteria have lipolytic intracellular enzymes and these are responsible for the release of large fatty acid levels in mature cheeses, that is, bacterially matured cheeses.[7,8] LAB-linked lipolytic enzymes are intracellular and released by lysis into the cheese matrix.[13] The optimal pH of an enzyme is pH 7.0–8.5 and the optimal temperature is 35°C and is mainly active in short-chain fatty acids.[5,7,8,26] The reaction of sulfhydryl groups (*S-methyl thioacetate, thioethyl-2-methylpropanoate* and *S-methyl thiobutyrate*) with free fatty acids leads to develop eggy and garlic flavors.[33] The creamy or buttery, sweet flavors impart due to nonalactones, which are produced by intramolecular transesterification of hydroxyacids.[25,38]

Formation of carboxylic acids is because of deamination of branched-chain amino acids. The dehydrogenases and NAD+ catalyzed deamination in Swiss cheese; oxidases catalyze deamination in surface-ripened and Camembert cheeses.[38,39] Carboxylic acids produced include pentanoic acid and butyric acid, they impart rancid flavor and fatty flavor.[25] Finally, β-oxidation and succeeding decarboxylation of free fatty acids forms alkan-2-ones or methyl ketones, specially heptanone and nonanone. The *G. candidum, P. camemberti, and P. roqueforti* have triggered this reaction.[38] In blue cheese, alkan-2-ones are responsible for moldy flavor.[30]

9.2.3 LACTATE METABOLISM

Lactate, a significant substrate for several reactions occurs throughout the ripening of cheese, produced by the growth of starter microorganisms during lactose metabolism.[4] The fundamental biochemical glycolytic and proteolytic reactions during the processing of cheese (Figure 9.4). The composition of NSLAB has a significant impact on L-lactate racemization, that is, racemization in raw milk cheese is almost certainly faster than in pasteurized milk cheese.[38]. Lactate racemization may be concerned with the oxidation by lactate dehydrogenase of L-lactate to produce pyruvate. Pyruvate is further reduced by D-lactate dehydrogenase to D-lactate. The Ca-DL-lactate solubility is lesser as compared to Ca-L-lactate; thus, racemization is necessary, so racemization contributes to the formation of white Ca-DL-lactate crystals. These crystals are primarily visible on cheese cut surfaces.[11,49]

These are nontoxic, but since they are moldy and contain foreign particles, they might cause product rejection.[11] Higher lactose levels promote NSLAB development, leading to the formation of crystals[45,48] and thus leading to a better release of casein-bound Ca or a decrease in Ca-Lactate solubility.[38] Lactic acid bacteria (LAB) further oxidize lactate into formate, ethanol, acetate, and CO_2.[20]

Clostridium tyrobutyricum anaerobic lactate metabolism to hydrogen and butyrate causes late gas blowing defect, which allows the appearance of cracks during ripening in cheese and advancement of off-flavors.[22,38] The big problem in brine-salted cheese is late gas blowing.[36] Higher NaCl levels are not prone to the late blowing of gas in dry salted cheese varieties like Cheddar. Secondly, the adjunct NSLAB and bacterial culture contribute primarily to ripening by which most cheese flavors are formed. The variability of aroma and flavor of the cheese in various forms is attributed to the introduction of different secondary probiotics.[2] In curd, residual citrate is metabolized in numerous flavorings compounds (such as acetoin, acetate, diacetyl, 2-butanone, and 2,3-butanidiol) by certain citrate-positive LAB (such as *S. diacetyl lactis and L. lactis and L. cremoris*).[30]

9.3 PROTEOLYSIS IN CHEESE

Proteolysis is the major complicated primary biochemical reaction which occurs during cheese ripening. It adds softness to texture of the cheese by casein hydrolysis in cheese curd by reducing the cheese curd's water activity. It also directly affects the cheese flavor by producing smaller peptides and amino acids.[18,31] Coagulants, starter, and nonstarter microorganisms are peptidases and proteinases responsible for proteolysis during ripening. To fasten the ripening process, exogenous proteinases and peptidases are

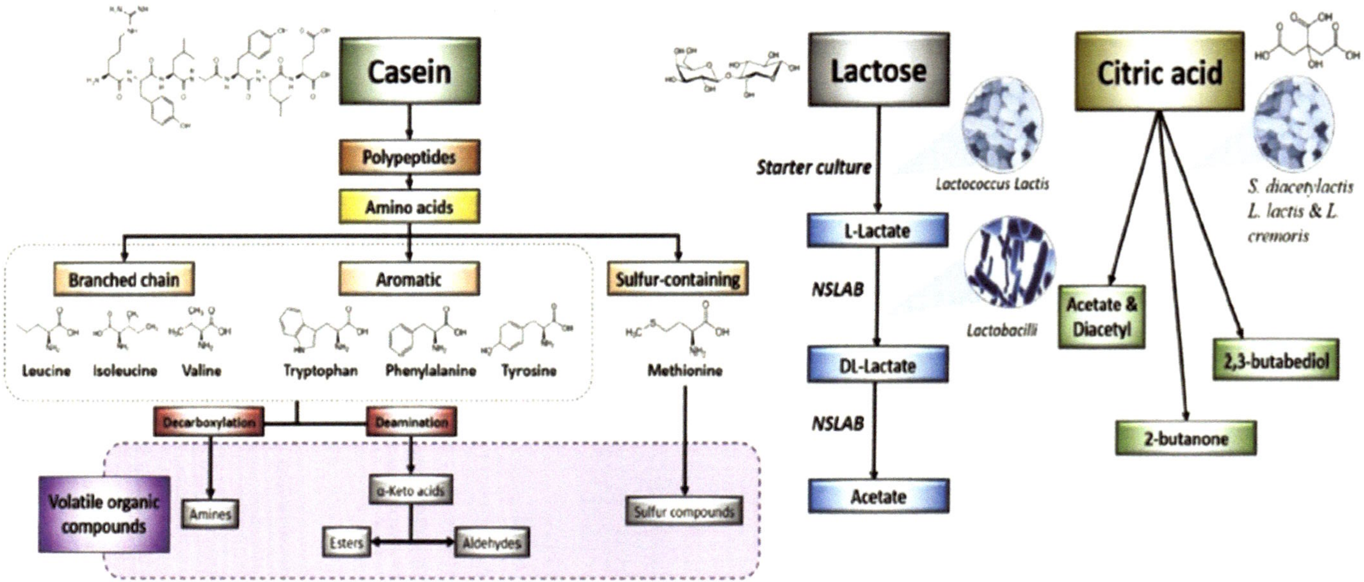

FIGURE 9.4 Biochemical reactions involved cheese flavor formation.

Source: Reprinted with permission from Ref. 35. © 2019 Elsevier.

also added into milk. In most cheese types, the major source of proteolytic enzymes is the residual coagulant, often chymosin, which during whey removal remains stuck in curd.[41]

In the cheese curd, approximately 30% of added coagulant activity retains based on several variables, that is, enzyme form, cooking temperature and whey pH.[52] Chymosin is unique to β-casein at seven sites, most sites are present near the β-casein hydrophobic C-terminal and function on these sites at Leu192-Tyr193. This results in short, hydrophobic, bitter peptide.[53] Chymosin primarily acts on Phe23-Phe24 against αs1-casein,[40] leading to the release of smaller peptides (αs1-CN f1–23) that are hydrolyzed by starter proteases further. Chymosin degrades αS1-casein primarily at Leu101-Lys102 at different locations and is further hydrolyzed during the ripening process. It is understood that αs2-casein is less vulnerable to chymosin than αs1-casein. The chymosin activity on αs2-casein is limited to its hydrophobic area (sequences 90-120 and 160-207).[24,40,42]

However, several sites of para κ-casein are susceptible to chymosin.[28] Milk contains various proteolytic enzymes. Trypsin-like serine proteinase from blood has plasmin and milk indigenous proteinase (optimum pH of 7.5 and temperature of 37°C). Plasmin deteriorates fibrin clots during blood clotting therefore it's activity must be regulated. Plasminogen (the precursor of plasmin) is converted to plasmin by the plasminogen activators (PAs). Plasmin and plasminogen activator inhibitors are also components of this system.[24,40,42]

PAs are predominantly allied with casein micelle, whereas "plasmin and plasminogen activator inhibitors" are present in serum. The plasmin activity is specific to the peptide, that is, Lys-X; to a lesser degree, Arg-X. The order of casein degradation is β-casein >αs1-casein >αs2-casein; K-casein is resistant to proteinase activity.[3] The specificity of plasmin on various casein fractions, that is, β-, αS_2- and αS_1-caseins, is known,[14,52] among them β-casein is most important. The plasmin-based degradation of β-casein is shown in Figure 9.5. The most plasmin vulnerable casein is αS_2 Casein. In high-temperature cheese varieties like Swiss cheeses, plasmin action is most important.[15,16]

Plasminogen and plasmin activator inhibitors are inactivated at high temperatures (~55°C); thereby lead to activation of plasmin.[12] Plasmin is important in the mold ripened cheese.[52] Aspartyl proteinase and cathepsin D are other indigenous proteinases.[34] The mammalian proteinase derived from procathepsin D is cathepsin D (Kelly).[2] The specificity of cathepsin D for caseins, primarily αs1-caseinin, is like that of chymosin.[40]

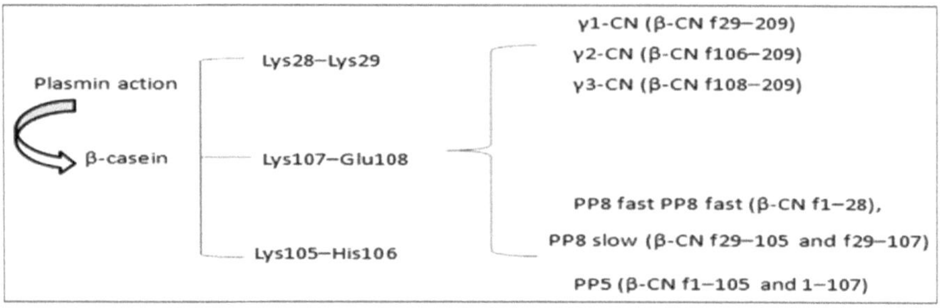

FIGURE 9.5 Plasmin degradation of β-casein at different sites.[17]

9.3.1 ACID METABOLISM OF AMINO ACIDS

Amino acids are the precursor to volatile aroma compounds that are further degraded into compounds of other flavors, such as, amines, alcohol, aldehyde, and ammonia. The compounds produced by amino acid metabolism contribute to cheese flavor. The amount and composition of amino acids are known as the ripening index in the cheese.[20] The increased amount of free amino acids in cheese, however, does not affect the flavor of the cheese, as flavor occurs due to the transformation of free amino acid to aroma compound.[54]

9.3.2 PRODUCTION OF VOLATILE FLAVOR COMPOUNDS VIA SECONDARY EVENTS DURING CHEESE RIPENING PROCESS

9.3.2.1 FREE FATTY ACID METABOLISM

The particular type of cheese flavor is primarily the contribution of SCFA. Free fatty acids (FFA) contribute to flavor of the cheese via performing like a precursor for the release of volatile aroma imparting compounds through a sequence of reactions. Even though SCFA directly contributes to cheese flavor, FFA also contributes to the flavor of cheese by serving as a precursor to the development of volatile flavoring compounds through a sequence of reactions collectively called fatty acid metabolism.[8,14,15,16] Figure 9.6 presents the generation of flavorings compounds through the metabolism of fatty acids.

Esters are produced in several types of cheese and released by FFA and alcohol reactions.[43] Lactones are produced from hydroxy-acid via intermolecular esterification. The "γ and δ lactones" have been identified in ripened cheeses. The release of lactones during the ripening process is dependent on the level of precursor compound, that is, hydroxy acids. In mammary glands hydroxy acids are formed through the reduction of ketones.[8,9]

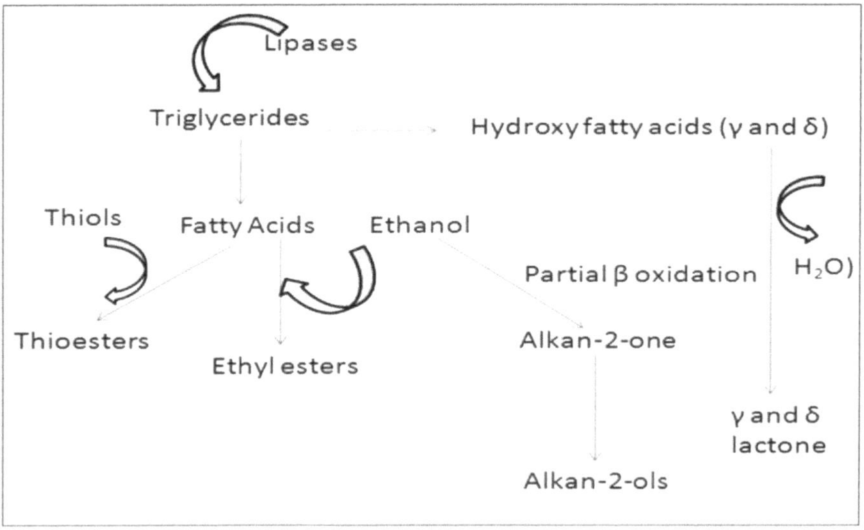

FIGURE 9.6 Generation of flavoring compounds through the metabolism of fatty acids in cheese during ripening.[38]

9.4 SUMMARY

Cheese ripening involves several interrelated events which result in the production of flavor compounds and the development of textural attributes. The biochemical events through which milk constituents (i.e., fat, protein, lactose, and lactate) are converted into flavor compounds that have not been studied in detail. Despite available knowledge of the cheese ripening process of different cheese varieties, it is not possible to get the best cheese quality from daily production. Therefore, it remains a challenging issue for productive research study in this area.

KEYWORDS

- **cheese**
- **glycolysis**
- **lipolysis**
- **proteolysis**
- **ripening**

REFERENCES

1. Agarwal, S.; Sharma, K.; Swanson, B. G.; Yüksel, G.; Clark, S. Nonstarter Lactic Acid Bacteria Biofilms and Calcium Lactate Crystals in Cheddar Cheese. *J. Dairy Sci.* **2006,** *89*(5), 1452–1466.
2. Andiç, S.; Tunçtürk, Y.; Boran, G. Changes in Volatile Compounds of Cheese. In *Processing and Impact on Active Components in Food*, 2015; pp 231–239.
3. Bastian, E. D.; Brown, R. J. Plasmin in Milk and Dairy Products: An Update. *Int. Dairy J.* **1996,** *6*(5), 435–457.
4. Bintsis, T. Lactic Acid Bacteria as Starter Cultures: An Update in Their Metabolism and Genetics. *AIMS Microbiol.* **2018,** *4*(4), 665.
5. Chich, J. F.; Marchesseau, K.; Gripon, J. C. Intracellular Esterase from *Lactococcus Lactis* Subsp. Lactis NCDO 763: Purification and Characterization. *Int. Dairy J.* **1997,** *7*(2–3), 169–174.
6. Clark, S.; Costello, M.; Drake, M.; Bodyfelt, F, Eds.; *The Sensory Evaluation of Dairy Products*; Springer Nature: Cham, 2009; p 551.
7. Collins, Y. F.; McSweeney, P. L. Wilkinson, M. G. Lipolysis and Catabolism of Fatty Acids in Cheese. *Cheese Chem. Phys. Microbiol.* **2004**, *1*, 373–389.
8. Collins, Y. F.; McSweeney, P. L.; Wilkinson, M. G. Lipolysis and Free Fatty Acid Catabolism in Cheese: A Review of Current Knowledge. *Int. Dairy J.* **2003,** *13*(11), 841–866.
9. Curtin, Á. C.; McSweeney, P. L. H. Catabolism of Amino Acids in Cheese During Ripening. *Cheese: Chem. Phys. Microbiol.* **2004,** *1*, 435–454.
10. Driessen, F. M. Inactivation of Lipases and Proteinases (Indigenous and Bacterial). *Bullet. Int. Dairy Fed.* **1989,** *238*, 71–93.
11. Dybing, S. T.; Wiegand, J. A.; Brudvig, S. A.; Huang, E. A.; Chandan, R. C. Effect of Processing Variables on the Formation of Calcium Lactate Crystals on Cheddar Cheese. *J. Dairy Sci.* **1988,** *71*(7), 1701–1710.
12. Farkye, N. Y.; Fox, P. F. Observations on Plasmin Activity in Cheese. *J. Dairy Res.* **1990,** *57*(3), 413–418.
13. Fernández, L.; Beerthuyzen, M. M.; Brown, J.; Siezen, R. J.; Coolbear, T.; Holland, R.; Kuipers, O. P. Cloning, Characterization, Controlled Overexpression, and Inactivation of the Major Tributyrin Esterase Gene of *Lactococcus Lactis*. *Appl. Environ. Microbiol.* **2000,** *66*(4), 1360–1368.

14. Fox, P. F. 1989. Proteolysis During Cheese Manufacture and Ripening. *J. Dairy Sci.* **1989,** *72*, 1379–1400.
15. Fox, P. F.; McSweeney, P. L. H. Proteolysis in Cheese During Ripening. *Food Rev. Int.* **1996,** *12*, 457–509.
16. Fox, P. F.; Guinee, T. P.; Cogan, T. M.; McSweeney, L. H. M., Eds.; *Fundamentals of Cheese Science*, 2nd ed.; Springer Nature: Cham, 2017; p 814.
17. Fox, P. F.; Law, B. A.; McSweeney, P. L. H.; Wallace, J. Biochemistry of Cheese Ripening. *Cheese Chem. Phys. Microbiol.* **1999,** *1*, 347–360.
18. Fox, P. F.; Law, J. Enzymology of Cheese Ripening. *Food Biotechnol.* **1991,** *5*(3), 239–262.
19. Fox, P. F.; Lucey, J. A.; Cogan, T. M. Glycolysis and Related Reactions During Cheese Manufacture and Ripening. *CRC Crit. Rev. Food Sci. Nutr.* **1990,** *29*(4), 237–253.
20. Fox, P. F.; McSweeney, P. L. H. In *Biochemistry of Cheese Ripening. Chemistry, Biochemistry and Control of Cheese Flavor*, Proceedings of 4th Cheese Symposium; Teagasc: Moorepark, 1995; pp 135–159.
21. Fox, P. F.; McSweeney, P. L. H. Rennets: Their Role in Milk Coagulation and Cheese Ripening. In *Microbiology and Biochemistry of Cheese and Fermented Milk*, 2nd ed.; Law, B. A. Ed.; Chapman & Hall: London, 1997; pp 1–49.
22. Fox, P. F.; Uniacke-Lowe, T.; McSweeney, P. L. H.; O'Mahony, J. A. Enzymology of Milk and Milk Products. In *Dairy Chemistry and Biochemistry*; Springer Nature: Cham, 2015; pp 377–414.
23. Fox, P. F.; Wallace, J. M. Formation of Flavor Compounds in Cheese. *Adv. Appl. Microbiol.* **1997,** *45*, 17–86.
24. Fox, P. F.; Wallace, J. M.; Morgan, S.; Lynch, C, M.; Niland, E. J.; Tobin. J. Acceleration of Cheese Ripening. *Antonie Van Leeuvenhoek* (Antonie Van Leeuwenhoek), **1996**, *70*, 271–297.
25. Gan, H. H.; Yan, B.; Linforth, R. S.; Fisk, I. D. Development and Validation of an APCI-MS/GC–MS Approach for the Classification and Prediction of Cheddar Cheese Maturity. *Food Chem.* **2016,** *190*, 442–447.
26. Gobbetti, M.; Fox, P. F.; Smacchi, E.; Stepaniak, L.; Damiani, P. Purification and Characterization of a Lipase from *Lactobacillus Plantarum*, 2739. *J. Food Biochem.* **1996,** *20*(1), 227–246.
27. Grappin, R.; Rank, T. C.; Olson, N. F. Primary Proteolysis of Cheese Proteins During Ripening - A Review. *J. Dairy Sci.* **1985,** *68*(3), 531–540.
28. Green, M. L.; Foster, P. M. D. Comparison of the Rates of Proteolysis During Ripening of Cheddar Cheeses Made with Calf Rennet and Swine Pepsin as Coagulants. *J. Dairy Res.* **1974,** *41*(2), 269–282.
29. Gripon, J. C. Mould Ripened Cheeses. In *Cheese: Chemistry, Physics and Microbiology*, 2nd ed.; Major Cheese Groups; Fox, P. F., Ed.; Chapman & Hall: London, 1993; Vol. 2; pp 111–136.
30. Hassan, F. A.; Abd El-Gawad, M. A. M.; Enab, A. K. Flavor Compounds in Cheese. *Res. Precis. Instrum. Mach.* **2013,** *2*(2), 15–29.
31. Hayaloglu, A.; McSweeney, P.; Özer, B.; Akdemir-Evrendilek, G. Primary Biochemical Events During Cheese Ripening. *Dairy Microbiol. Biochem. Recent Dev.* **2014,** *2014*, 134–166.
32. Ianni, A.; Bennato, F.; Martino, C.; Grotta, L.; Martino, G. Volatile Flavor Compounds in Cheese as Affected by Ruminant Diet. *Molecules* **2020,** *25*(3), 461–470.

33. Iwasawa, A.; Suzuki-Iwashima, A.; Iida, F.; Shiota, M. Effects of Flavor and Texture on the Desirability of Cheddar Cheese During Ripening. *Food Sci. Technol. Res.* ***2014,*** *20*(1), 23–29.
34. Kelly, A. L.; McSweeney, P. L. H. Indigenous Proteolytic Enzymes in Milk. *Adv. Dairy Chem. Proteins* **2003,** *2003*, 495–521.
35. Khattab, A. R.; Guirguis, H. A.; Tawfik, S. M.; Farag, M. A. Cheese Ripening: A Review on Modern Technologies Towards Flavor Enhancement, Process Acceleration and Improved Quality Assessment. *Trends in Food Sci. Technol.* **2019,** *88*, 343–360.
36. Kleter, G.; Lammers, W. L.; Vos, E. A. The Influence of PH and Concentration of Lactic Acid and NaCl on the Growth of *Clostridium Tyrobutyricum* in Whey and Cheese, Part II: Experiments in Cheese. *Neth. Milk Dairy J.* **1984,** *38*(1), 31–41.
37. Lucey, J. A.; Johnson, M. E.; Horne, D. S. Invited Review: Perspectives on the Basis of the Rheology and Texture Properties of Cheese. *J. Dairy Sci.* **2003,** *86*(9), 2725–2743.
38. McSweeney, P. L. H. Biochemistry of Cheese Ripening. *Int. J. Dairy Technol.* **2004,** *57*(2–3), 127–144.
39. McSweeney, P. L. H. Biochemistry of Cheese Ripening: Introduction and Overview. *Cheese Chem. Phys. Microbiol.* **2004,** *1*, 347–360.
40. McSweeney, P. L. H.; Fox, P. F. Metabolism of Residual Lactose and of Lactate and Citrate. *Cheese Chem. Phys. Microbiol.* **2004,** *1*, 361–371.
41. McSweeney, P. L.; Fox, P. F.; Olson, N. F. Proteolysis of Bovine Caseins by Cathepsin D: Preliminary Observations and Comparison with Chymosin. *Int. Dairy J.* **1995**, *5*(4), 321–336.
42. McSweeney, P. L. H.; Sousa, M. J. Biochemical Pathways for the Production of Flavor Compounds in Cheeses During Ripening - A Review. *Le Laite* **2000,** *80*(3), 293–324.
43. Meinhart, E.; Schreier, P. *Study of Flavor Compounds from Parmigiano* Reggiano Cheese. *Milchwissenschaft* **1986,** *41*(11), 689–691.
44. Olivecrona, T.; Vilaro, S.; Olivecrona, G. Lipases in Milk. In *Advanced Dairy Chemistry-1: Proteins*, 3rd ed.; Fox, P. F., McSweeney, P. L. H., Eds.; Kluwer: New York, 2003; pp 473–494.
45. Pearce, K. N.; Creamer, L. K.; Gilles, J. Calcium Lactate Deposits on Rindless Cheddar Cheese. *N. Z. J. Dairy Sci. Technol.* **1973,** *8*, 3–7.
46. Piveteau, P. Metabolism of Lactate and Sugars by Dairy Propionibacteria: A Review. *Le Lait* **1999,** *79*(1), 23–41.
47. Soodam, K.; Ong, L.; Powell, I. B.; Kentish, S. E.; Gras, S. L. Effect of Rennet on the Composition, Proteolysis and Microstructure of Reduced-Fat Cheddar Cheese During Ripening. *Dairy Sci. Technol.* **2015,** *95*(5), 665–686.
48. Sutherland, B. J.; Jameson, G. W. Composition of Hard Cheese Manufactured by Ultrafiltration. *Aust. J. Dairy Technol.* **1981,** *36*(4), 136.
49. Thomas, T. D.; Crow, V. L. Mechanism of D (-)-Lactic Acid Formation in Cheddar Cheese. *N. Z. J. Dairy Sci. Technol.* **1983,** *18*, 131–141.
50. Turner, K. W.; Thomas, T. D. Lactose Fermentation in Cheddar Cheese and the Effect of Salt. *N. N. Z. J. Dairy Sci. Technol.* **1980,** *15*, 265–276.
51. Turner, K. W.; Morris, H. A.; Martley, F. G. Swiss-Type Cheese, Part II: The Role of Thermophilic Lactobacilli in Sugar Fermentation. *N. Z. J. Dairy Sci. Technol.* **1983,** *18*, 117–124.

52. Upadhyay, V. K.; McSweeney, P. L. H.; Magboul, A. A. A.; Fox, P. F. Proteolysis in Cheese During Ripening. *Cheese Chem. Phys. Microbiol.* **2004,** *1*(3), 391–434.
53. Visser, S. Specificity of Chymosin (Rennin) in its Action on Bovine Beta-Casein. *Neth. Milk Dairy J.* **1977,** *31*, 16–30.
54. Yvon, M.; Rijnen, L. Cheese Flavor Formation by Amino Acid Catabolism. *Int. Dairy J.* **2001,** *11*(4–7), 185–201.

PART III

THERAPEUTIC CHARACTERISTICS OF MILK AND MILK PRODUCTS

CHAPTER 10

THERAPEUTIC AND NUTRITIONAL PROPERTIES OF FERMENTED MILK PRODUCTS

DIVYANG SOLANKI, SUVARTAN G. RANVIR, HEENA PARMAR, and SUBROTA HATI

ABSTRACT

Fermented dairy products are highly nutritious and therapeutic foods in Indian cultures and in other Asian countries. Lactic culture's role is crucial in the fermentation process and in the development of rheology of fermented milk products. Some yeasts or molds have also been inoculated with lactic strains in the medium, which gives unique characteristics to the fermented products. Along with the progress of technology, the fermentation process has taken place from area to area. The manufacture of multiple fermented dairy products of good workability is enabled from chosen microorganisms widely known as bacteria. The world's most manufactured fermented dairy foods include: the Bulgarian butter milk, kefir, curd (*dahi*), yogurt, acidophilus, koumiss, and several cheeses. Several types of lactic cultures, single of blended microorganism are used to produce desirable semifinished and finished products, while undesirable products may be generated in the presence of natural bacteria or milk contaminants. Thus, various chemical/biochemical modifications occur during fermentation. These changes depend upon processing steps, milk quality, microbes, and the forms of product. The taste, texture, and consistency of the product are influenced by compounds generated during the fermentation, such as, acetic acid, diacetyl, dioxide, ethyl alcohol, exopolysaccharides, propionic acid, and bacteriocins

The Chemistry of Milk and Milk Products: Physicochemical Properties, Therapeutic Characteristics, and Processing Methods. Megh R. Goyal, Suvartan Ranvir, & Junaid Ahmad Malik (Eds.)

by the spoilage microorganisms. In addition to these, fermentation causes predigestion of the nutrients that can be easily metabolized by the human body. Gastrointestinal diseases are also treated with fermented dairy products because of their hypocholesterolemic, anticarcinogenic, and immune-enhancing effects. Researchers have developed fermented dairy products with multifunctional healing benefits. This will lead to the development of fermented dairy products as biomedicine.

10.1 INTRODUCTION

The term "Fermentation" stems from Latin word *fermentare*, described as leaven. Food fermentation method is embedded throughout the world with different cultural values; fermentation being spread from village fermenting to marketed and industrial processors for the consumer market. The fermentation is related to "food manufacturing by different bacteria and fungi (different microorganisms and enzymes) is the result of it."[40] In addition to food preservation, fermentation can be used as a mechanism to promote eating habits, and to produce a unique and healthy foods.[79] Formulated fermented foods and specific manufacturing methods differ between cultures depending on the tastes and preferences, sources of food, raw products, the latest technical advancement, and eventually, the environment.[72] Starting from food and beverage preservation to preparation of a ready-to-eat product globally, fermentation is an important beverage and food production process.

All in all, humans eat nearly 5000 different fermented foods. The main goal is to meet the needs of a certain city, most of which are ethnic and developed on a pilot scale. The production of fermented foodstuffs is extended to include meat, vegetables, milk, grains, and fish products through increasing technologies and prospects of fermented foodstuffs.[82] Mostly, the rural residents preferred fermented foods as compared to non-fermented because of the pleasant taste, color, and texture.[52]

It has been reported that fermented foods provide health benefits to the consumers. The scientists have thoroughly investigated these benefits, such as, reduction in blood glucose levels,[65] antihypertensive activity,[44] antithrombotic activity, and antidiarrheal.[35] Further, these health benefits were searched for the specific bioactive molecules or components from the fermented foods, which show the bio-functional activity. Research studies have been conducted on the components of the fermented foods and to identify certain amino acids, vitamins, minerals, and phytochemicals.[86]

The fermented dairy foods are also most popular foods for providing nutrient and health benefits in a single product. Cultured dairy products are manufactured by the activity of acceptable and harmless microorganisms in the milk.[29] Bioactive molecules and microbial chemicals are produced during fermentation. Fermented dairy products are functional and provide medium for the addition of other components or elements, which make the end-product more functional and health promoting. The improved nutritional profile and bioavailability of food-nutrients were closely correlated with fermentation. Fermentation is an activity of lactic acid bacteria (LAB), which boosts the dietary properties of foods and excludes harmful or antinutritional compounds from foods (such as, galactose and lactose) that are specifically connected to certain health conditions such as, galactose and lactose intolerance.[87]

Deliberately added lactic strains in the product have pivotal role in the liberation of bioactive compounds during the fermentation. These strains produce lactic acids from lactose as well as improve several bioactive components in the food matrix.[31] Production of lactic acid or other compound from the process of fermentation depends upon the types of bacterial strains inoculated, medium, and growth conditions during the fermentation. Generally, these are the common lactic cultures used during the fermentation, which includes "*Streptococcus thermophilus* in company of Bifidobacteria (*Bifidobacterium longum*, *Bifidobacterium breve* C50, *Bifidobacterium animalis*, and *Bifidobacterium lactis,* and) or strains of Lactobacillus which includes *Lactobacillus rhamnosus, Lactobacillus acidophilus, Lactobacillus casei,* and *Lactobacillus johnsonii.*"[31]

Considering that cultured dairy products are now considered as functional foods, this chapter depicts the role of starters, and bio-functionality of fermented dairy foods, and physicochemical changes during the fermentation. This chapter also provides composition of some of these products, and origin of some health-promoting products, commercially available functional dairy-based products, and world scenario of fermented dairy foods.

10.2 FERMENTED DAIRY PRODUCTS: CATEGORIES, COMPONENTS, AND FORMS

It is almost impossible to locate the true origin of fermented milks. However, the root of fermented milk is expected to be in Western Asia and continue further east. New options for addressing the complex atmosphere have been established in the eastern region. Milk fermentation enabled people to store and make surplus milk available when driving long distances.[52] In the traditional methods, a left portion of the fermented liquid from previous days was used as the inoculation starter, but thanks to modern advances and technology, well-developed freeze-dried starters have been used for fermented milk and dairy items.[67]

The fermentation process is now being customized to manufacture milk-based foods having diverse tastes, medical benefits, and different flavors. Fermented milk as well as other associated dairy products are well-known and available worldwide in different forms in the multiple countries.[52]

10.2.1 CATEGORIES OF DAIRY-FERMENTED MILK

There are three different groups in which fermented milks are classified[51,91]: mesophilic sour-milks, alcoholic milks, and thermophilic sour-milks.

- In case of mesophilic, the products are cultured buttermilk, Ymer, Skyr, and Viili (growth temperature 20–30°C).
- The alcoholic milks include gioddu and kefir (growth temperature 15–25°C, generation of carbon dioxide and alcohol).
- Thermophilic sour-milks include matzoon (growth temperature 42–45°C, formation of lactic acid).

Not only cow and buffalo milk but various milks are used for the manufacturing of fermented dairy foods. This preparation is affected by the agricultural conditions, different species of milk used (such as, goat milk, cow milk, sheep milk, buffalo milk, camel milk, milk from yak, and reindeer).[57] Other factors affecting the cultured dairy products include type of milk used, type of starter used, adjunct cultures, and parameters observed during fermentation (i.e. incubation time, incubation temperature, and storage vessel).[111]

10.2.2 CONSTITUENTS OF CULTURED DAIRY PRODUCTS

Oligosaccharides, proteins, vitamins, peptides fatty acids, and organic acids are vital biogenic compounds.[22]

- **Lactose-carbohydrate:** Lactose is an important carbohydrate in milks that changes to lactic acid after fermentation. It is converted into "glucose and galactose under the action of β-galactosidase" in intestinal brush border. Depending on the strains, presence of oligosaccharides, and polysaccharides observed in the end-product. Some stains are also producing exopolysaccharides (EPS) (homo and hetero polysaccharides).[22,25]

- **Lipids:** Microbes from the fermented milk and milk products can digest fatty acids and produce aroma.[102] A common fermented milk product like yogurt, contain triglycerides constitute 95% of total lipids. Saturated fats constitute to 72%, 25% of monosaturated fats, while polyunsaturated fats accounts 3% of the fats of yoghurt and reported some extent amount in conjugated linoleic acid.[25]
- **Proteins:** Fermented milk products contains valuable proteins such as, α-s_1 and α-s_2; β-casein and κ-casein and whey proteins (α-lactoalbumin, β-lactoglobulin, immunoglobulins, enzymes, lactoferrin, growth factors, and glucomacropeptide). Proteolytic activity of lactic culture is attributed to release of specific peptides.[25] This proteolytic activity of LAB affects digestive process and ultimately improve biological value of the food product and digestibility of the product.[102] Proteinase from LAB releases bioactive oligopeptides from caseins (α-caseins and β-caseins) and from whey proteins. The bio-functionality of the peptides was determined as ACE inhibitor, antithrombotic, apoptosis modulation, antioxidant, antihypertensive, opioid, antiopioid, and immunomodulatory activity.[35]
- **Vitamins and minerals:** Cultured dairy foods are enriched with bioavailable minerals and vitamins.[25] Fermented dairy products contribute vitamins (such as, A, D B1, B2, B3, B5, B6, B9, and B12 vitamin). Minerals (such as, calcium, magnesium, phosphorus, potassium, potassium iodide, and zinc) are also found in fermented milk products.[102] Fermentation process and acidity in cultured dairy products raises the bioavailability of micronutrients as compared with the raw, milk which ultimately improves the vitamin content.[25]
- **Prebiotics:** Is a component, which is selected based on the types of fermentation to develop specific change in the gastrointestinal microflora and to improve activity leading to health benefits for the host.[35] In the body, microbiota found in the gastrointestinal tract produces prebiotic friendly environment which modify intestinal health thorough the creation of biologically active compound,[20] mainly fatty acids with short chain (propionate, butyrate, lactate, acetate)[16] produced during the fermentation[108] and poly-unsaturated fatty acids[20] showing an efflux from gut into integral circulation. The most important function of prebiotics is to support and enhance the growth of good bacteria "probiotics" or as helping agent for health-promoting microflora in

the gut.[35] Sources of prebiotics are lactose, EPS,[6] sialyllactose, galactooligosaccharides, and polydextrose that can extensively impart the synthesis and absorption of vitamin B.[4]

- **Probiotics:** Is defined as live microorganisms, which when administered in adequate amounts confer health benefit on the host.[27] Probiotics are widely used for stoppage of antibiotics-associated diarrhea, healing inhibiting agent against *Helicobacter pylori* infection, remedy for traveler's diarrhea, curing agent for constipation, therapy for pediatric acute diarrhea, prevention against allergy, cure of irritable bowel syndrome, treatment for and bacterial vaginosis and vaginitis, remedy for inflammatory bowel disease, cure of healing agent for dental infections, cure of acute diarrhea in adults, *Clostridium difficile* infection, sepsis, and obesity.[56]
- **Yeasts:** Are largely associated with traditionally fermented milk products. Grain form of yeasts is a unique feature of some traditionally prepared fermented milk products. There is some yeast utilized in fermented milk, especially most famous one is *Saccharomyces boulardii.*[13] Reports are also available for the bio-functional property of *S. boulardii*, as it heals the diarrheal symptoms corresponding with *C. difficile.*[13]

10.3 VARIOUS FERMENTED MILK AND CULTURED DAIRY PRODUCTS

Fermented milk and milk products are produced through the fermentation carried out in the presence of lactic cultures, which are safe for human consumption. These are lactic cultures, which remain active, abundant, and viable in the fermented dairy product.[92] Every single fermented dairy product is prepared using specific LAB and growth conditions specific to the inoculated strains/lactic culture.[3] Table 10.1 indicates fermented milk products with specific lactic cultures.[33,91]

10.3.1 CHEESE

Cheese is a milk-based product containing of lactic cultures which are safe in nature and produces lactic acid, encloses bacteria that provides aroma and taste, and includes suitable enzymes which are inducing coagulation.[3] Cheese is a common tittle of a variety of fermented milk product produced by across the world with different forms, flavors, and textures. Variety of technological aspects were studied with the types of milk (goat, buffalo, cow, sheep), type of coagulation (enzymatic, acidic), agents for coagulation (microbial, rennet, and vegetable coagulant), manufacturing, and the ripening conditions.[43,45]

TABLE 10.1 Fermented Milk With Specific Lactic Cultures.

Milk type	Microbes associated	Fermented milk product
Cow, buffalo, goat, and sheep milk	*Priopionibacterium shermanii; Penicillium roqueforti; L. lactis* subsp. *Cremoris; L. lactis* subsp. *Lactis*	Cheese
Buffalo's/cow's milk	*Streptococcus lactis; S. thermophiles; S. cremoris; L. lactis; L. delbrueckii* subsp.; *L bulgaricus; L. plantarum*	Curd
	S. cremoris; S. lactis subsp. *Diacetylactis*	Cultured butter milk
	S. thermophiles; L. bulgaricus	Shrikhand
Cow milk	*L. delbueckii* subsp. *Bulgaricus*	Bulgarian butter milk
	L. acidophilus	Acidophilus milk
	S. thermophiles; L. bulgaricus; L. acidophilus;	Yogurt
Goats and sheep milk	*L. bulgaricus; S. lactis; S. thermophiles*	Leben
Mare, camel, and ass milk	*Saccharomyces; Micrococci; L. acidophilus L. bulgaricus*	Kumiss
Sheep, cow, goat, and mixed milk	*Saccharomyces; Micrococci; S. lactis Leuconostoc* sp.	Kefir

Source: Reprinted with permission from Ref. [91]. © 2011 Elsevier.

Major steps of cheese preparation include culture addition/rennet addition and formation of curd, cutting, stirring, washing, heating, and pressing.[28] Probiotic cheese is another category in the group of cheese families. It offers special marketing opportunity as a probiotic carrier. Usage of lactic cultures particularly probiotics improve the functional properties of cheese and it would increase value of these products as being functional foods.[30] Many bio-functional properties of cheese have been reported in the literature, which includes regulation of blood pressure, building of muscle, prevention of diabetes, controlling low-density lipoprotein cholesterol, obesity, tooth decay, and cancer.[103]

10.3.2 YOGURT

It is a traditionally produced semisolid product prepared from either milk and/or cream using the process of fermentation. This can be carried out using two specific lactic cultures (strains), which holds symbiotic relationship; *S. thermophilus* and *Lactobacillus bulgaricus*. In other countries, some other microorganism is also used for yogurt preparation which includes Bifidobacterium spp and *L. acidophilus*.[8] In this symbiotic relationship, *Lb. delbrueckii* subsp. *bulgaricus* utilizes casein and produces certain amino acids which facilitate the growth of *S. thermophilus*, whereas *S. thermophilus* removes oxygen and produces formic acid (lower down pH) which supports the multiplication of *Lb. delbrueckii* subsp. *bulgaricus*. In this product, culture inoculated as rods: cocci ratio (1:1) for better product development.[73]

Apart from milk solids, other ingredients are added into yogurt, which includes sweeteners, sugars, vegetables, fruits, flavoring agents, NaCl, coloring agents, preservatives, and stabilizers.[106] Worldwide, various types of yogurts are produced and consumed. It includes, plain yogurt, low-lactose, flavored (like sundae), and carbonated, stirred, or Swiss-style yogurt (contains fruit, flavors, and other bulky ingredients). Liquid yogurts are one of the types of yogurts that are famous in Europe, Japan, and Canada, which are homogeneous and pourable in nature.[53]

10.3.3 CULTURED BUTTERMILK AND SOUR CREAM

Buttermilk is generated during the butter manufacturing as a byproduct. After cream has been churned into butter, the liquid left is called buttermilk. It is an excellent source of fat, lactose, protein, phospholipids, and minerals and possesses functional properties as a natural emulsifier. It can be used in baked goods, which requires tangy flavors and smooth textures. Low-fat/skimmed milk is used for the preparation of cultured butter milk using the harmless microorganisms. Mesophilic and multiple strain lactic cultures are used for making cultured buttermilk and sour cream which produces acid, aroma, and flavor in the end-product.[52]

Citrate fermenting cultures (such as, *Leuconostoc lactis* subsp. *lactis* var. *diacetylactis* or *Leuconostoc mesenteroides* subsp. *cremoris* combined with *Lc. lactis* subsp. *lactis or Lc. lactis* subsp. *Cremoris*) are utilized in manufacturing of cultured butter milk and sour cream. Cultures utilized for this type of fermentation must have a low diacetyl reductase activity. Flavor improvement in the product is supported by the Leuconostocs (but not lactococci) with the production of ethanol from acetaldehyde. Diacetyl to acetaldehyde balanced ratio is much more important for the unique flavor in cultured buttermilk (ranges between 3.2:1 and 4.4:1).[54] Sour cream and cultured buttermilk hold almost similar characteristics, such as, the culture or strain used for the sour cream is mesophilic in nature with the capacity to produce acid, flavor, and body that to similar to cultured buttermilk. The flavoring compounds and the incubation parameters are almost similar in both the products.

10.3.4 KEFIR

Kefir confers the health benefits. Kefir word may be acquired from the Turkish word "keyif" meaning "feeling good" after eating.[97] Kefir is a product with a distinct flavor and effervescent attributed to the yeast used in the product manufacturing.[76] Kefir is thought to have originated in the Caucasus Mountains. It is a conventional product, which is mostly ingested in Eastern Europe, Southwest Asia, and Russia,[97] Central Asia and Middle East, particularly in Turkey, Ukraine, Russia, Czech Republic, and Poland. Kefir is different from other fermented milk products as this contains kefir grains and more population of yeast in end-product. Cow milk was traditionally used in kefir manufacturing; simultaneously other milks are also used, such as, buffalo, sheep, goat, or soy milk.[49]

In order to make kefir, kefir granules (2–3%) are combined with several lactic cultures from the genera *Acetobacter*, *Lactococcus*, and *Leuconostoc*, as well as *Lactobacillus kefiri* and *Lactobacillus kefiranofaciens*, which have a specific close relationship with the product. The granules are composed of *Kluyveromyces marxianus* (lactose fermenting yeast) and *Saccharomyces cerevisiae*, Saccharomyces exiguous, and *Saccharomyces unisporus*, which are lactose-free fermenting yeasts found in the kefir granules.[13]

10.3.5 KOUMISS

Koumiss is a popular dairy product in Russia having acidic-alcoholic characteristics that is prepared from Mare milk.[43] It is a fermented product having gray color and sharp alcoholic and acidic taste with effervescent. Koumiss provides nutrition and therapeutic effects upon consumption. The word Koumiss is derived from Kumanese tribe, and it has been consumed by the people of Central Asia (Aryans of central Asia) since 2000 B.C.[105] Chunks are used for the preparation of Kumiss as an inoculum, that are composed of bacteria and yeasts. Turdusk (also known as a "burduk" or "saba") a leather sac (25–30 L) is used to keep the milk. It takes about 3–8 h for the fermentation. The microflora identified from the product includes *Lb. casei*, *Lb. delbrueckii* subsp. *bulgaricus*, *Lactococcus lactis* subsp. *lactis, S. unisporus,* and *Kluyveromyces fragilis.*[105] Moreover, yeasts (such as, *Candida* spp. and *Torula* spp.) along with the lactic cultures (such as, *Lb. lactis* and *Lb. acidophilus*) are also utilized in the manufacturing of Koumiss.[94,105]

10.3.6 YAKULT

Yakult is a Japanese fermented milk product. *Lb. casei* strain Shirota (LcS) is the main lactic culture used in the manufacturing of this product. It was isolated by Minoru Shirota in Japan. LcS is reported to be a resistant against bile acids and gastric acids. This strain remains active up to reach the lower intestine of the body after ingestion. It is developed upon the proposition that daily oral ingestion of LAB helps in improving health of intestine and inhibits occurrence of severe diseases. Therefore, it extends the lifespan of human beings. According to the manufacturer, a single bottle of this milk contains 20 billion cells of living LcS, which helps to improve intestinal health.[52]

10.3.7 FILMJOK AND LANGFIL

Filmjolk is a Swedish fermented milk product produced from cow milk using *Leuconostoc mesenteroides* and *L. lactis* species.[43] As compared to yogurt, it is less sour fermented milk product. Traditionally, it is prepared by adding former active batch of *filmjolk* into pasteurized milk and is further incubated for 1–2 days at room temperature or more preferably in a cool cellar. *Langfil* is also a type of *filmjolk* that has a sort of elastic texture due to the development of EPS by *L. lactis* subsp. subsp. *cremoris* and *L. lactis* subsp. *lactis* species.[101]

10.3.8 ACIDOPHILUS MILK

Acidophilus milk is an nutritive product having higher population of *Lb. acidophilus.*[83] Acidophilus milk and other associated cultured dairy products like acidophilus yogurts are popular in the United States. Skim milk/whole milk is heated at 95°C for 1 h, then is inoculated @ 2–5% milk and is followed by incubation at 37°C till it is coagulated. Acidophilus milk is more suitable for the patients of the lactose intolerance because the enzyme of *L. acidophilus* hydrolyzes the lactose.[33]

10.3.9 VIILI

Viili is a convention product that may have originated from Sweden but largely consumed in Finland. It is also known as *viilia*. Now, it is a national treasure in Finland.[48] The term Viili may be derived from the Swedish word fil.[48] From Sweden, the technique and skill to produce Viili has been passed to other countries. Mostly, Viili is consumed in Denmark, Sweden, Finland, and in less extent in Norway. Almost similar fermented milk products with sweet taste and consistency are popular in different countries, such as, *tatmjolk and langfil* in Sweden; *skyr* in Iceland; *taette* in Norway; *ymer* in Denmark; and *viilipiima*, and *pitkapiima, piima* in Finland.[96]

The sharp pleasant taste and good diacetyl aroma are closely related to the stringy texture of the product (Viili), which can be easily cut through the spoon. The lactic strains taking part in process of fermentation are *L. lactis* subsp. *cremoris* and *L. lactis* subsp. *Lactis,* which liberate the EPS. Viili exhibits a velvety layer on its surface alike to that of Camembert and Brie cheeses. This is associated with the presence of *Geotrichum candidum*, a filamentous fungal species that exhibits characteristics of both yeasts and molds. This species utilizes the lactate, which further reduces the acidity and produces characteristic moldy aroma of *Viili.*[37]

10.3.10 KISHK

Kishk is a product from Middle East, which is composed of fermented milk and wheat. Further, this product is also being used in Mexico[14] and Europe[10] with the changing of dietary patterns. It is prepared by mixing yogurt (strained yogurt) and parboiled wheat. This mixture is further incubated at different periods at optimum temperature. In this manufacturing process, milk is individually fermented in a different vessel and further concentrated and mixed with moistened wheat flour. Milk is fermented solely, and the paste produced is allowed to dry for reaching moisture content to 10–13%, followed by grounding as powder and stored as dried balls (brown color with rough surface and hard texture).[93] The *L. plantarum, L. casei, L. brevis*, and *Bacillus subtilis* and yeasts are considered as major fermenting microorganisms.

10.3.11 TARHANA (TRAHANAS)

Tarhana is consumed in Greece and Turkey, and it is developed using sheep milk yoghurt, wheat flour, and cooked vegetable, which includes onions and tomatoes, and spices like mint and salt. Further, the mixture is fermented for 1–7 days. This fermented mixture is allowed to dry and stored in the form of biscuits.[15] The product thus obtained and the fermentation process both are related to that of Kishk. Major microflora associated in fermentation of sheep milk is *L. bulgaricus and S. thermophiles.*[23] It provides overall nutrition as it provides dairy protein along with cereal protein in a single product.

10.3.12 RAABADI

It is a conventional product from India (Rajasthan, Punjab, and Haryana), which is manufactured by cooking blend of maize flour and buttermilk. There are two developed methods for the preparation of raabadi, which includes fermenting the mixture of moth bean flour and buttermilk before cooking or fermenting the mix after cooking.[9]

10.3.13 KUMUDU

Kumudu is a conventional product in the central African region. It is produced using the blend of fermented milk and sorghum flour (nongerminated/germinated) through sun drying of the mixture.[1]

10.3.14 DAHI (INDIAN YOGURT: CURD)

Dahi is famous Indian-originated fermented milk product and is also known as Indian yogurt. It is quite like yogurt based on its consistency and appearance. It is popular among the consumers of all age groups due to its distinctive flavor, nutritional, and therapeutic values. It is a part of Indian culinary dishes and utilized in various form in daily diet. *Dahi* has been used since the Vedic times, and its presence is described in Upanishads, Vedas, and various hymns.[107] It is mentioned that butter and *dahi* is utilized in the era (ca. 3000 BC) of Lord Krishna. *Dahi* is always used in the rituals as a part of *panchamrut* (five nectars). Therapeutic importance of dahi prepared from cow milk or buffalo milk has been cited in the Ayurveda (the traditional scientific system of Indian medicine), in its treatises Charaka Samhita and Sushruta Samhita.[72]

Dahi provides distinct taste to the diet and shows putrefactive effect and refreshing taste. Number of LAB are used for manufacturing of dahi according to type of product requirement. Mainly lactic cultures like *L. bulgaricus and Streptococcus,* along with probiotic cultures or aromatic cultures are also used. Mainly, mixed species of *Streptococci*, *Lactobacilli*, *Lactococcus*, and *Leuconostoc* have been isolated from naturally fermented dahi. Probiotic dahi is prepared by fermentation of milk with probiotics and lactic cultures. Dahi can be considered as a functional food with nutritive and therapeutic values.[81]

10.4 ROLE OF STARTERS IN FERMENTATION AND FERMENTED MILK PRODUCTS

Starter culture can be defined as the population of microorganisms, which are composed of huge number of cells of at least single strain, which can be added into the raw material to develop the desirable product using the process of fermentation. A group of bacteria called LAB plays important role to produce fermented dairy products. They have been proved to be safe to apply on the raw material. These bacteria can produce lactic acid, EPS, bacteriocins, ethanol, aroma compounds, acetic acid, and numerous enzymes having technological and functional values in the product development. Therefore, *LAB* can raise the shelf-life, microbial safety, sensory, and texture the finished product.

The Lactic cultures (bacteria), propionic-bacteria, surface-ripening bacteria, yeasts, and molds are utilized as the starter cultures for making different type of cultured dairy products. Major role of dairy starters includes production of acidity (lactic acid) in milk. In the fermented milk products, flavor and aroma are contributed to the *LAB*, which produces the compounds called acetaldehyde, diacetyl, and acetic acid during fermentation process.[34]

The flavor production in fermented products is mainly due to the three different pathways, such as, glycolysis (fermentation of sugars and production lactic acid, diacetyl, acetic acid, acetoin, and aldehyde), lipolysis (hydrolysis of fat), and proteolysis (hydrolysis of proteins-production of alcohols, aldehydes, acids, esters, sulfur compounds).[45,88,94] In hetero-fermentation, production of carbon dioxide is observed, which provides characteristic eye formation in the fermented milk product. In case of cheeses, occurrence of textural changes and flavor development flavor during ripening process is related with the enzymes originally isolated from bacterial/fungal cultures based on cheese type.

Starters have some bio-functional properties to improve, prevent, or modify the health status of the consumer. The health-promoting characteristics of fermented milk are due to the prebiotic EPS, probiotics, bioactive peptides, lipids, and oligosaccharides. Starters play important role in aroma and flavor production of the product, tolerance to bacteriophage, exopolysaccharide production, salt tolerance, vitamin production, production of bacteriocin, reduction of toxic or antinutritive factors, low calorie sugar production, temperature susceptibility, pH susceptibility, and nutraceutical production of the final product.[34] Based on the characteristics, starters are selected and used in fermentation of various dairy products.

10.5 BIO-FUNCTIONALITY OF FERMENTED MILK AND MILK PRODUCTS

Bio-functional role of fermented milk and associated products has been widely studied and mentioned in the literature. In general, fermented milk products are highly digestible than the unfermented ones. Thus, in some group of people, fermented milk products are consumed as weaning foods. Fermented milk components, especially probiotics, have shown potential to prevent gastrointestinal infection in humans. Either reducing the pH or through colonization resistance, they prevent such infections.

Several health benefits of probiotics have been identified, such as, healing lactose intolerance, improvement in symptoms of Rota virus-associated diarrhea prevention, diminished fecal mutagenicity, and immune improvement. Probiotics are also inhibiting agents of *H. pylori* in humans. Bacteriocins and bioactive peptides promote the health function of the fermented milk in the host.[34] Nowadays, fermented milk products gain novel name as a functional or fortified products. Various bio-functional roles of fermented milk and milk product are explained here in this chapter.

10.5.1 ANTIHYPERTENSIVE ACTIVITY

Antihypertensive function of fermented milk products is associated with release of peptides from intact milk protein sequences during the process of fermentation. Small peptides generated like "Val-Pro-Pro and Ile-Pro-Pro" show inhibition of ACE-activity. Casokinins and lactokinis are the types of bioactive peptides (hypotensive activity) generated due to action of proteolytic bacteria on milk proteins during milk fermentation.[22] In vivo studies have revealed the reduction in systolic blood pressure when spread (consisting bioactive peptides IPP and VPP) was offered to hypertensive subjects.[104]

Some studies have indicated ACE-inhibitory peptides in fermented camel milk and goat milk grown under optimal conditions.[68,90] Traditionally fermented camel milk is subjected for isolation of *Lactobacillus helveticus* 130B4, which further proved to release "Ala-Ile-Pro-Pro-Lys-Lys-Asn-Gln-Asp" (ACE-inhibitory peptide) from milk proteins.[74] *Lb. helveticus* was also examined to identify the release of ACE-inhibitory peptides from bovine milk proteins.[59]

It has been proved that some fermented milks show antihypertensive effect in vitro and in vivo studies, directly associated with bioactive peptides released in milk. The effective action of these peptides is validated as like synthetic ACE inhibitors. Thus, fermented milk can be considered as hypotensive agent.

10.5.2 ANTIBACTERIAL OR ANTIMICROBIAL ACTIVITY

Antimicrobial or antibacterial activity of cultured dairy products is attributed to liberation of bioactive peptides during the fermentation process. Milk fermented with *Lactobacillus helveticus* shows antimicrobial and immunomodulatory effects due to release of such peptides.[55] Water soluble extract (cell-free supernatant) obtained from milk, fermented with *Lactobacillus helveticus* LH-2 and further the peptide fraction F5 was subjected for the test in macrophages. This results in increased production of IL-6, TNF-α, and IL-1 via macrophage stimulation, as well as enhanced nitric oxide release and phagocytic activity.

The TNF-α induces the production of nitric oxide which is one of the cytotoxic agents that can kill bacteria, pathogens, and tumor cells. These findings imply that the F5 peptide fraction may influence macrophage activities.[99] Immuno-compromised hosts exhibited increased resistance against the infections of *Salmonella typhimurium* and *Streptococcus pneumoniae* after getting the supplemented diet with goat milk fermented with *Lactobacillus rhamnosus* CRL1505.[80]

Kefir has shown similar antibacterial effects to that of ampicillin, ceftriaxone, amoxicillin, cetoconazol, and azithromycin. Antimicrobial or antibacterial effects of kefir are associated with the microorganisms, which produce bactriocins. Bacteriocin ST8KF produced by *Lactobacillus plantarum* ST8KF has exhibited antimicrobial activity against *Listeria innocua* and *Enterococcus mundtii.*" Similarly, other microflora associated with kefir grains (*Lactobacillus anofaciens, Streptococcus thermophiles* stains, *L. acidophilus*) has exhibited antimicrobial activity against *S. typhimurium, Listeria monocytogenes, Escherichia coli, Salmonella enteritidis, Staphylococcus aureus, Pseudomonas aeruginosa, Yersinia enterocolitica, and Shigella flexneri.*[13]

10.5.3 ANTICARCINOGENIC ACTIVITY

Fermented milk and milk products possess the potential to prevent or control or act against certain types of cancers. Yogurt and *LAB* of the fermented milk have extensively studied for their effects to reduce cancer, intestinal inflammation, and related problems. It is determined that intake of cheese, yogurt, and buttermilk can prevent from breast cancer. The *LAB* have exhibited anticarcinogenic effects though the suppression of cancer already initiated or prevention of cancer initiation in animal models.

Mice model-based studies have proved anticarcinogenic effects of cultured dairy product (yogurt and fermented milk) with *L. acidophilus.* Preventive functions of probiotics are associated with the changes in the gut microflora, modulation in immune response, and checking the growth of the microbes, which work as procarcinogens in carcinogens.[19] Other potential mechanisms have been studied, which regulate antitumor effect through fecal enzyme alteration, regulation of mutagenicity of chemical mutagens, cellular uptake of mutagenic elements, and controlling tumors through better immune system.[21,36]

10.5.4 ACTIVITY AGAINST COLON CANCER

Murine cancer model fed with yogurt indicated inhibition of tumor growth through excessive secretion of IL-10, apoptosis, and reduction of pro-carcinogenic enzymes, which can overall regulate the reduction of inflammatory response.[19] Bioactive peptides derived from kefir shows the initiation of phagocytosis, activation of macrophages, and nitric oxide (NO) production.[85]

BALB/c Mice fed with yogurt showed a greater number of IFN-γ (+) cells than the control group dimethylhydrazine DMH-yogurt group. It was suggested that yogurt had supported in the raised quantity of immune cells in gut and cytokines like IL-10 could regulate these cells. It was found that mice fed with yogurt have shown regulation of immune response and production of greater number of cells IL-10(+).[70] Supernatant of kefir has shown preventive effect on the DNA damage stimulated by carcinogen agents was shown at all the concentrations used in human colon adenocarcinoma cells.[75,85]

10.5.5 EFFECTS ON OTHER TYPES OF CANCER

Fermented milks play significant role against the carcinogens and inhibit different variety of cancer cells. *L. kefiri* has shown rising apoptosis of human myeloid leukemia cells resistant to many drugs in vitro with the stimulation of caspase 3 with a dose-dependent way.[85] On the other hand, Kefir has no impact on healthy cells, but it suppresses cell proliferation in leukemia patients.[75]

10.5.6 EFFECTS ON BREAST CANCER

Diet enriched with cultured dairy products has shown inhibition of breast tumors too.[18] Fermented milks are enriched with peptides that are reported to promote the immune system and check growth of the tumors.[55] A mouse model of breast cancer has shown reduction in tumor growth rate and raised lymphocyte proliferation after fed with *L. acidophilus*, which was isolated from yogurt.[69] Cultured dairy products have shown to initiate apoptosis and proved to reduce the tumor growth in breast cancer cells,[85] which may be appropriate in treatment or prevention of breast cancer.

10.5.7 ANTIOXIDANT ACTIVITY

Bioactive peptides with antioxidant effects are isolated from fermented milk; and free radical scavenging activity of lactic cultures of cultured dairy products are mainly associated with the antioxidative effect of cultured dairy foods. Fermented goat milk has shown antioxidative effect by some of the known mechanisms, which include raising the serum melatonin levels, total antioxidant capacity, controlling the biomolecular oxidative damage, reduction in glutathione peroxidase 1 expression in the duodenal mucosa.[60,61]

Bacteria with probiotic potential have been proved to trap reactive oxygen species from the food matrix. A study conducted using the intracellular extracts of Lactobacillus sp. exhibited the curing effect of vitamin-E deficiency in rats. Lactic cultures like *Lactobacillus delbrueckii* subsp. *bulgaricus* and *S. thermophiles* have been proved to inhibit the lipid peroxidation through free radical scavenging activity or scavenging reactive oxygen species.

Clinical studies are being conducted using goat milk fermented with *Lactobacillus fermentum* ME-3. Comparative analysis of sour milk and unfermented milk revealed that sour milk improved total antioxidant activity as compared to unfermented one. It was determined that sour milk has reduced oxidized low-density lipoprotein, lower down glutathione redox ratio, reduced lipoprotein peroxides, extended oxidative resistance of lipoprotein fraction, and further raised total antioxidant activity."

Factors responsible in antiatherogenic effect of fermented milk have been identified as bioactive peptides liberated from α-casein, α-lactalbumin and β-lactoglobulin, and antioxidative function of some lactobacillus in the human GI tract.[22] Antioxidant activity of Kefir may be due to EPS. EPS may improve oxidative status of host may be through regulating the microorganisms to selectively inhibit or remove the microflora, which induce oxidative stress.[110]

10.5.8 HYPOCHOLESTEROLEMIC ACTIVITY

Hypocholesterolemic effects of probiotics have been proved clinically in many studies. Regular administration of some selected probiotics shows hypocholesterolemic effects, due to their unique characteristic to break cholesterol and regulating its reabsorption in the digestive tract. Different studies were conducted using *lactobacilli, bifidobacteria*, and other milk bacteria in vitro and in vivo state and proved that these bacteria can assimilate cholesterol, by incorporation into cellular membranes, precipitation, and deconjugation of cholesterol with bile acids.[22]

Yogurt consists of *Bifidobacterium pseudocatenulatum*-G4 and *Bifidobacterium longum* -BB536 and it has shown beneficial properties in decreasing risk of cardiovascular health issues.[5] Further, the probiotics *Bifidobacterium longum* BB536 and *Bifidobacterium pseudocatenulatum* G4 increased bile acid excretion and reduced cholesterol in plasma.[5] In a study to compare the ordinary yogurt and probiotic yogurt, it was determined that consumption of yogurt containing *Bifidobacterium lactis* and *L. acidophilus* had significantly reduced the cholesterol level in serum as compared to ordinary yogurt.[7]

Koumiss originated lactic culture *Lactobacillus fermentum* SM-7 had exhibited in vitro and in vivo cholesterol-reducing activity.[66] Fermented camel milk product especially *Garissa* consisting bacteria *Bifidobacterium lactis* (BB-12) has demonstrated in vivo hypocholesterolaemic effect in rat[24] and reduction of cholesterol observed in liver.[2]

10.5.9 ANTIALLERGIC ACTIVITY

Prevalence of higher level of Bifidobacterium and lactobacilli (group 1) in the intestine of children has reduced the occurrence of allergic diseases. Further, it was demonstrated that kefir and kefiran showed antiallergic activity.[33]

External complement of Bifidobacterium improves the infant's intestinal microflora, which has reduced the levels of Bacteroides, which further reduced the occurrence of allergic reaction due to food.[13] Imbalance in the Th1/Th2 cells is a major mechanism of food allergy. This imbalance raised the production of IgE antibodies. Mice fed with mixture of lactobacilli have shown lower production of antibodies and specific IgE to ovalbumin. Reduction in the symptoms of allergy were observed through the treatment with probiotics.[13]

10.5.10 EFFECTS ON BONE RESORPTION

Positive effects of fermented milk have been determined on the health of children and adolescents through intervention trials through improvement in bone density, and bone mineral content. Studies revealed that cheese supplements are more favorable for cortical bone accumulation than supplying calcium in the form of tablets.[77] The consumption of yogurt or soft cheese fortified with or without Calcium and vitamin D has shown reduction in serum bone resorption markers (CTX and TRAP 5b) and PTH.[12] It was concluded that fermented milk products whether fortified or not could improve the IGF-I and work as promoting agents in the accumulation of bone minerals in children and adolescents. Improvement was observed in balance of calcium and resorption in adults.[77]

Positive association of yogurt has shown significant improvement in bone formation, increasing bone mass density, prevent whit hip fracture risk, and preventing bone loss.[11] Bone formation is also supported by the bioactive peptides. Bioactive peptides VPP derived from milk fermented with *L. helveticus* LBK-16H had improved bone formation in vitro.[63,64] After 12 weeks of intervention, ovariectomized rats fed with fermented milk containing VPP showed minimum bone loss.[63,65]

10.5.11 GUT MODULATION BY FERMENTED MILK

The products generated from lactic fermentation influence composition of intestinal microbiota. It was evident that greater population of *bifidobacteria* was observed in the feces of the group, who obtained cell-free concentrated whey from *Bifidobacterium breve* C50 fermented milk as compared to the control group consuming diluted milk. Beneficial effects of these components were also observed in the intestinal microbiota.[31] Studies conducted on children who were fed with formula-containing fermented milks had higher population of *Bifidobacteria* with less adult species as compared to the children, who obtained a standard formula not fermented milk.[62]

Improvement in the barrier function and immune priming was observed in children due to breast feeding and it showed higher colonization of *Bifidobacterium breve* and *Bifidobacterium longum infantis*. Colonization of some specific species of *Bifidobacterium* has proved to improve immune system and to help in the proper maturation of intestinal microbiota.[31] In a randomized controlled trials, episodes acute diarrhea in healthy infants was reduced after feeding fermented milk-based formula and it showed reduction in the dehydration, and needs of oral rehydration and medical visits as compared to the other groups.[100]

Fermented milk containing bacteria having probiotic properties can modify intestinal microbiota when reached to intestinal transit or they can promote growth of health beneficial microbes. It was evident that the consumption of kefir and kefiran had raised the counts of *Lactobacillus* and *Bifidobacterium* in an animal model and showed the reduction of harmful microbes like *Clostridium perfringens*. C57BL/6 mice fed with kefir had shown reduction in the severity of the infection caused by *Giardia intestinalis* because of modulation of immune system.[87]

10.5.12 ANTIOBESITY ACTIVITY

Fermented dairy products formulated with low calorific value are important foods for the individual with obesity. The antiobesity action of fermented milk is mediated by different of pathways. Particularly, they support to extend or check the maturation of preadipocytes into mature adipocytes inside adipose tissues; and they hinder the fat depository mechanism in adipose cells. Despite this, some studies suggest that probiotics included in dairy products might reduce the body's brown fat accumulation. Entirely, fat reduction observed through any mechanism is supportive in reducing the waist circumference of the body. It was also proposed that lower deposition of fat around the abdominal portion of the body could reduce the chances of occurrence and associated heart diseases.[109]

10.5.13 FERMENTATION AND GENERATION OF BIOACTIVE PEPTIDES

Bioactive peptides are hydrolyzed sequences derived from native protein, which exerts positive physiological impact on humans. These are inactive amino acids in the intact protein sequences but get to active form due to action of microbial fermentation or enzymatic action or through food processing or through gastrointestinal enzymes during digestion. Peptides can be produced using fermentation carried through bacteria, yeasts, or fungi, which act on the protein and generate bioactive peptides.

The degree of protein hydrolysis depends upon the microbial strain, protein source, and the fermentation time. These bioactive peptides have several health-beneficial effects, such as, ACE-inhibition, antimicrobial, anticancer, antiobesity, anticariogenic, hypocholesterolemic, mineral binding, and opioid activities.[26] Commercially produced bioactive peptides enriched fermented milks are also available in the market conferring the health beneficial effects to the consumer, for example, Calpis milk and evolus milk.[41,42]

10.6 PHYSICOCHEMICAL CHANGES IN THE FERMENTED DAIRY PRODUCTS

Fermentation is a process of conversion of lactose into lactic acid and many biochemical changes. During fermentation process of milk and milk products, many changes were observed in final product, such as, change in pH, titratable acidity, fat content, protein content, total solids, and survival of lactic cultures.

10.6.1 PROCESSING CONDITIONS AFFECTING FERMENTED DAIRY PRODUCTS

The quality of final fermented milk products depends upon various processing stages involved in manufacturing process. Quality of milk is very much important to produce good quality fermented product. Milk should be clean, fresh, should not carry predeveloped acidity or off-flavors, lower microbial load or free from mastitis.[71] Addition of milk powders improves the consistency of the product. Skim milk powders, whey protein hydrolysates, or whet protein concentrate are used for preparation of fermented products.[89] It was also indicated that there is an increase in the total solids on addition of external protein supplements and it enhances the growth *of L. acidophilus* and *S. thermophiles.*[50] Furthermore, the type and quantity of proteins in use have a key impact on the final product's texture and rheology.[17]

After homogenization of milk, the number of fat globules is increased, and this improves the whiteness of the milk. Homogenization also reduces the risk of free whey separation on to surface of set fermented dairy products (called as syneresis) and increases the firmness of finished product.[95] Pathogenic and spoilage microorganisms in raw milk are killed by heat treatment, thus starter cultures are less competitive and growth medium is improved because of this. Heating of milk expels oxygen and decreases the redox potential, which encourages growth of starter bacteria.[46] Although, heating milk at 100°C decreases the syneresis and enhances yogurt coagulum smoothness.[78]

Heat treatment of milk has a significant impact on chemical structure and nature of coagulum. Protein coagulation is development of three-dimensional network that entraps fat globules and serum with dissolved components. The resulting semisolid gel has a smooth texture and a custard-like consistency.[47] Cooling is one of most common process that is used for controlling the starter culture metabolic activity. Air incorporation occurs during the stage of cooling resulting in reduced whey syneresis and viscosity.[58]

10.6.2 CHANGES DURING STORAGE OF MILK PRODUCT

Yogurt prepared commercial lactic starters *Str. thermophilus* and *L. delbrueckii* ssp. *Bulgaricus* (control) and with the same starter *St Lb* with added *L. salivarius* LS (Lsal StLb) as an adjunct was studied for the storage changes at 4°C until 4 weeks. Author noted non-significant changes in LAB populations during storage period. Fat content, proteins, and total dry solids content remained constant, while pH value of the yogurt was decreased up to 4 weeks of storage. Firmness of the yogurt with commercial lactic cultures showed gradual increase from 1st day to 4th week of the storage, while the yogurt with adjunct culture showed increasing trends but reduction on 4th week of the storage period. Lactic acid content of the control yogurt (only with commercial starter) has shown highest value on 3rd week, while reduction on 4th week of the storage. On the other hand, yogurt with added adjunct culture has shown gradual increase up to the storage period.[39]

10.6.3 CHANGES DUE TO ADDED INGREDIENTS

Addition of external ingredients affects the consistency and characteristics of fermented milk products. Addition of hydrocolloids in the reconstituted sweetened yogurt had reduced pH of the final product. It has also reduced the whey expulsion rate of the reconstituted sweetened yogurt with hydrocolloids. Samples of sweetened yogurt prepared using drying method showed lower survival rate of starter bacteria than the sweetened yoghurt prepared with hydrocolloids (hydrocolloids provides protective effects to the starter cultures).[84]

10.7 FUTURE PROSPECTS OF FERMENTED MILK AND MILK PRODUCTS

With the application of probiotics, there is huge scope in prevention of risk of contamination, risk of intoxication for producing certain toxins during fermentation process, bacteremia, and endocarditis or various lifestyle diseases like obesity, diabetes, hypertension, etc. Various traditional fermented foods are popular in every country around the globe. But lack of well characterized potential starter cultures with GRAS status are the need for popularizing such fermented foods. These fermented foods are being consumed since long. But scientific evidence to establish the human health claim is very important.

Currently, there is a scarcity of clinical data demonstrating the benefits of probiotic therapy or functional fermented foods in the management of several lifestyle diseases. More clinical information is required (1) to screen for optimal dosage regimens, (2) to characterize the required properties of probiotics with a view to strain selection, (3) to develop novel technology for manufacturing of traditional fermented foods, and (4) to extend our knowledge of safety aspects by conducting the study on animal models or in human subjects.

10.8 SUMMARY

Fermentation is a shelf-life extending preserving process. History of fermentation is not yet determined but the fermented milk and milk products become vital constituents of daily human diet. Added novel lactic starts for the desired fermentation adds bio-functionality in the fermented milk. Bio-functional activities of fermented milk and milk products are related to the EPS, bacteriocins, probiotics, and bioactive peptides formed in the dairy food matrix. Future research work in the field of bio-functionality fermented milk and milk products will quench the thrust of many individuals who loves to eat food as medicine in their daily diet. However, the type of fermentation and selection of starters is more important with the environmental conditions during the process of fermentation of milk. Surely, fermented dairy foods consisting of probiotics would be the most preferable selection of the consumer with the surety of nutrition and therapeutic effects on consumption.

KEYWORDS

- **bacteriocins**
- **bio-functional properties**
- **bioactive peptides**
- **fermentation**
- **fermented camel milk**
- **kefir**
- **lactic acid bacteria**

REFERENCES

1. Achi, O. K.; Ukwuru, M. Cereal-Based Fermented Foods of Africa as Functional Foods. *Int. J. Microbiol. Appl.* **2015,** *2*(4), 71–83.
2. Ali, A. A.; Alyan, A. A.; Bahobail, A. S. Effect of Fermented Camel Milk and Cow Milk Containing *Bifidobacteria* Enriched Diet in Rats Fed on Cholesterol Level. *Res. J. Agric. Sci.* **2013,** *3,* 342–346.
3. *Alimentarius, Codex. Milk and Milk Products*, 2nd ed.; World Health Organization, and Food and Agriculture Organization: Rome, Italy, 2011; p 248.

4. Allen, J. M.; Jaggers, R. M.; Solden, L. M.; Loman, B. R.; Davies, R. H.; Mackos, A. R.; Bailey, M. T. Dietary Oligosaccharides Attenuate Stress-Induced Disruptions in Immune Reactivity and Microbial B-Vitamin Metabolism. *Front. Immunol.* **2019,** *10,* 1774.
5. Al-Sheraji, S. H.; Ismail, A.; Manap, M. Y.; Mustafa, S.; Yusof, R. M.; Hassan, F. A. Hypocholesterolemia Effect of Yoghurt Containing *Bifidobacterium pseudocatenulatum* G4 or *Bifidobacterium longum* BB536. *Food Chem.* **2012,** *135*(2), 356–361.
6. Aryana, K. J.; Olson, D. W. 100-year Review: Yogurt and Other Cultured Dairy Products. *J. Dairy Sci.* **2017,** *100*(12), 9987–10013.
7. Ataie-Jafari, A.; Larijani, B.; Alavi Majd, H.; Tahbaz, F. Cholesterol-lowering Effect of Probiotic Yogurt in Comparison with Ordinary Yogurt in Mildly to Moderately Hypercholesterolemic Subjects. *Ann. Nutr. Metab.* **2009,** *54*(1), 22–27.
8. Bamforth, C. W.; Cook, D. J. Yoghurt and Other Fermented Milk Products. In F*ood, Fermentation and Micro-organisms*; John Wiley & Sons Ltd.: New York, 2019; pp 183–186.
9. Basu, S.; Tomar, S.K.; Hati, S. Isolation and Identification of Phytate Degrading *Lactobacilli* from Indian Cereal-Based Fermented Milk Product: Raabadi. *Int. J. Food Ferment. Technol.* **2015,** *4*(1/2), 35–40.
10. Berghofer, E. Use of non-European fermented foods in Austrian market. *Zeitschrift fur Ernahrungswissenschaft* (*J. Nutr. Sci.*) **1987,** *11*(1), 14–22.
11. Bian, S.; Hu, J.; Zhang, K.; Wang, Y.; Yu, M.; Ma, J. Dairy Product Consumption and Risk of Hip Fracture: A Systematic Review and Meta-Analysis. *BMC Public Health* **2018,** *18*(1), 165–170.
12. Bonjour, J. P.; Benoit, V.; Payen, F.; Kraenzlin, M. Consumption of Yogurts Fortified in Vitamin D and Calcium Reduces Serum Parathyroid Hormone and Markers of Bone Resorption: A Double-Blind Randomized Controlled Trial in Institutionalized Elderly Women. *J. Clin. Endocr. Metab.* **2013,** *98*(7), 2915–2921.
13. Bourrie, B. C. T.; Willing, B. P.; Cotter, P. D. The Microbiota and Health Promoting Characteristics of the Fermented Beverage Kefir. *Front. Microbiol.* **2016,** *7*, 1–17.
14. Cadena, M. A.; Robinson, R.K. The Acceptability of Yoghurt-Cereal Mixtures to A Rural Community in Mexico. *J. Nutr. Ecol. Food Res.* **1979,** *8*(3), 169–174.
15. Campbell-Platt, G. Fermented Foods: A World Perspective. *Food Res. Int.* **1994,** *27*(3), 253–257.
16. Chen, M.; Ye, X.; Shen, D.; Ma, C. Modulatory Effects of Gut Microbiota on Constipation: The Commercial Beverage Yakult Shapes Stool Consistency. *J. Neurogastroenterol. Motil.* **2019,** *25*(3), 475–477.
17. Damin, M. R.; Alcântara, M. R.; Nunes, A. P.; Oliveira, M. N. Effects of Milk Supplementation With Skim Milk Powder, Whey Protein Concentrate, and Sodium Caseinate on Acidification Kinetics, Rheological Properties, and Structure of Nonfat Stirred Yogurt. *LWT Food Sci. Technol.* **2009,** *42*, 1744–1750.
18. De Moreno de LeBlanc, A.; Matar, C.; Farnworth, E.; Perdigon, G. Study of Immune Cells Involved in the Antitumor Effect of Kefir in a Murine Breast Cancer Model. *J. Dairy Sci.* **2007,** *90,* 1920–1928.
19. De Moreno De Leblanc, A.; Perdigón, G. The Application of Probiotic Fermented Milks in Cancer and Intestinal Inflammation. *Proc. Nutr. Soc.* **2010,** *69*(3), 421–428.
20. Druart, C.; Bindels, L. B.; Schmaltz, R.; Neyrinck, A. M.; Cani, P. D.; Walter, J. Delzenne, N. M. Ability of The Gut Microbiota to Produce PUFA-Derived Bacterial Metabolites: Proof of Concept in Germ-Free Versus Conventionalized Mice. *Mol. Nutr. Food Res.* **2015,** *59*(8), 1603–1613.

21. Ebel, B.; Lemetais, G.; Beney, L.; Cachon, R.; Sokol, H.; Langella, P.; Gervais, P. Impact of Probiotics on Risk Factors for Cardiovascular Diseases: A Review. *Crit. Rev. Food Sci. Nutr.* **2014,** *54*(2), 175–189.
22. Ebringer, L.; Ferencik, M.; Krajcovic, J. Beneficial Health Effects of Milk and Fermented Dairy Products – Review. *Folia Microbiol.* **2008,** *53*(5), 378–394.
23. Economidou, P. L.; Steinkraus, K. H. Greek Trahanas. In *Handbook of Indigenous Fermented Foods*; Steinkraus, K. H., Ed.;; Marcel Dekker: New York, 1993; pp 299–304.
24. Elayan, A. A.; Sulieman, A. E.; Saleh F. A. The Hypocholesterolemic Effect of Gariss and Gariss Containing Bifidobacteria in Rats Fed on a Cholesterol-Enriched Diet. *Asian J. Biochem.* **2008,** *3,* 43–47.
25. Fernandez, M. A.; Picard-Deland, É.; Le Barz, M.; Daniel, N.; Marette, A. Yogurtand Health. In *Fermented Foods in Health and Disease Prevention*; Frías, J., Martínez-Villaluenga, C., Peñas, E., Eds.; Elsevier Science and Technology Books: Québec, Canada, 2016; pp 305–338.
26. FitzGerald, R. J.; Murray, B. A. Bioactive Peptides and Lactic Fermentations. *Int. J. Dairy Technol.* **2006,** *59,* 118–125.
27. Food and Agricultural Organization (FAO) of the United Nations and World Health Organization (WHO). Health and Nutritional Properties of Probiotics in Food Including Powder Milk With Live Lactic Acid Bacteria; World Health Organization: Rome, 2001; pp 1–20.
28. Fox, P. F.; Guinee, T. P.; Cogan, T. M.; McSweeney, P. L. Eds.; *Fundamentals of Cheese Science*. Springer: New York, 2017; pp 121–183.
29. García-Burgos, M.; Moreno-Fernández, J.; Alférez, M. J.; Díaz-Castro, J.; López-Aliaga, I. New Perspectives in Fermented Dairy Products and Their Health Relevance. *J. Funct. Foods* **2020,** *72,* E-article 104059.
30. Gardiner, G.; Stanton, C.; Lynch, P. B.; Collins, J. K.; Fitzgerald, G.; Ross, R. P. Evaluation of Cheddar Cheese as a Food Carrier for Delivery of a Probiotic Strain to the Gastrointestinal Tract. *J. Dairy Sci.* **1999,** *82*(7), 1379–1387.
31. Granier, A.; Goulet, O.; Hoarau, C. Fermentation Products: Immunological Effects on Human and Animal Models. *Pediatr. Res.* **2013,** *74,* 238–244.
32. Hamet, M. F.; Londero, A.; Medrano, M.; Vercammen, E.; Van Hoorde, K.; Garrote, G. L.; Huys, G; Vandamme, P.; Abraham, A. G., Application of Culture-Dependent and Culture-Independent Methods for the Identification of *Lactobacillus Kefir anofaciens* in Microbial Consortia Present in Kefir Grains. *Food Microbiol.* **2013,** *36*(2), 327–34.
33. Hati, S.; Das, S.; Mandal. S. Technological Advancement of Functional Fermented Dairy Beverages. In *Engineering Tools in the Beverage Industry, the Science of Beverages*; Woodhead Publishing: London, 2019; vol 3, pp 101–136.
34. Hati, S.; Mandal, S.; Prajapati, J. B. Novel Starters for Value Added Fermented Dairy Products. *Curr. Res. Nutr. Food Sci.* **2013,** *1*, 83–91.
35. Hill, D.; Sugrue, I.; Arendt, E.; Hill, C.; Stanton, C.; Ross, R. P. Recent Advances in Microbial Fermentation for Dairy and Health. *F1000 Res.* **2017,** *6,* 751–755.
36. Hosono, A.; Kashina, T.; Kada, T. Antimutagenic Properties of Lactic Acid-Cultured Milk on Chemical and Fecal Mutagens. *J. Dairy Sci.* **1986,** *69*(9), 2237–2242.
37. Kahala, M.; Mäki, M.; Lehtovaara, A.; Tapanainen, J.M.; Katiska, R.; Juuruskorpi, M.; Juhola, J.; Joutsjoki, V. Characterization of Starter Lactic Acid Bacteria from the Finnish Fermented Milk Product Viili. *J. Appl. Microbiol.* **2008,** *105*(6), 1929–1938.

38. Kamiya, S; Owasawara, M.; Arakawa, M.; Hagimori, M. The Effect of Lactic Acid Bacteria Fermented Soybean Milk Products on Carrageenan-Induced Tail Thrombosis in Rats. *Biosci. Microbiota Food Health* **2013,** *32,* 101–105.
39. Karsheva, M.; Paskov, V.; Tropcheva, R.; Georgieva, R.; Danova, S. Physicochemical Parameters and Rheological Properties of Yoghurts During the Storage. *J. Chem. Technol. Metall.* **2013,** *48*(5), 483–488.
40. Katz, S. E. *The Art of Fermentation an in-Depth Exploration of Essential Concepts and Processes from Around the World*; Chelsea Green Publishing: White River Junction, Vermont, USA, 2012; p 498.
41. Korhonen H.; Pihlanto A. Food-derived Bioactive Peptides: Opportunities for Designing Future Foods. *Curr. Pharm. Des.* **2003,** *9*(16), 1297–1308.
42. Korhonen, H.; Pihlanto, A. Bioactive Peptides: Production and Functionality. *Int. Dairy J.* **2006,** *16*(9), 945–960.
43. Kosikowski, F.V.; Mistry, V.V. Cheese and Fermented Milk Foods. In *Origin and Principles. Fermented Milks*; 2nd ed.; Blackie Academic and Professional: New York, 1997; vol 1; pp 57-74.
44. Koyama, M.; Hattori, S.; Amano, Y.; Watanabe, M.; Nakamura, K. Blood pressure-lowering Peptides From Neo-Fermented Buckwheat Sprouts: A New Approach to Estimating ACE-Inhibitory Activity. *PLoS One* **2014,** *9,* E-article 105802
45. Law, B. A. Cheese Ripening and Cheese Flavor Technology. In *Technology of Cheese Making*; Sheffield Academic Press Ltd: Sheffield, 1999; pp 163–192.
46. Law, B. A. Ed.; *Microbiology and Biochemistry of Cheese and Fermented Milk*, 3rd ed. Blackie Academic and Professional: London, 1999; p 242.
47. Lee, W. J.; J. A. Lucey. Formation and physical properties of yoghurt. *Asian-Australasian J. Anim. Sci.* **2010,** *23,* 1127–1136.
48. Leporanta, K. Viili and Långfil - Exotic Fermented Products from Scandinavia. *Valio Foods Funct.* **2003,** *2,* 3–5.
49. Liu, J. R.; Wang, S. Y.; Chen, M. J.; Chen, H. L.; Yueh, P. Y.; Lin, C. W. Hypocholesterolaemic Effects of Milk-Kefir and Soymilk-Kefir in Cholesterol-Fed Hamsters. *Br. J. Nutr.* **2006,** *95*(5), 939–946.
50. Lucey, J. A.; Tamehana, M.; Singh, H.; Munro, P. A. Comparison of the Formation, Rheological Properties, and Microstructure of Acid Skim Milk Gels Made with a Bacterial Culture or Glucono-Delta-Lactone. *Food Res. Int.* **1998**, *31,* 147–155.
51. Macori, G.; Cotter, P.D. Novel Insights into the Microbiology of Fermented Dairy Foods. *Curr. Opin. Biotechnol.* **2018,** *49,* 172–178.
52. Makwana, M.; Hati, S. Fermented Beverages and Their Health Benefits. In *Fermented Beverages*; Woodhead Publishing: London, 2019; pp. 1–29.
53. Mann, B.; Athira, S.; Sharma, R.; Bajaj, R. Bioactive Peptides in Yogurt. In *Yogurt in Health and Disease Prevention*; Academic Press: New York, 2017; pp 411–426.
54. Marth, E. H.; Steele, J. E. Eds.; *Applied Dairy Microbiology*; CRC Press: Boca Raton, FL, 2001; p 736.
55. Matar, C.; Valdez, J. C.; Medina, M.; Rachid, M.; Perdigon, G. Immunomodulating Effects of Milks Fermented by *Lactobacillus Helveticus* and its Non-Proteolytic Variant. *J. Dairy Res.* **2001,** *68*(4), 601–609.
56. McFarland, L. V. From Yaks to Yogurt: The History, Development, and Current Use of Probiotics. *Clin. Infect. Dis.* **2015,** *60*(2), 85–90.

57. McKevith, B.; Shortt, C. Other relevant products. In *Fermented Milks*; McKevith, B., Shortt, C., Eds.; Yakult UK Ltd: Acton, London, 2013; pp 2383–2389.
58. Mehta, B.; Maricê, O. Fermented Dairy Products. In *Fermentation: Effects on Food Properties*; Mehta, B., Kamal-Eldin, A., Iwanski, R. Z., Eds.; CRC Press, Taylor and Francis Group: Boca Raton, FL, USA, 2012; pp 259–283.
59. Minervini, F.; Algaron, F.; Rizzello, C. G.; Fox, P. F.; Monnet, V.; Gobbetti, M. Angiotensin-I-Converting-Enzyme-Inhibitory and Antibacterial Peptides From *Lactobacillus Helveticus* PR4 Proteinase-Hydrolyzed Caseins of Milk From Six Species. *Appl. Environ. Microbiol.* **2003,** *69,* 5297–5305.
60. Moreno-Fernandez, J.; Diaz-Castro, J.; Alférez, M. J. M.; Nestares, T.; Ochoa, J. J.; Sánchez-Alcover, A.; López-Aliaga, I. Fermented Goat Milk Consumption Improves Melatonin Levels and Influences Positively the Antioxidant Status during Nutritional Ferropenic Anemia Recovery. *Food Funct.* **2016,** *7*(2), 834–842.
61. Moreno-Montoro, M.; Olalla-Herrera, M.; Rufián-Henares, J. Á.; Martínez, R. G.; Miralles, B.; Bergillos, T.; Jauregi, P. Antioxidant, ACE-Inhibitory and Antimicrobial Activity of Fermented Goat Milk: Activity and Physicochemical Property Relationship of the Peptide Components. *Food Funct.* **2017,** *8*(8), 2783–2791.
62. Mullié, C.; Yazourh, A.; Thibault, H.; Odou, M. F.; Singer, E.; Kalach, N.; Kremp, O.; Romond, M. B. Increased Poliovirus-Specific Intestinal Antibody Response Coincides With Promotion of *Bifidobacterium longum-Infantis* and *Bifidobacterium breve* in Infants: A Randomized, Double-Blind, Placebo-Controlled Trial. *Int. Pediatr. Res. Foundation* **2004,** *56*(5), 791–795.
63. Narva, M.; Rissanen, J.; Halleen, J.; Vapaatalo, H.; Väänänen, K.; Korpela, R. Effects of Bioactive Peptide, Valyl-Prolyl-Proline (VPP), and *Lactobacillus Helveticus* Fermented Milk Containing VPP on Bone Loss in Ovariectomized Rats. *Ann. Nutr. Metabol.* **2007,** *51,* 65–74.
64. Narva, M.; Halleen, J.; Väänänen, K.; Korpela, R. Effects of *Lactobacillus Helveticus* Fermented Milk on Bone Cells *in vitro*. *Life Sci.* **2004,** *75,* 1727–1734.
65. Oh, M. R.; Park, S. H.; Kim, S. Y.; Back, H. I.; Kim, M. G.; Jeon, J. Y.; Ha, K. C.; Na, W. T.; Cha, Y. S.; Park, B. H.; Park, T. S.; Chae, S. W. Postprandial Glucose-Lowering Effects of Fermented Red Ginseng in Subjects With Impaired Fasting Glucose or Type 2 Diabetes: A Randomized, Double Blind, Placebo-Controlled Clinical Trial. *BMC Complement. Altern. Med.* **2014,** *14,* 237–245.
66. Pan, D. D.; Zeng, X. Q.; and Yan, Y. T. Characterization of *Lactobacillus* Fermentum SM-7 Isolated From *Koumiss*, a Potential Probiotic Bacterium with Cholesterol-Lowering Effects. *J. Sci. Food Agric.* **2011,** *91*(3), 512–518.
67. Panesar, P. S. Fermented Dairy Products: Starter Cultures and Potential Nutritional Benefits. *Food Nutr. Sci.* **2011,** *2,* 47–51.
68. Parmar, H.; Hati, S.; Panchal, G.; Sakure, A. A. Purification and Production of Novel Angiotensin I-Converting Enzyme (ACE) Inhibitory Bioactive Peptides Derived From Fermented Goat Milk. *Int. J. Pept. Res. Ther.* **2020,** *26*(2), 997–1011.
69. Pei, R.; Martin, D. A.; Di-Marco, D. M.; Bolling, B. W. Evidence for the Effects of Yogurt on Gut Health and Obesity. *Crit. Rev. Food Sci. Nutr.* **2017,** *57*(8), 1569–1583.
70. Perdigon, G.; de LeBlanc, A. D. M.; Valdez, J.; Rachid, M. Role of Yoghurt in the Prevention of Colon Cancer. *Eur. J. Clin. Nutr.* **2002,** *56*(S3), S65–S68.
71. Prajapati, J. B. *Fundamentals of Dairy Microbiology*; Akta Prakashal: Nadiad, Gujarat, India, 1995; pp 4–45.

72. Prajapati, J. B.; Nair, B. M. The History of Fermented Foods. In *Handbook of Fermented Functional Foods*; Farnworth, E. R., Eds.; CRC Press., Taylor and Francis Group, LLC: Boca Raton, FL, 2008; chapter 1, pp 1–22.
73. Puniya, A. K.; Kumar, S.; Puniya, M.; Malik, R. Fermented Milk and Dairy Products: An Overview. In *Fermented Milk and Dairy Products*; CRC Press/Taylor and Francis: Boca Raton, FL, 2015; pp 3–24.
74. Quan, S.; Tsuda, H.; Miyamoto, T. Angiotensin I-Converting Enzyme Inhibitory Peptides in Skim Milk Fermented with *Lactobacillus Helveticus* 130B4 From Camel Milk in Inner Mongolia, China. *J. Sci. Food Agric.* **2008,** *88,* 2688–2692.
75. Rafie, N.; Hamedani, S. G.; Ghiasv, R.; Miraghajani, M. Kefir and Cancer: Systematic Review of Literature. *Arch. Iranian Med.* **2015,** *18*(12), 852–857.
76. Rattray, F. P.; O'Connell, M. J. Fermented Milks Kefir. In *Encyclopedia of Dairy Sciences*; 2nd ed.; Fukay, J.W., Ed.; Academic Press: San Diego, CA, 2011; pp 518–524.
77. Rizzoli, R.; Biver, E. Effects of Fermented Milk Products on Bone. Ca*lcif. Tissue Int.* **2018,** *102*(4), 489–500.
78. Robinson, R. K.; Tamime A. Y. Microbiology of Fermented Milks. In *Dairy Microbiology: The Microbiology of Milk Products*, 2nd ed.; Robinson, R. K., Ed.; Elsevier Applied Science Publishers: London, 1990; vol. 2, pp 291–343.
79. Rodgers, S. Novel Applications of Live Bacteria in Food Services: Probiotics and Protective Cultures. *Trends Food Sci. Technol.* **2008,** *19,* 188–197.
80. Salva, S.; Nuñez, M.; Villena, J.; Ramón, A.; Font, G.; Alvarez, S. Development of A Fermented Goats' Milk Containing *Lactobacillus Rhamnosus*: *In Vivo* Study of Health Benefits. *J. Sci. Food Agric.* **2011,** *91*(13), 2355–2362.
81. Sarkar, S. Innovations in Indian Fermented Milk Products: A Review. *Food Biotechnol.* **2008,** *22,* 78–97.
82. Sarkar, S.; Mariappan, S. Usage of Traditional Fermented Products by Indian Rural Folks and IPR. *Indian J. Tradit. Knowl.* **2007,** *6(*1), 111–120.
83. Senan, S.; Prajaparti J. B. *Acidophilus* Milks. In *Fermented Milk and Dairy Products*; CRC Press/Taylor and Francis: Boca Raton – FL, 2015; pp 227–259.
84. Seth, D.; Mishra, H. N.; Deka, S. C. Effect of Hydrocolloids on the Physicochemical and Rheological Properties of Reconstituted Sweetened Yoghurt Powder. *J. Sci. Food Agric.* **2018,** *98*(5), 1696–1702.
85. Sharifi, M.; Moridnia, A.; Mortazavi, D.; Salehi, M.; Bagheri, M.; Sheikhi, A. Kefir: A Powerful Probiotics With Anticancer Properties. *Med. Oncol.* **2017,** *34*(11), 1–7.
86. Sheikh, I. C.; Fang, T. J.; Wu, T. K.; Chen, R. Y. Effects of Fermentation on Antioxidant Properties and Phytochemical Composition of Soy Germ. *J. Sci. Food Agric.* **2014,** *94,* 3163–3170.
87. Shiby, V. K.; Mishra, H. N. Fermented Milks and Milk Products as Functional Foods: A Review. *Crit. Rev. Food Sci. Nutr.* **2013,** *53*(5), 482–496.
88. Smit, G.; Smit, B. A.; Engels, W. J. Flavor Formation by Lactic Acid Bacteria and Biochemical Flavor Profiling of Cheese Products. *FEMS Microbiol. Rev.* **2005,** *29,* 591–610.
89. Sodini, I.; Lucas, A.; Tissier, J. P.; Corrieu, G. Physical Properties and Microstructure of Yoghurts Supplemented with Milk Protein Hydrolysates. *Int. Dairy J.* **2005,** *15,* 29–35.
90. Solanki, D.; Hati, S.; Sakure, A. *In Silico* and *in Vitro* Analysis of Novel Angiotensin I-Converting Enzyme (ACE) Inhibitory Bioactive Peptides Derived From Fermented Camel Milk *(Camelus dromedarius). Int. J. Pept. Res. Ther.* **2017,** *23*(4), 441–459.

91. Surono, I. S.; Hosono, A. Types and Standards of Identity. In *Fermented Milks Encyclopedia of Dairy Sciences*; Academic Press: New York, 2011; chapter 7, pp 470–476,
92. Tamang, J. P.; Watanabe, K.; Holzapfel, W. H. Review: Diversity of Microorganisms in Global Fermented Foods and Beverages. *Front. Microbiol.* **2016,** *7*, E-article 00377.
93. Tamang, J. P.; Thapa, S.; Tamang, N.; Rai, B. Indigenous Fermented Food Beverages of Darjeeling Hills, and Sikkim: A Process and Product Characterization. *J. Hill Res.* **1996,** *9*(2), 401–411.
94. Tamime, A. Y. Yoghurt-based Products. In *Encyclopedia of Food Science, Food Technology and Nutrition*; Macrae, R., Robinson, R. K., Saddler, M. J., Eds.; Academic Press: London, 1993; vol 7, pp 4972–4977.
95. Tamime, A. Y. Microbiology of Starter Cultures. In *Diary Microbiology Handbook*; 3rd ed.; Robinson, R. K., Ed.; John Wiley & Sons Ltd.: Danvers, MA, 2002; pp. 261–347.
96. Tamime, A. Y. Production of Kefir, Koumiss and Other Related Products. In *Fermented Milk*; Tamime, A. Y., Ed.; Blackwell Science Ltd.: Oxford, 2006; pp 174–2
97. Tamime, A. Y.; Muir, D. D.; Wszolek, M. Kefir, Koumiss and Kishk. *Int. Dairy J.* **1999,** *64,* 32–33.
98. Tamime, A. Y.; Robinson, R. K. *Yoghurt Science and Technology*, 3rd ed.; Woodhead Publishing Ltd.: Cambridge, 2007; p 808.
99. Tellez, A.; Corredig, M.; Brovko, L. Y.; Griffiths, M. W. Characterization of Immune-Active Peptides Obtained from Milk Fermented by *Lactobacillus Helveticus*. *J. Dairy Res.* **2010,** *77*(2), 129–136.
100. Thibault, H.; Aubert-Jacquin, C.; Goulet, O. Effects of Long-term Consumption of A Fermented Infant Formula (With *Bifidobacterium breve C50* and *Streptococcus Thermophilus 065*) on Acute Diarrhea in Healthy Infants. *J. Pediatr. Gastroenterol. Nutr.* **2004,** *39*(2), 147–152.
101. Toba, T.; Yoshioka, E.; Itoh, T. Acidophilucin A, a New Heat-Labile Bacteriocin Produced by *Lactobacillus Acidophilus LAPT 1060*. *Lett. Appl. Microbiol.* **1991,** *12*(4), 106–108.
102. Tojo-Sierra, T.; Leis-Trabazo, R.; Barros-Velázquez, J. Probiotics in Infant Nutrition: Fermented Milk Products. *Ann. Pediatr. Monograph* **2006,** *4(*1), 54–66.
103. Tunick, M. H.; Van-Hekken, D. L. Dairy Products and Health: Recent Insights. *J. Agric. Food Chem.* **2015,** *63*(43), 9381–9388.
104. Turpeinen, A. M.; Kumpu, M.; Rönnback, M.; Seppo, L.; Kautiainen, H.; Jauhiainen, T.; Korpela, R. Antihypertensive and Cholesterol-Lowering Effects of a Spread Containing Bioactive Peptides IPP and VPP and Plant Sterols. *J. Funct. Foods* **2009,** *1*(3), 260–265.
105. Uniacke-Lowe, T. Studies on Equine Milk and Comparative studies. PhD Thesis; University College Cork, Cork, 2011; p 218.
106. Weerathilake, W. A. D. V.; Rasika, D. M. D.; Ruwanmali, J. K. U.; Munasinghe, M. A. D. D. The E Volution, Processing, Varieties and Health Benefits of Yogurt. *Int. J. Sci. Res. Publ.* **2014,** *4*(4), 1–10.
107. Yegna-Narayan, A. K. Dairying in ancient India. *Indian Dairyman* **1953,** *5,* 77–83.
108. Zaiss, M. M.; Jones, R. M.; Schett, G., Pacifici, R. The Gut-Bone Axis: How Bacterial Metabolites Bridge the Distance. *J. Clin. Invest.* **2019,** *129*(8), 3018–3028.
109. Zemel, M. B. Dairy Augmentation of Total and Central Fat Loss in Obese Subjects. *Int. J. Obes.* **2005,** *29,* 391–397.

110. Zhang, J.; Zhao, X.; Jiang, Y.; Zhao, W.; Guo, T.; Cao, Y.; Yang, Z. Antioxidant Status and Gut Microbiota Change in an Aging Mouse Model as Influenced by Exopolysaccharide Produced by *Lactobacillus Plantarum* YW11 Isolated from Tibetan Kefir. *J. Dairy Sci.* **2017,** *100,* 6025–6041.
111. Zhong, Z.; Hou, Q.; Kwok, L.; Yu, Z.; Zheng, Y.; Sun, Z.; Menghe, B.; Zhang, H. Bacterial Microbiota Compositions of Naturally Fermented Milk are Shaped by Both Geographic Origin and Sample Type. *J. Dairy Sci.* **2016,** *99*(10), 7832–7841.

CHAPTER 11

THERAPEUTIC CHARACTERISTICS OF MILK-DERIVED BIOACTIVE PEPTIDES: AN OVERVIEW

M. A. SYAMA, V. K. AMMU, and S. ATHIRA

ABSTRACT

The significance of protein in daily diet is gaining huge attention in the light of scientific evidences, which prove its vital functionalities in maintaining human health. Dietary proteins can liberate a wide range of multifunctional biologically active peptides upon gastrointestinal digestion or by microbial fermentation or hydrolysis by proteolytic enzymes. Milk proteins are chief sources of bioactive peptides production. These bioactive peptides from milk have the potential to regulate nervous, immune, cardiovascular, and gastrointestinal systems. Hence, production of distinct dietary peptides to improve the health condition or to prevent risk of diseases has been a matter of concern for industrial and scientific experts. This chapter documents the production, regulatory functionalities, bioavailability, and commercial use of milk-derived bioactive peptides.

11.1 INTRODUCTION

Milk is a wide repository of bioactive components, which have potential role in immunological, gastrointestinal, neurological, and nutritional responses. These bioactive components mainly fall in the category of milk proteins and peptides secreted by mammary glands.[20] Within the parent protein sequence, bioactive peptides that are inactive will be liberated by proteolytic action

The Chemistry of Milk and Milk Products: Physicochemical Properties, Therapeutic Characteristics, and Processing Methods. Megh R. Goyal, Suvartan Ranvir, & Junaid Ahmad Malik (Eds.)

during processing, fermentation, and digestion of milk in gut. Upon liberation, bioactive peptides act as hormone-like regulatory compounds.[46]

The dietary protein caters the body with adequate amount of requisite amino acids and organic nitrogen. A healthy adult has to consume 0.6 g of dietary protein and/or amino acids per kg of body weight, so that an adequate protein status can be maintained.[73] Several researchers have studied the immunomodulatory, antihypertensive, mineral utilizing, or opiate properties of numerous peptides.[62,64] Milk protein are considered as significant source of bioactive peptides. The major fraction of bovine milk protein comprises of casein, α-la, β-lg, immunoglobulins, lactoferrin, serum albumin, and proteose-peptides. Bioactive peptides are obtained either by gastrointestinal digestion (in vivo) from different proteins or by enzymatic hydrolysis (in vitro). Peptides obtained as a result of enzymatic hydrolysis are purified by different separation techniques and further analyzed for bioactivities.[5]

The potential health benefits of bioactive peptide mostly depend on sequence of inherent amino acid and its composition. The amino acid residue in active sequences can vary between 2 and 20; and several peptides exhibit multifunctional properties.[39] Opioid peptide was the first bioactive peptide found in milk followed by immunomodulatory peptides.[27] Numerous bioactive peptides from milk exhibit multifunctional role, for example, α- and β-lactorphin have both opioid and Angiotensin Converting Enzyme (ACE) inhibitory activity; immunomodulatory and ACE-inhibitory activities shown by α_{s1}-CN fraction 194–199; β-CN fraction 60–70 showing opioid, immunomodulatory, and ACE-inhibitory activity.[29]

This chapter explores the bioactive peptides in milk, their production, different functionalities, bioavailability, and prospective applications.

11.2 FORMATION OF MILK-DERIVED BIOACTIVE PEPTIDES

Dietary protein plays a key role in energy supply and essential amino acids are important for growth of living organisms. Compared with other protein sources, bovine milk protein is defined by high digestibility, higher biological value, and balanced amino acid profile. Besides wide array of nutritive properties, milk proteins, and peptides exhibit numerous biological properties.[32] The amino acid residues of active sequences show different functionalities, such as, opioid, opioid antagonist, antihypertensive, immunomodulatory, antithrombotic, antifungal, antibacterial, metal binding, and transporting.[8,60]

In the parent protein's amino acid sequence, inert bioactive peptides may be released by several means. Figure 11.1 depicts the bioactive peptide formation from bovine milk proteins. Gastrointestinal milk digestion, milk fermentation either by hydrolysis by protease enzymes or proteolytic starter cultures may enable the liberation of bioactive peptides from milk proteins.[30]

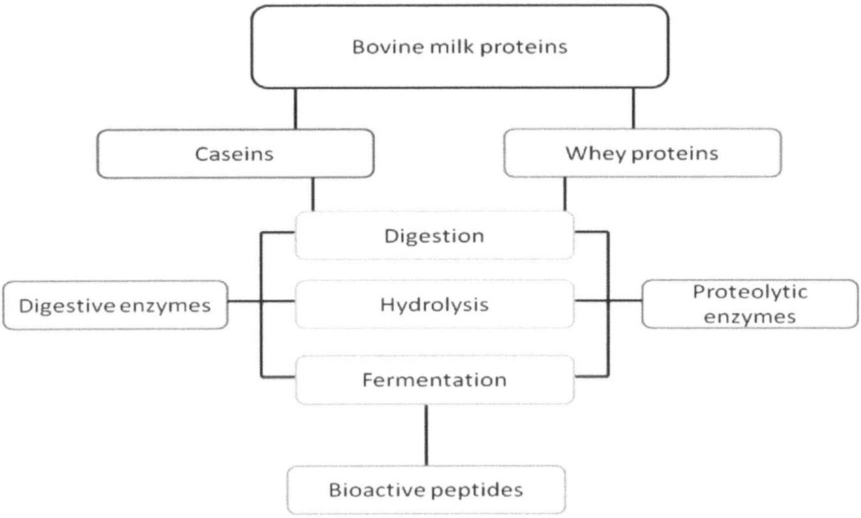

FIGURE 11.1 Schematic portrayal of bioactive peptide formation from milk proteins.

It is possible to generate biologically active peptides by hydrolytic action of digestive enzymes on milk proteins. The most eminent enzymes are digestive enzyme (such as, pepsin) and pancreatic enzymes (such as, chymotrypsin, trypsin), and carboxypeptidases. Bioactive peptides from other sources were also produced by enzymes derived from fungi and bacteria. The extent to which bioactive peptides show efficient physiological activity is directly related to their ability to keep up the integral state while these transport to target of action in body.[68,52]

Enzymatic hydrolysis of casein and whey protein release opioid peptides, immunomodulatory peptides, calcium-binding phosphopeptides, antihypertensive, and antibacterial peptides.[12,39,71] Various bioactive peptides were released by simulating the gastrointestinal digestion by proteolytic enzymes, such as, thermolysin, alcalase, subtilisin, and consecutive treatment with trypsin and pepsin.[17,42,58,66] Although whey peptide (Ala-Leu-Pro-Met-His-Ile-Arg) from tryptic digest of β-lg showed intensive antihypertensive activity[9,34,44]; yet another studies[48] showed that peptides from casein hydrolysate had higher ACE-inhibitory activity than the peptides from whey protein hydrolysate. A study conducted to evaluate the impact of nine different proteolytic enzymes, that are commercially available with casein hydrolysate on ACE-inhibitory activity, revealed highest activity per peptide of protease isolated from *Aspergillus oryzae.*[42]

Several studies have been carried out to monitor bioactive peptides liberated due to hydrolysis of milk proteins by the action microbial produce proteinase.[12] Most widely used microbial proteases are those obtained from lactic acid bacteria (LAB), *Bifidobacterium*, and *Bacillus* spp.[29]

The advantages of microbial proteases over those from other sources are wide diversity, less cultivation cost due to short maturation time and less nutritional requirement, and most of the microbial proteases are expressed on cell membrane; thus, production of microbial proteases are economical and requires minimal manpower. The proteases produced by the action of LAB differ in specificity and enzymatic activity. Hence, a bioactive peptide with desirable activity can be produced by selecting appropriate strain of LAB based on their proteolytic potential.[19] The proteolytic system of LAB contains transport proteins, cell envelope proteinases, and intracellular peptidases.[25] Table 11.1 depicts the milk-derived bioactive peptides liberated by microbial proteolysis. The strains of *Lactobacillus helveticus* can release antihypertensive peptides, such as, ACE-inhibitory tripeptides (i.e., Ile-Pro-Pro and Val-Pro-Pro).[1,22]

TABLE 11.1 Milk-Derived Bioactive Peptides Released by Microbial Proteolysis.

Protein precursor	Microorganism	Bioactivity	Reference
Skim milk hydrolysate	*Lb. helveticus* ICM 1004	ACE-inhibitory	49
Whey proteins	*Lb. helveticus* CPN 4	ACE- inhibitory	70
β-CN	*Lb. delbrueckii* subsp. *Bulgaricus*	ACE- inhibitory	2
β-CN	*Lb. rhamnosus*	Antioxidative, ACE-inhibitory	21
β-CN, κ-CN	*Lb. helveticus, Saccharomyces cerevisiae*	ACE- inhibitory, Antihypertensive	47
κ-CN	*Lb. delbrueckii* subsp. *bulgaricus* IFO13953	Antioxidative	[31]

Donkor et al.[7] evaluated proteolytic activity of probiotic microorganisms, such as, *Bifidobacterium lactis*, *Lactobacillus acidophilus*, *Lactobacillus casei*, and several LAB cultures. All these cultures produced peptides with

ACE-inhibitory activity while growing with *Bifidobacterium longum* strain and strongest activity was shown by *L. acidophilus*. The isolated strains of *Enterococcus faecalis* from raw milk when used for milk fermentation resulted in liberation of several ACE-inhibitory peptides.[57] Fermentation of caseins and cheese whey with commercial starter cultures used for production of sour milk, ropy milk, and yoghurt can liberate potential ACE-inhibitory peptides.[53]

11.3 PHYSIOLOGICAL FUNCTIONALITY OF BIOACTIVE PEPTIDES EXTRACTED FROM MILK

Certain protein fragments have beneficial effects on our body due to bioactive peptides.[26] Depending upon their sequence of amino acid, the site of action in body system differs for various bioactive peptides, that is, nervous, cardiovascular, immune, and digestive system.

11.3.1 IMPACT ON REGULATION OF NERVOUS SYSTEM

Enzymatic hydrolysis of casein fractions has been found to release peptides with opioid property. Opioid peptides have stereo-specifically reversible opiate-like effects by naloxone and attraction for opiate receptors. The peptide that binds to opioid receptor and thereby changes its proportion in an active form is called opioid agonist; and peptide, which reduces the action of agonist, is called opioid antagonist. Naloxone is a pure opioid antagonist. Unlike other opioid receptor antagonists, it has no concomitant agonist properties. In 1975, peptides having affinity for opiate receptors were isolated from brain and named enkephalins. Majority of opioid peptides originate from the precursor proteins-prodynorphin, proopiomelanocortin, and proenkephalin.[56]

The N-terminal of several peptides has same sequence, that is, Tyr-Gly-Gly-Phe. Biochemical, pharmacological, and behavioral studies have revealed that three opioid receptors exist (μ, δ, and κ) in mammalian peripheral and central nervous system. Substances with opioid-like effect can bind to μ, δ, and κ receptors with preferential affinities.

The receptors of opioid peptide are situated in endocrine, gastrointestinal tract, and nervous system of mammals. The endogenous ligands of opioid peptides can have an effect on exogenous opioids and opioid antagonists. Once the opioid peptides have been orally administrated, these may exert an influence on absorption process in gut mainly by reducing the transit time by affecting smooth muscles and affecting electrolytes intestinal transport.

Oral administration of milk or casein showed the presence of β-casomorphins or their precursors in human small intestine, in duodenal chyme of minipigs, and in plasma of new borne calves.[37–39] Opioid peptide derived from α_{s1}- (fraction 91–100) exhibited stress relieving property in human trials and in animal models; and this peptide has application as an ingredient in soft drinks and confectionery. Table 11.2 indicates milk-derived opioid peptides.

TABLE 11.2 Milk Derived Opioid Peptides.

Name of opioid peptide	Release protease	Physiological function	Reference
Casoxin D	Pepsin-chymotrypsin	Opioid antagonist	72
Lactoferroxin A	Pepsin	Opioid antagonist	69
Serorphin	Pepsin	Opioid agonist	65
β-Lactorphin	Synthetic or trypsin	Opioid agonist	43

11.3.2 IMPACT ON IMMUNE SYSTEM REGULATION

Milk-derived immunomodulatory peptides influence both cell proliferation responses and immune system. Peptides liberated from casein, α-la, β-lg, and milk protein hydrolysate exert an improved immune cell function, as they influence in the antibody synthesis, cytokine regulation, and proliferation of lymphocyte.[18] Milk-derived immunomodulatory peptides may reduce allergic responses in atopic individuals and intensify immunity in gastrointestinal tract.[28] This is how immunomodulatory peptides control immune system development in newborns. Whey protein isolates hydrolysis by enzymes can liberate immunomodulatory peptides and which can find application in infant formula with desired immunomodulatory properties.[40]

Examples of milk-derived immunomodulatory peptides are given in Table 11.3. Microbial infection by *Klebsiella pneumoniae* was resisted by immunomodulatory peptide derived from casein.[41] The anti-cancerous properties shown by different fermented milks may be attributed by immuno-peptides released during milk fermentation. These cytomodulatory peptides liberated from fractions of casein may restrain cancerous cell outgrowth or stimulate activity of neonatal intestinal cells and immunocompetent cells.[39] Phagocytic function of human macrophages may be enhanced by the immuno peptides liberated from casein hydrolysates.[10]

TABLE 11.3 Milk-Derived Immunomodulatory Peptides.

Name of peptide	Precursor protein	Release protease	Reference
Immunopeptide	β-CN (fraction 191–193)	Trypsin or chymosin	41
Immunopeptide	β-CN (fraction 63–68)	Trypsin or chymosin	41
α_{s1}-casokinin-6	α_{s1}-CN	Trypsin	24
β-casomorphin-5	β-CN	Trypsin	6
β-casomorphin-7	β-CN	Trypsin	35

10.3.3 IMPACT ON REGULATION OF GASTROINTESTINAL SYSTEM

Dietary proteins may exert regulatory effect on digestive enzymes and absorption of nutrients in intestinal tract, before it gets hydrolyzed to amino acids.[61] Among the bioactive peptides from milk protein, casein-derived peptides have been identified with potent mineral binding functions. Bioactive peptides from casein bind minerals through specific or nonspecific binding sites. Several studies were carried out on caseino phospopeptides, which are considered to be the most important mineral-binding peptides. At intestinal pH, the sequence Ser(P)-Ser(P)-Ser(P)-Glu-Glu possessed by most of the phosphopeptides in whole bovine casein keeps the minerals in solution. Enzymatic digestion of milk protein releases several phosphopeptides, which possess high anionic property. Thus, it withstands subsequent proteolysis and leads to development of soluble complexes with calcium and restricts production of calcium phosphate, which is insoluble.[3]

The extent to which phosphopeptide interact with colloidal calcium phosphate depends on their phosphoserine residue content.[16] Difference in calcium binding activities of phosphopeptide fractions may be because of the variation in composition of amino acid surrounding the phosphorylated region. The absorption of bound minerals may be more throughout the intestinal epithelium and intestinal microbial growth can be decreased due to entrapment of iron by peptides thereby reducing intraluminal availability.[67]

Antimicrobial peptides are cationic, positively charged, hydrophobic peptides liberated from milk proteins, which restrict the development of most of the spoilage and few pathogenic microorganisms.[20] Most studied antimicrobial peptide is lactoferricin, which is obtained from human and bovine lactoferrin. Peptides with antibacterial property have been isolated from α_{s1}-CN and α_{s2}-CN and they exhibit antibacterial property against Gram-negative and Gram-positive bacteria. Caseino macropeptides reduce stomach contractions and prohibit gastric secretions. The satiety hormone, cholecystokinin release is stimulated by caseino macropeptides. Even though the physiological function of milk-derived antimicrobial peptide is not yet clear, yet when formed in vivo during digestion of milk might have regulatory effect on intestinal microflora.[61]

11.3.4 IMPACT ON REGULATION OF CARDIOVASCULAR SYSTEM

Peptides with antithrombotic, antihypertensive, and antioxidative properties have great influence on regulation of cardiovascular system. Mechanisms behind milk clotting (interaction of chymosin with κ-CN) were comparable to blood clotting (interaction of thrombin with fibrinogen). Any changes in the equilibrium created by antithrombotic components result in accumulation of platelets on a surface, thus formation of thrombosis occurs. Further, the accumulated platelets have the ability to synthesize prostaglandin H2, which then transform to a potent platelet aggregation stimulator called thromboxan A2. The platelets will bind with other proteins and fibrinogen and result in clotting.[14] A fragment of κ-casein, casopiastrin released by trypsin hydrolysate show antithrombotic activity by restricting fibrinogen binding.[23]

Several studies have reported that bioactive peptides obtained from lactotransferrin and casein restricted the platelet function.[10,36] Antithrombotic peptides were also obtained from κ-caseinoglycopeptides. The κ-caseinoglycopeptides of bovine as well as human origin detected in newborns after administration of cow milk-based formula and breastfeeding showed antithrombotic property.[10] A reduction in collagen- and thrombin-induced aggregation of platelets was caused by peptides obtained from sheep κ-CN (fraction 106–171), or κ-caseinoglycopeptide.[54] Lactoferrin obtained from human and sheep pepsin digest may restrict thrombin-induced platelet aggregation.[55]

Antihypertensive peptides can restrict the action of ACE or they can be called as ACE inhibitors. ACE (chemically known as peptidyl dipeptide hydrolase) is a multifunctional enzyme, which basically cleaves dipeptides from a substrate's carboxy-terminal end. Bradykinine and renin-angiotensin systems control the peripheral blood pressure. The activities of ACE include conversion of angiotensin I to II, thereby increasing aldersterone and blood pressure levels as well as inactivates the depressing property of bradykinin. Enzymatic digestion of different milk proteins liberates many ACE-inhibitory peptides and peptides with antihypertensive effect. The mechanism behind antihypertensive functionality of these peptides is achieved through improving production of nitric oxide from endothelium,[63] liberation endothekin-I by endothelial cells,[34] and enhancement of bradykinin activity.[51]

In order to exhibit antihypertensive effect, the peptide has to take up from intestine in an intact form; and also it should be resistant to the plasma peptidase degradation. This fact has been demonstrated with the help of human intestinal Caco-2 cell studies.[59] During production of sour milk with *L. helveticus* CP790, antihypertensive peptides have been formed from α_{s1}-CN and β-CN.[33] During production of Japanese sour fermented milk known Calpis with *L. helveticus* and *Saccharomyces cerevisiae* as starter cultures released two potent ACE-inhibitory peptides, that is, Ile-Pro-Pro and Val-Pro-Pro.[47]

Blood pressure reduction property has been shown by tripeptides formed during preparation of fermented milk, Evolus. The hypocholesterolemia effect possessing peptide has been identified (Ile-Ile-Ala-Glu-Lys) from β-lg. Even though in vitro Caco-2 cell study revealed cholesterol absorption by this peptide, yet the exact underlying mechanism is not clear.[45]

11.4 BIOAVAILABILITY

Potential of bioactive peptides to generate intended physiological outcome in target organism when administrated orally can be termed as bioavailability of bioactive peptides. This can be achieved only when the bioactive peptides reach the target site intact by withstanding gastrointestinal digestion and absorption.[68] The major factors determining the bioavailability of peptides include: molecular weight, hydrophobicity, protease resistance, charge, hydrogen bonding, and specific residues present.[50]

Peptides, which possess proline and hydroxyl proline, are relatively resistant to digestive enzymes.[4] The intestinal transit and gastric emptying significantly influence the time elapsed by a peptide in gastrointestinal tract and its absorption. Another factor, which determines the transport of peptide, is physiological pH. Because of the higher affinity of bioactive peptides for tissues, certain bioactive peptides of milk origin possess ACE-inhibitory activity as compared with relative synthetic ACE-inhibitors, but possess high in vivo activity.[15] In general, a clear correlation between in vivo and in vitro bioactive peptide activities is difficult to establish. Upon digestion the amount of peptide liberated and its beneficial effect on human well-being is difficult to forecast. Theoretically, yield of milk-derived opioid peptides (such as, β-casomorphin from β-CN) and α-lactorphin from α-la was estimated as 2% and 6%, respectively.[38]

11.5 COMMERCIAL APPLICATIONS

Bioactive peptides are promising ingredients in wide range of health-promoting functional foods. The eminent source for those ingredients is milk proteins. The large-scale manufacturing of such ingredients may be achieved by membrane separation technique, which enhances the peptides with required molecular weight range.[27] Casein or whey protein hydrolysate-based bioactive peptide-containing ingredients were industrially produced by ultrafiltration and nanofiltration technologies. Such ingredients find application in pastilles, fruit, and dairy-based drinks, chewing gum, and confectionery. Products containing antihypertensive, anticarcinogenic, stress relieving, mineral binding, and satiety-inducing peptide[29] are available in commercial products (Table 11.4).

TABLE 11.4 Milk-Derived Bioactive Peptide Containing Commercial Products.

Product	Brand name	Bioactive peptide present	Health claim
Calcium enriched fermented milk drink	Evolus	β-CN and κ-CN fragments	Blood pressure reducing
Chewing gum	Recaldent	Calcium casein peptone calcium phosphate	Anticariogenic
Flavored milk	PRODIET	α_{S1}-CN fragments	Stress relieving
Hydrolyzed whey protein isolate	BioZate	β-Lg fragments	Blood pressure reducing
Sour milk	Calpis	β-CN and κ-CN fragments	Blood pressure reducing
Whey protein isolate	BioPURE-GMP	κ-CN fragments	Influence blood clotting, antibacterial

11.6 SUMMARY

Novelties in research and scientific approaches have led a path for new dimension to explore the potential functionalities of various bioactive peptides obtained from milk. The array of health benefits rendered by these peptides offers its application in treating lifestyle disease conditions and as a promising functional ingredient in different health-promoting foods and pharmaceutical industry. Preparation of milk-derived bioactive peptides with desired function can be augment to industrial level by controlled fermentation. By considering the role of bioactive peptides derived milk regulating nervous, immune, gastrointestinal, and cardiovascular system, application of these peptides as ingredient in different food system need to be thoroughly studied.

KEYWORDS

- **bioactive peptides**
- **bioavailability**
- **fermented milk**
- **milk proteins**
- **regulatory functionality**

REFERENCES

1. Aihara, K.; Kajimoto, O.; Hirata, H.; Takahashi, R.; Nakamura, Y. Effect of Powdered Fermented Milk With *Lactobacillus Helveticus* on Subjects With High-Normal Blood Pressure or Mild Hypertension. *J. Am. Coll. Nutri.* **2005,** *4*(4), 257–265.
2. Ashar, M. N.; Chand, R. Fermented Milk Containing ACE-Inhibitory Peptides Reduces Blood Pressure in Middle Aged Hypertensive Subjects. *Milchwissenschaft* (Dairy Sci.) **2004,** *59,* 363–366.
3. Berrocal, R.; Chanton, S.; Juilleart, M. A.; Pavillard, B.; Scherz, J. C.; Jost, R. Tryptic Phosphopeptides From Whole Casein, Part II: Physicochemical Properties Related to the Solubilization of Calcium. *J. Dairy Res.* **1989,** *56*, 335–341.
4. Cardillo, G.; Gentilucci, L.; Tolomelli, A.; Calienni, M.; Qasem, A. R.; Spampinato, S. Stability Against Enzymatic Hydrolysis of Endomorphin -1 Analogues Containing Beta-Proline. *Org. Biomol. Chem.* **2003,** *1*, 1498–1502.
5. Clare, D. A.; Swaisgood, H. E. Bioactive Milk Peptides: Prospectus. *J. Dairy Sci.* **2000,** *83*, 1187–1195.
6. Cieslinska, A.; Kostyra, E.; Kostyra, H.; Olenski, K.; Fiedorowicz, E.; Kaminski, S. Milk From Cows of Different B-Casein Genotypes as a Source of B-Casomorphin-7. *Int. J. Food Sci. Nutr.* **2012,** *63*, 426–430.
7. Donkor, O.; Henriksson, A.; Vasiljevic, T.; Shah, N. P. Proteolytic Activity of Dairy Lactic Acid Bacteria and Probiotics as Determinant of Growth and *In-Vitro* Angiotensin-Converting Enzyme Inhibitory Activity in Fermented Milk. *Lait* **2007,** *86*, 21–38.
8. Ebringer, L.; Ferencik, M.; Krajcovic, J. Beneficial Health Effects of Milk and Fermented Dairy Products. *Folia Microbiol.* **2008,** *53*(5), 378–394.
9. Ferreira, I. M. P. L. V. O.; Pinho, O.; Mota, M. V.; Tavares, P.; Pereira, A.; Goncalves, M. P.; Torres, D.; Rocha, C.; Teixeira, J. A. Preparation of Ingredients Containing an ACE-Inhibitory Peptide by Tryptic Hydrolysis of Whey Protein Concentrates. *Int. Dairy J.* **2007,** *17*, 481–487.

10. Fiat, A. M.; Migliore-Samour, D.; Jolles, P.; Drouet, L.; Sollier, C. B.; Caen, J. Biologically Active Peptides From Milk With Emphasis on Two Examples Concerning Antithrombotic and Immunomodulating Activities. *J. Dairy Sci.* **1993**, *76*, 301–310.
11. FitzGerald, R. J.; Meisel, H. Milk Protein Hydrolysates and Bioactive Peptides. In *Advanced Dairy Chemistry*; Fox, P. F., McSweeney, P. L. H., Eds.; Kluwer Academic/Plenum Publishers: New York, USA, 2003; vol. 1; pp 675–698.
12. FitzGerald, R. J.; Murray, B. A.; Walsh, D. J. Hypotensive peptides From Milk Proteins. *J. Nutr.* **2004,** *13*(4), 980S–988S.
13. FitzGerald, R. J.; Murray, B. A. Bioactive Peptides and Lactic Fermentations. *Int. J. Dairy Technol.* **2006,** *59*, 118–125.
14. Fosset, S.; Tome, D. Dietary Protein-Derived Peptides With Antithrombotic Activity. Bull. *Int. Dairy Fed.* **1998,** *353*, 11–15.
15. Fujita, H.; Yoshikawa, M. LKPNM: Prodrug-Type ACE-Inhibitory Peptide Derived from Fish Protein. *Immunopharmacology* **1999,** *44*, 123–127.
16. Gagnaire, V.; Molle, D.; Herrouin, M.; Leonil, J. Peptides Identified During Emmental Cheese Ripening; Origin and Proteolytic Systems Involved. *J. Agric. Food Chem.* **2001,** *49*, 4402–4413.
17. Gauthier, S. F.; Pouliot, Y.; Saint-Sauveur, D. Immunomodulatory Peptides Obtained by The Enzymatic Hydrolysis of Whey Proteins. *Int. Dairy J.* **2006,** *16*, 1315–1323.
18. Gill, H. S.; Doull, F.; Rutherford, K. J.; Cross, M. L. Immunoregulatory Peptides in Bovine Milk. *Br. J. Nutr.* **2000,** *84*(1), S111–S117.
19. Gobbetti, M.; Smacchi, E.; Corsetti, A. The Proteolytic System of *Lactobacillus* CB1: Purification and Characterization of a Proteinase, A Dipeptidase, and an Aminopeptidase. *Appl. Environ. Microbiol.* **1996,** *62*(9), 3220–3226.
20. Haque, E.; Chand, R. Antihypertensive and Antimicrobial Bioactive Peptides From Milk Proteins. *Eur. Food Res. Technol.* **2008,** *227*, 7–15.
21. Hernandez-Ledesma, B.; Amigo, L.; Ramos, M.; Recio, I. Angiotensin Converting Enzyme Inhibitory Activity in Commercial Fermented Products. Formation of Peptides Under Simulated Gastrointestinal Digestion. *J. Agric. Food Chem.* **2004,** *52*, 1504–1510.
22. Hirota, T.; Ohki, K.; Kawagishi, R.; Kajimoto, Y.; Mizuno, S.; Nakamura, Y.; Kitakaze, M. Casein Hydrolysate Containing the Antihypertensive Tripeptides Val-Pro-Pro and Ile-Pro-Pro Improves Vascular Endothelial Function Independent of Blood Pressure-Lowering Effects: Contribution of the Inhibitory Action of Angiotensin-I-Converting Enzyme. *Hypertens. Res.* **2007,** *30*, 489–496.
23. Jolles, P.; Levy-Toledano, S.; Fiat, A. M.; Soria, C.; Gillesen, D.; Thomaidis, A.; Dunn, F. W.; Caen. J. Analogy Between Fibrinogen and Casein: Effect of an Undecapeptide Isolated From *K*-Casein on Platelet Function. *Eur. J. Biochem.* **1986,** *158*, 379–384.
24. Karaki, H.; Doi, K.; Sugano, S.; Uchiwa, H.; Sugai, R.; Murakami, U.; Takemoto, S. Antihypertensive Effect of Tryptic Hydrolysate of Milk Casein in Spontaneously Hypertensive Rats. *Comp. Biochem. Physiol.* **1990,** *96*, 367–371.
25. Khalid, N. M.; Marth, E. H. *Lactobacilli*-Their Enzymes and Role in Ripening and Spoilage of Cheese: A Review. *J. Dairy Sci.* **1990,** *73*(10), 2669–2684.
26. Kitts, D. D.; Weiler, K. Bioactive Proteins and Peptides From Food Sources: Applications of Bioprocesses Used in Isolation and Recovery. *Curr. Pharm. Des.* **2003,** *9*, 1309–1323.
27. Korhonen, H. Technology Options for New Nutritional Concepts. *Int. J. Dairy Technol.* **2002,** *55*, 79–88.

28. Korhonen, H.; Pihlanto, A. Food-Derived Bioactive Peptides-Opportunities for Designing Future Foods. *Curr. Pharm. Des.* **2003,** *9*, 1297–1308.
29. Korhonen, H.; Pihlanto, A. Bioactive Peptides: Production and Functionality. *Int. Dairy J.* **2006,** *16*(9), 945–960.
30. Korhonen, H. Milk Derived Bioactive Peptides: From Science to Application. *J. Funct. Foods* **2009,** *1*, 177–187.
31. Kudoh, Y.; Matsuda, S.; Igoshi, K.; Oki, T. Antioxidative Peptide From Milk Fermented With *Lactobacillus delbrueckii* subsp. *bulgaricus* IFO13953. *J. Jpn Soc. Food Sci.* **2001,** *48*, 44–50.
32. Litwinczuk, Z.; Krol, J.; Brodziak, A.; Barlowska, J. Changes of Protein Content and Its Fractions in Bovine Milk From Different Cow Breeds Subject to Somatic Cell Count. *J. Dairy Sci.* **2011,** *94*(12), 684–691.
33. Maeno, M.; Yamamoto, N.; Takano, T. Identification of an Antihypertensive Peptide From Casein Hydrolysate Produced by A Proteinase From *Lactobacillus heleveticus* CP790. *J. Dairy Sci.* **1996,** *79*, 1316–1321.
34. Maes, W.; Camp, V.; Vermeirssen, J.; Hemeryck, V.; Ketelslegers, M.; Schrezenmeier, J. M.; Oostveldt, P.; Huyghebaert, A. Influence of the Lactokinin Ala-Leu- Pro-Met-His-Ile-Arg (ALPMHIR) on the Release of Endothelin-1 by Endothelial Cells. *Regul. Pept.* **2004,** *118*, 105–109.
35. Maruyama, S.; Nakagomi, K.; Tomizuka, N.; Suzuki, H. Angiotensin I-converting Enzyme Inhibitor Derived From an Enzymatic Hydrolysate of Casein. II. Isolation and Bradykinin-Potentiating Activity on the Uterus and The Ileum of Rats. *Agric. Biol. Chem.* **1985,** *49*, 1405–1409.
36. Mazoyer, E., S.; Levy-Toledano.; Rendu, F.; Hermant, L.; Lu, H.; Fiat, A. M.; Jolles, P.; Caen. J. KRDS, A New Peptide Derived from Lactotransferrin, Inhibits Platelet Aggregation and Release Reactions. *Eur. J. Biochem.* **1990,** *194*, 43–49.
37. Meisel, H. Overview on Milk Protein-Derived Peptides. *Int. Dairy J.* **1998,** *8*, 363–373.
38. Meisel, H.; FitzGerald, R. J. Opioid Peptides Encrypted in Intact Milk Protein Sequences. *Br. J. Nutr.* **2000,** *84*(1), 27–31.
39. Meisel, H.; FitzGerald, R. J. Biofunctional Peptides From Milk Proteins: Mineral Binding and Cytomodulatory Effects. *Curr. Pharm. Des.* **2003,** *9*, 1289–1295.
40. Mercier, A.; Gauthier, S. F.; Fliss, I. Immunomodulating Effects of Whey Proteins and their Enzymatic Digests. *Int. Dairy J.* **2004,** *14*, 175–183.
41. Migliore-Samour, D.; Floch, F.; Jolles, P. Biologically Active Casein Peptides Implicated in Immunomodulation. *J. Dairy Res.* **1989,** *56*, 357–362.
42. Mizuno, S.; Matsuura, K.; Gotou, T.; Nishimura, S.; Kajimoto, O.; Yabunc, M.; Kajimoto, Y.; Yamamoto, N. Antihypertensive Effect of Casein Hydrolysate in a Placebo Controlled Study in Subjects With High Normal Blood Pressure and Mild Hypertension. *Br. J. Nutr.* **2005,** *94*, 84–91.
43. Mullally, M. M.; Meisel, H.; FitzGerald, R. J. Synthetic Peptides Corresponding to Alpha-Lactalbumin and Beta-Lactoglobulin Sequences With Angiotension-I-Converting Enzyme Inhibitory Activity. *Biol. Chem.* **1996,** *377*, 259–260.
44. Mullally, M. M.; Meisel, H.; FitzGerald, R. J. Identification of a Novel Angiotensin-I Converting Enzyme Inhibitory Peptide Corresponding to A Tryptic Fragment of Bovine β-lactoglobulin. *FEBS Lett.* **1997,** *402*, 99–101.
45. Nagaoka, S.; Futamura, Y.; Miwa, K.; Awano, T.; Yamauchi, K.; Kanamaru, Y. Identification of Novel Hypocholesterolemic Peptides Derived From Bovine Milk β-lactoglobulin. *Biochem. Biophys. Res. Commun.* **2001,** *218*, 11–17.

46. Nagpal, R.; Behare, P.; Rana, R.; Kumar, A.; Kumar, M.; Arora, S.; Morotta, F.; Jain, S.; Yadav, H. Bioactive Peptides Derived From Milk Proteins and Their Health Beneficial Potentials: An Update. *Food Funct.* **2011,** *2*, 18–27.
47. Nakamura, Y.; Yamamoto, M.; Sakai, K.; Okubo, A.; Yamazaki, S.; Takano, T. Purification and Characterization of Angiotensin-I-converting Enzyme Inhibitors From Sour Milk. *J. Dairy Sci.* **1995,** *78*, 777–783.
48. Otte, J.; Shalaby, S. M.; Zakora, M.; Pripp, A. H.; El-Shabrawy, S. A. Angiotensin-Converting Enzyme Inhibitory Activity of Milk Protein Hydrolysates: Effect of Substrate, Enzyme and Time of Hydrolysis. *Int. Dairy J.* **2007,** *17*, 488–503.
49. Pan, D.; Luo, Y.; Tanokura, M. Antihypertensive Peptides From Skimmed Milk Hydrolysate Digested by Cell-Free Extract of *Lactobacillus helveticus* JCM1004. *Food Chem.* **2004,** *91*, 123–129.
50. Pauletti, G. M.; Gangwar, S.; Knipp, G. T.; Nerurkar, M. M.; Okumu, F. W.; Tamura, K.; Siahaan, T. J.; Borchardt, R.T. Structural Requirements for Intestinal Absorption of Peptide Drugs. *J. Control. Release* **1996,** *41*, 3–17.
51. Perpetuo, E. A.; Juliano, L.; Lebrun, I. Biochemical and Pharmacological Aspects of Two Bradykinin-Potentiating Peptides From Tryptic Hydrolysis of Casein. *J. Protein Chem.* **2003,** *22*, 601–606.
52. Picariello G.; Ferranti, P.; Fierroa, O.; Mamonea, G.; Cairaa, S.; Di Luccia, A.; Monica, S.; Addeon, F. Peptides Surviving the Simulated Gastrointestinal Digestion of Milk Proteins: Biological and Toxicological Implications. *J. Chromatogr. B* **2010,** *878*, 295–308.
53. Pihlanto-Leppala, A.; Rokka, T.; Korhonen, H. Angiotensin I-converting Enzyme Inhibitory Peptides Derived From Bovine Milk Proteins. *Int. Dairy J.* **1998,** *8*, 325–331.
54. Qian, Z. Y.; Jolles, P.; Migliore-Samour, D.; Fiat. A. M. Isolation and Characterization of Sheep Lactoferrin, An Inhibitor of Platelet Aggregation and Comparison With Human Lactoferrin. *Biochim. Biophys. Acta* **1995,** *1243*, 25–32.
55. Qian, Z. Y.; Jolles, P.; Migliore-Samour, D.; Schoentgen, F.; Fiat. A. M. Sheep Kappa-Casein Peptides Inhibit Platelet Aggregation. *Biochim. Biophys. Acta* **1995,** *1244*, 411–417.
56. Quirion, R.; Weiss, A. S. Peptide-E and Other Proenkephalin-Derived Peptides are Potent Kappa Opiate Receptor Agonists. *Peptides* **1983,** *4*, 445–449.
57. Quiros, A.; Ramos, M.; Muguerza, B.; Delgado, M.; Miguel, M.; Alexaindre, A.; Recio, I. Identification of Novel Antihypertensive Peptides in Milk Fermented With *Enterococcus faecalis*. *Int. Dairy J.* **2007,** *17*, 33–41.
58. Roufik, S.; Gauthier, S. F.; Turgeon, S. L. *In-vitro* Digestibility of Bioactive Peptides Derived from Bovine β-lactoglobulin. *Int. Dairy J.* **2006,** *16*, 294–302.
59. Satake, M.; Enjoh, M.; Nakamura, Y.; Takano, T.; Kawamura, Y.; Arai, S. Transepithelial Transport of a Bioactive Tripeptide, Val-Pro-Pro, in Human Intestinal Caco-2 Cell Monolayers. *Biosci. Biotechnol. Biochem.* **2002,** *66*, 378–384.
60. Seppo, L.; Jauhiainen, T.; Poussa T.; Korpela R. A Fermented Milk High in Bioactive Peptides has a Blood Pressure Lowering Effect in Hypertensive Subjects. *Am. J. Clin. Nutr.* **2003,** *77*, 326–330.
61. Shimizu, M. Food-Derived Peptides and Intestinal Functions. *BioFactors* **2004,** *21*, 43–47.
62. Singh, T. K.; Fox, P. F.; Healy, A. Isolation and Identification of Further Peptides in the Diafiltration Retentate of the Water-Soluble Fraction of Cheddar Cheese. *J. Dairy Res.* **1997,** *64*, 433–443.
63. Sipola, M.; Finckenberg, P.; Korpela, R.; Vapaatalo, H.; Nurminen, M. L. Effect of Long-Term Intake of Milk Products on Blood Pressure in Hypertensive Rats. *J. Dairy Res.* **2002,** *69*, 103–111.

64. Smacchi, E.; Gobbetti, M. Peptides from Several Italian Cheeses Inhibitory to Proteolytic Enzymes of Lactic Acid Bacteria (LAB): *Pseudomonas fluorescens* ATCC 948 and to the Angiotensin-I-Converting Enzyme. *Enzyme Microbiol. Technol.* **1998,** *22*, 687–694.
65. Tani, F. A.; Shiota, H.; Chiba, Yoshikawa, M. Serophin, an Opioid Peptide Derived From Serum Albumin. In *β-Casomorphins and Related Peptides: Recent Developments*; Brantl, V., Teschemacher, H., Eds.; Wiley: Weinheim, Germany, 1994; vol 1; pp 49–53.
66. Teschemacher, H. Opioid Receptor Ligands Derived From Food Proteins. *Curr. Pharm. Design* **2003,** *9*, 1331–1344.
67. Vegarud, G. E.; Langsurd, T.; Svenning, C. Mineral-Binding Milk Proteins and Peptides: Occurrence, Biochemical and Technological Characteristics. *Br. J. Nutr.* **2000,** *84*, S91–S98.
68. Vermeirssen, V.; Vancamp, J.; Verstraete, W. Bioavailability of Angiotensin I Converting Enzyme Inhibitory Peptides. *Br. J. Nutr.* **2004,** 92, 357–366.
69. Yamamoto, N. Antihypertensive Peptides Derived From Food Proteins. *Biopolymers* **1997,** *43*, 129–134.
70. Yamamoto, N.; Maeno, M.; Takano, T. Purification and Characterization of an Antihypertensive Peptide From a Yogurt-Like Product Fermented by *Lactobacillus helveticus* CPN4. *J. Dairy Sci.* **1999,** *82*, 1388–1393.
71. Yamamoto, N.; Ejiri, M.; Mizuno, S. Biogenic Peptides and Their Potential Use. *Curr. Pharm. Des.* **2003,** *9*, 1345–1355.
72. Yoshikawa, M.; Tani, F.; Shiota, H.; Suganuma, H.; Usui, H.; Kurahashi, K.; Chiba, H. Casoxin, D. An Opioid Antagonist Ileum-Contracting/ Vasorelaxing Peptide Derived from Human α_{s1}-casein. In β-Casomorphins and Related Peptides: Recent Developments; Brantl, V., Teschemacher, H., Eds.; VCH: Weinheim, Germany, 1994; pp 43–48.
73. Young, V. R.; Pellett, P. L. How to Evaluate Dietary Proteins. In *Milk Proteins: Nutritional, Clinical, Functional and Technological Aspects*; Barth, C. A., Schlimme, E., Eds.; Springer: New York, New York, 1988; pp 7–36.

CHAPTER 12

POTENTIAL ASPECTS OF WHEY PROTEINS IN DAIRY PRODUCTS: CHEMISTRY, BIO-FUNCTIONAL CHARACTERISTICS, AND THEIR APPLICATIONS

DRISHTI KADIAN, CHANDNI DULARIA, and CHANDER MOHAN

ABSTRACT

Whey proteins (WP) are high quality protein with clear flavor and are soluble at low pH, which makes WP suitable as a protein source in different food products. Nowadays, membrane technology is being effectively used to fractionize different whey proteins. High heat treatment applied to whey proteins cause denaturation, ranking according to the increasing order of susceptibility to heat denaturation in milk is beta-lactoglobulin, alpha-lactalbumin, immunoglobulins, and bovine serum albumin. Utilization of whey-derived proteins in foods has direct positive affect on health and techno-functional properties. Alteration in whey proteins can result in enhancement of its functional properties, such as gel forming, emulsifiers, foaming. Various functional properties, imparted by WP are extensively applied in food industries with respect to alter texture and quality in numerous food categories. They are rich in essential amino acids making them unique in promoting and preventing muscle synthesis as well as mass loss of muscle tissue, respectively. These properties make them ideal to be used in sports nutrition, infant formula, and in different beverages. This chapter will review the chemistry associated with whey proteins and major functional applications of whey proteins in food sectors.

The Chemistry of Milk and Milk Products: Physicochemical Properties, Therapeutic Characteristics, and Processing Methods. Megh R. Goyal, Suvartan Ranvir, & Junaid Ahmad Malik (Eds.)

12.1 INTRODUCTION

Whey is a byproduct obtained when milk is acid or rennet coagulated during the manufacturing of cheese or casein. The coagulation process separates casein curd, acid, or mineral from milk by the action of chymosin and the remaining liquid is called whey.[100] Worldwide production of cheese whey (CW) is anticipated to be over 10^8 tons/year.[19] The annual surplus of CW is around 13×10^6 tons, whereas the estimated production is 40×10^6 tons/year according to the European Union (EU). The market value of whey protein was 8.7 billion US dollar in 2019; and by the year 2027 it is projected to inflate at a compound annual growth rate of 9.8%.[90] In the past, whey was considered as a waste product, but now, whey products contribute in variety of ingredients due to their unique functional and nutritional attributes.

Earlier attempts were made to dry liquid whey basically to preserve it and to minimize the pollution. Further, ideas were generated to utilize whey in the best possible way, some of them was to dry by roller dryer, spray dryer or in combination. Development of multistage vacuum evaporators was able to improve its functionality. The only application of whey was in animal feed or utilizing the lactose property for infant food or browning sugar in confectionery industries. Purification of whey protein from rest of the liquid was the main concern at that time until membrane technology came into the picture.

The pros of membrane technology are: (1) it can minimize the whey-protein denaturation caused by various thermal treatments; (2) protein from whey can be retrieved safely by effective use of microfiltration (MF) or ultrafiltration (UF). Commercially available products from whey proteins are:

- *lactose* widely applicable in infant formula, confectionery, and bakery industry.
- *whey powder* (sweet or acid) has found its applicability in milk solid substitute; however, whey powder contain high amounts of ash content. Ash can be removed by ion exchange, diafiltration (DF) or electrodialysis (ED) to produce demineralized whey powder, which can be used in yogurt, infant formula, and in many more products.
- *whey protein concentrate* (WPC) and isolate (WPI) possess high functional and nutritional properties. Different varieties of WPC are available in the market differentiated according to their protein content, namely, WPC34, WPC60, and WPC80. To get WPI of 90% protein, this WPC80 of 80% protein content can be further concentrated by MF

to remove extra fat (Figure 12.1). They find their vast applications in sports and adult protein supplements.

- *fractionation products* whey protein enriched fractions like alpha-lactalbumin, beta-lactoglobulin, lactoferrin, lactoperoxidase, and glycomacropeptide have specific high end application.

Membrane technology separates and fractionizes whey proteins from liquid whey. This technology allows filtration of protein molecules through molecular sieving of 150 μm. The porous filter membrane allows low molecular weight molecules (permeate) to pass through and other materials of high molecular weight get retained in the retentate. The material of filter is size selective and is made from zirconium oxide, cellulose acetate, ceramic, polysulfone, etc.; and filter arrangement is habitually placed in spiral wound stainless steel housing.[43] After going through these processes, whey is spray dried with less than 5% moisture content in the final product. The whey-processing industries have five types of filtration techniques, which are employed individually or in combination, and these are:

- electrodialysis (ED)
- reverse-osmosis (RO)
- nanofiltration (NF)
- UF and
- MF.

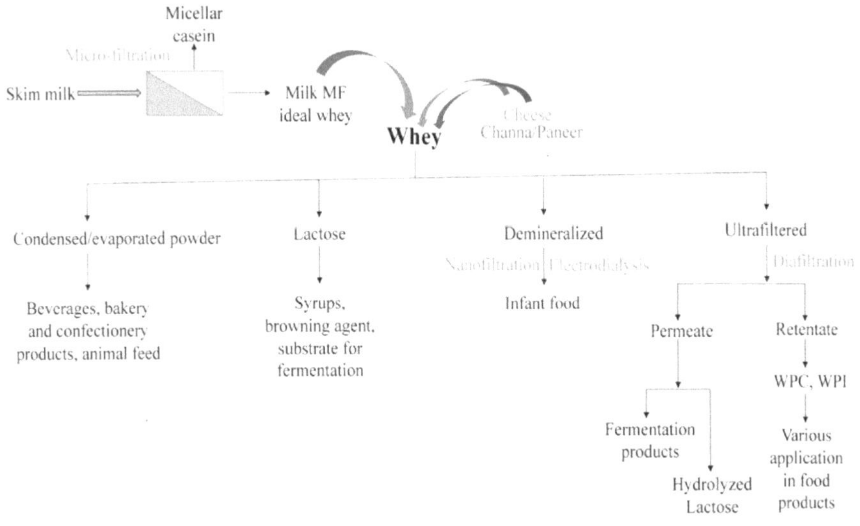

FIGURE 12.1 Flowchart for processing liquid whey proteins.

Separation of whey by UF of molecular weight cutoff (MWCO) of 10 kDa is usually preferred by industries. MWCO in the range 3–100 kDa can be used depending on what proteins size is desired. The temperature of the feed should be kept <55°C, inlet pressure of around 300 kPa, and preferred membrane pore size is 250 nm.[94] Permeate feed has lactose, soluble minerals, and water, whereas insoluble minerals, fat, and proteins are retained in the retentate.

The other technique associated with UF is DF in which water is added in the retentate before proceeding for second UF. This process allows further removal of lactose and salts. MF uses 200 kDa MWCO and the pore size of membrane can vary from 0.05 to 10 μm for selective removal of microbial flora. ED is a electrochemical process having chamber of ion permeable walls in which uninterrupted current is passed through the whey.

Another membrane technique is NF or ultraosmosis that selectively separates lactose and operates at 1.5 and 3 MPa pressure. It is frequently used in removing salts and minerals from whey. RO membrane is also used to remove minerals from whey at high pressure ranges from 2.7 to 10 MPa. The MWCO used by this membrane is 150 kDa that may cause fouling problem in the membrane. It removes around two-third of water from whey, leaving a concentrated solution that can be easily dried down to powder.[26]

This chapter explores potential aspects of whey proteins in food and dairy sectors. Its chemical composition, functional properties along with applications of its ingredients in food sector are also reviewed.

12.2 CHEMICAL COMPOSITION AND PROPERTIES OF WHEY PROTEINS

The major whey protein fractions are immunoglobulins (IgS), α-lactalbumin (α-La), β-lactoglobulin (β-Lg), bovine serum albumin (BSA), minor whey proteins (e.g., lactoperoxidase, lactoferrin, lysozyme, proteose peptone 3 (PP3), osteopontin, cathepsin D, and glycomacropetide). Chemical properties of the following whey fractions have been discussed in this section.

12.2.1 β-LACTOGLOBULIN (β-LG OR BETA-LG)

In ruminant animals, β-Lg is found in the highest amount as compared to other whey protein fractions.[79] β-Lg represents 10–12% and nearly 50% of the total proteins and total whey proteins in bovine milk, respectively.[25,77] There are two types of beta-Lg, that is, A and B that contains two bonds containing disulfide linkages at Cys 66-160 and Cys106-119 and one sulfhydryl group that is free at Cys121, hidden within the protein matrix.[68,80] Though, in some species, the sulfydryl group that is free is absent in beta-Lg.[76]

12.2.2 α-LACTALBUMIN (ALPHA-LA OR β-LA)

The second most abundant WP is α-La after β-Lg and around 1.2–1.5 gL^{-1} is present in bovine milk.[40] It contributes 3.5% of the whole protein in milk and 20% of the complete whey proteins.[77] It consists of four intramolecular disulfide bonds and there is no free sulfyhydryl group present in α-La structure. It is having a bilobal structure alike lysozyme. One of the lobes contain β-sheets and unordered structures, while another contains α-helices.[13] Both lobes are parted by a branch closed at one end.[12]. The concentration of α-La is decreased with end of lactation.[17]

12.2.3 BOVINE SERUM ALBUMIN (BSA)

BSA is present in very low amount compared to other major WP fractions, that is, β-Lg and α-La. In ruminant animals (such as cow), it consists of 0.4 g/L of BSA, having 1.5% of the whole protein while 8% of the complete whey proteins.[40] In camel milk, the concentration of BSA is found higher compared to bovine animals.[72] BSA does not contain phosphorus, although it has one free sulfhydryl group and 17 disulfides. It contain low percentage of methionine, glycine, and tryptophan, while high in amino acids (AA).[85]

12.2.4 IMMUNOGLOBULINS (IGS)

Igs are found in colostrum that is responsible for the immunological activities, such as defence against viruses, bacteria, and other toxins. There are various categories of Igs, such as IgG (monomeric form), IgA (dimer form), and IgM (pentamer form). IgG is further subdivided into IgG1 and IgG2 in colostrum and milk.[52,69] The total amount of Igs is up to 200 g/L in colostrum while 0.6–1.0 g/L in bovine milk. The concentrations of various categories of Igs are: IgG of 60 and 0.47 g/L, IgA of 3.5 g/L and 0.05–0.4 g/L, and IgM of 5 g/L and 0.04–1.0 g/L in colostrum and bovine milk, respectively.[69]

12.2.5 MINOR WHEY PROTEINS

12.2.5.1 LACTOFERRIN

Lactoferrin also known as iron-binding glycoprotein. It has only one polypeptide chain organized in two highly homologous lobes associated with an α-helix structure. Ferric iron-binding site is present at each lobe. Lactoferrin lacks sulfhydryl groups but possesses 16 intramolecular disulfide bonds. In human milk, lactoferrin concentration is higher (1.5–2.0 g/L) compared to bovine milk (0.02–0.5 g/L).[23] However, in colostrum and mastitis milk lactoferrin concentration increases.[34,57]

12.2.5.2 LACTOPEROXIDASE

Lactoperoxidase is the second most abundantly soluble enzyme in bovine milk that is linked with the milk fat globule membrane. The concentration of lactoperoxidase is nearly 0.03 g/L equivalent to 0.5% of the complete whey proteins.[15] Due to their antibacterial properties, it is incorporated in cosmetics, toothpaste, mouthwashes, fish feed, and calf starters, etc.[30]

12.2.5.3 LYSOZYME

Muramidase is the other term used for lysozyme. They are living organisms that play an essential role in the defence mechanism. The usual concentration of lysozyme is <3 mg/L and lies in the range of 0.7–6 mg/L in bovine milk,[39,98] while in buffalo milk it is 3.85 mg/L. However, the concentration of lysozyme in human milk is significantly higher compared to bovine and buffalo milk, that is,100–120 mg/L.[1]

The properties of major and minor proteins are summarized in Table 12.1 and chemical constituents of the whey protein products are listed in Table 12.2.

12.3 EFFECT OF PROCESSING PARAMETERS ON CHEMICAL PROPERTIES

12.3.1 TEMPERATURE (HEATING/FREEZING)

Functional properties exhibited by whey proteins are greatly influenced by high temperature leading to conversion of native proteins to denaturated

TABLE 12.1 Summary of Major and Minor Whey Protein With Their Properties.

Whey protein fraction	No. of amino acid residues	Mol. wt.	pI	Sub-units	Isoionic points	Properties	Denaturation temperature	Reference
BSA	583	66,465.8 Da	4.7–4.9	2	5.13	• Heart-shaped structure • Single chain globular nonglycoprotein • Ligand-binding capacity • Binds fatty acids, metal ions, and calcium ions (~3 per protein molecule) • Cancer-inhibiting properties • Maintains blood pH	55–85°C	9,11,18,40, 48, 49, 56, 65, 66
Igs	>500	IgG ~160 kDa IgA~370 kDa IgM 900–1000 kDa	5.1–8.3	Two in heavy and light chain	-	• Provides adequate passive immunity against microbial infections • Heat stable • Reduces risk of GI tract disorders like traveler's diarrohea	–	9,11,49,71
Lactoferrin	689	~80,0000 Da	8.7	–	–	• Glycosylated metalloprotein • Binds two atoms of Fe per molecule • Heat stable • Antibacterial • Reduces risk of GI • Resistant to proteolytic enzymes	–	33,34,57

TABLE 12.2 *(Continued)*

Whey protein fraction	No. of amino acid residues	Mol. wt.	pI	Sub-units	Isoionic points	Properties	Denaturation temperature	Reference
Lactoperoxidase	612	78,030 Da	9.6	–	–	• Glycoprotein • At pH 6.7 it is heat-stable at 72°C for 15s (pasteurization HTST) • Antibacterial characteristic	–	7,15,30,38, 53,60,77
Lysozyme	154 (in bovine milk) 130 (in human milk)	18,000 Da (in bovine milk) 15,000 Da (in human milk)	–	–	–	• At pH 3–4 it is heat-stable but not at pH >7 • Heat stable at 80°C for 15 s (pasteurization HTST) • Antibacterial properties • Extends shelf-life of milk	–	21,39
α-La	123	14,178 Da	4.4	1	–	• Small compact protein • Calcium metallo-protein • Regulates lactose synthesis • Helps in manufacturing of aqueous phase of milk • Heat-stable	64°C	9,11,40,47, 49, 93
β-Lg	162	~18,300 Da	5.2	2	Variant A: 5.35 Variant B: 5.41	• Highly structured globular protein • Water-soluble protein • Most heat stable	70–73°C	9,10,35,40, 49

TABLE 12.2 Chemical Constituents of Whey Protein Products

Whey products		Protein, %	Fat, %	Moisture, %	Lactose, %	Ash, %	Reference
Whey	**Sweet whey**	0.9	0.3	93	4.9	0.6	42
	Acid whey	0.55–0.75	0.04	93.5	4.2–4.9	0.8	16
Whey powder		11.0–14.5	1.0–1.5	3.5–5.0	63.0–75.0	8.2–8.8	6
Demineralized whey		11.0–15.0	3.0–4.0	0.5–1.8	70.0–80.0	1.0–7.0	
Whey protein concentrate powder (WPC)	**34**	34–36	3–4.5	3–4.5	48–52	6.5–8.0	6
	50	50–52	5–6	3.5–4.5	33–37	7.5–8.5	
	80	80–82	4–8	3.5–4.5	4–8	3–4	
Whey protein isolate (WPI)		90–92	0.5–1.0	4.5	0.5–1.0	2–3	

whey proteins. During denaturation at moderate heating temperature (i.e., between 60 and 70°C), structural transformation, usually unfolding of the proteins takes place. Depending on compositional factors at temperature >70°C, protein aggregation takes place followed by precipitation. In the unfolding stage, molecular interactions, such as hydrogen and hydrophobic bonding occurs, while in the aggregation step disulfide linking takes place that is mediated by the calcium ions.[5,28,74] The whey protein fraction ranking according to the increasing order of susceptibility to heat denaturation in milk is Igs, β-Lg, α-La, and BSA.[31] Though, the overall rate of whey protein denaturation is decreased when the total solids content is increased. This phenomenon is related to the decrease in the denaturation of β-Lg whey fraction while increase in α-La with increase in total solid content.[46,76]

During the freezing process, the viscosity of whey proteins is increased due to the formation of ice crystals with increasing concentration of freezing that causes the changes or modifications in the structure of soluble protein resulting in aggregation.[73,91]

12.3.2 pH (ACIDIC/ALKALINE)

Whey proteins are aggregated at a pH range of 4.8–5.3 (which is the isoelectric pH range of α-La and β-Lg). At pH of 3, repulsive force occurs between the proteins, which is present in solution form due to high net negative charge in β-Lg. During the application of heat, the interactions between proteins are inhibited due to these repulsive forces that leads to high heat stability and clarity at 7% protein concentrations.[14]

As the pH is increased upto four, the net charge is declined along with the repulsive forces. Due to these actions, whey proteins are attracted in the absence of heat thus leading to increase in turbidity of whey protein solution. Protein aggregation along with precipitation occurs when the heating temperature is more than the denaturation temperature of β-Lg, that is, 78°C. Therefore, to make heat stability of whey proteins better, some other ingredients are added.[14]

The net charge becomes zero when the pH reaches to 5.2, when protein will precipitate out of solution and there is increase in the protein–protein interactions. At pH of 7, protein attains a net negative charge that is similar to net positive charge at pH of 4.2.[14]

12.3.3 ENZYMES

Functional properties and bioactive enhancement of whey proteins can be done by modifying the whey proteins using hydrolysis or covalent cross-linking.[36] Transglutaminase (TGase) enzyme is used for modifying whey proteins by incorporating amine via cross-linking and deamination. After enzymatic treatment, the functional properties are changed due to the release of high molecular weight polymers.[54] Because of the globular structure of whey proteins, their susceptibility is less with respect to reactions involved with TGase. Therefore, enzyme-specific sites (such as reactive glutamyl) and lysine molecules can be exposed to the enzymatic reactions for unfolding of the protein molecules.[24]

Use of TGase enzyme can modify the thermal stability of whey proteins via cross-linking covalent reactions among protein species.[3,96,102] Because of the disruption of the hydrophobic–hydrophilic equilibrium on the protein surface and some of the loss of positive charges from the lysine residues, the TGase-treatment on whey protein may affect susceptibility to pH change. Hence, it is not recommended to use TGase treated whey proteins in acidic beverages.[3]

12.4 EFFECT OF FUNCTIONAL PROPERTIES OF WHEY ON FOOD PRODUCTS

Whey-based food/dairy ingredients offer remarkable functional and biological advantages. Utilization of whey proteins in many food preparation has direct affect on functional and health properties. The versatile nature of whey proteins makes them very useful in food products. Although whey proteins has to undergo different treatments during their manufacturing that may alter the native state and also affects the whey proteins stability.[29] Whey proteins possess various functional properties that play a beneficial role and act as an ingredient in various food applications.

12.4.1 EMULSIFICATION

Milk is found in simpler emulsion form, which is termed as natural emulsion and to more complex emulsions, such as sausages.[32] The distribution of oil and water phases are relative to each other in an emulsion. It is a complex mechanism (flocculation, aggregation, creaming); therefore, its stability depends on interaction between oil-water molecules. These molecules tend to separate into two phases (e.g., fat and water) that do not mix.[27] To ensure stability of emulsion and to prolong the storage time of products, addition of emulsifiers is a must.

Whey proteins form organized viscous and elastic films at oil–water interfaces due to their amphiphilic property; thus forming fine dispersed emulsion droplets. This viscous as well as elastic film formation depends on hydrophobic, noncovalent, covalent, electrostatic interactions, and sulfhydryl disulfide interchange reactions between the adsorbed protein molecules.[86] Mainly, adsorption of β-LG and α-LA to oil–water interfaces gives stable emulsions. The study on emulsifying property of whey-protein concentrate was conducted at two different pH of 5.7 and 7.0. Hence, revealing elastic behavior of interfacial film of proteins at both pH, the emulsion stability at pH 5.7 was higher.[55] Addition of pectin may prevent protein–protein interaction due to the formation of huge protein aggregates at pH range of 4.8–5.4 that is pI of β-LG.[58] Stabilized emulsion was provided by whey protein under simulated gastric conditions, whereas at intestinal conditions the emulsion stability was not very good.[67]

12.4.2 WHIPPING AND FOAMING PROPERTIES

Whipping and foaming properties are determined by the proficiency of protein to unfold and to accommodate themselves on the air–water interface. The formation of protein condensed bubbles and interaction of proteins components occur via electrostatic, hydrophobic interaction, and hydrogen bondings. To provide foam structure, retention of air, and formation of strong matrix is a must through viscoelastic film formation. Several factors affect the foaming property of whey proteins, such as protein denaturation, heat treatment, protein solubility, pH, lipids, total solids, calcium content, and certain other reagents. Increment in total solid content improves the foaming ability of whey proteins, and addition of calcium has detrimental effect on WPC foaming property due to calcium-induced protein aggregation.[87]

The foaming properties of camel milk proteins at different temperatures were compared with bovine milk, and it was found that an increase in foamability occurred with increase in temperature from 70 to 100°C.[61] Camel milk lacks β-LG, whereas cow milk contains 72% of β-LG. During heat treatment, β-LG disconnects from a dimer structure to native monomer leading to thermal denaturation of proteins; thus forming disulphide aggregates. In case of camel milk, heating may increase the surface hydrophobicity due to denaturation and aggregation mechanism leading to decrease in electronegative charge and surface tension, therefore strengthening the foaming mechanism.[68]

12.4.3 SOLUBILITY OF WPC

The solubility of WPC mainly depends on the technology for manufacturing like membrane or gel filtration methods. The solubility reported for WPC manufactured by UF techniques is about 88–100%. Among all other properties of proteins, solubility is of prime importance. Influence of solubility on other properties like foaming, gelling, and emulsifying are highly significantly and it ultimately affects the product structure and functionality. Protein has surface hydrophobicity, which is due to protein–protein interaction, and hydrophilicity due to protein–water interaction that is directly related to solubility.

Heating and pH both have their negative effects on protein solubility. At pI, electrostatic forces are at minimum level, protein shows least solubility due to increase interaction between protein–protein. Interaction of water with protein molecules is less that makes favorable conditions for protein molecules to form aggregates and further gets precipitated.[83] If the pH values are above or below the pI, the net positive and negative charge contributes to protein solubility due to charge repulsion. Whereas, at severe acidic or basic pH protein reveals, its hydrophobic site cause unfolding of protein molecules.

Whey proteins are greatly affected by temperatures higher than 70°C for few minutes. Firstly, denaturation process occurs in Igs fraction followed by serum albumin. The β-LG is slightly less disturbed under same heating conditions; however, α-LA is the most impervious.[74] The effect of storage conditions on solubility (time of 2 weeks; temperature up to 60°C) was studied on WPC-35, and it was concluded that storage temperature above 30°C may induce denaturation to affect the protein solubility.[41]

12.4.4 GELATION

The ability of whey protein products to form gel when heated shows gelling property by solidifying fluid product. The physical reactions occurring are aggregation and denaturation of whey proteins. A three-dimensional network formation occurs under appropriate conditions, which upholds large amounts of water. At low concentration (3–5%) of protein, ionic strength and heating temperatures (55–70°C), translucent gel formation occurs; however, at higher concentration (10%) and elevated temperature (>90°C), more opaque gel forms.[89] Native whey proteins show low viscous behavior due to their small size. On the other hand, modification of whey protein can increase their viscosity and then can be widely used as gelling agent.[101] One of such modifications is heating of whey proteins, which results in polymerized whey protein (PWP) (Figure 12.2).

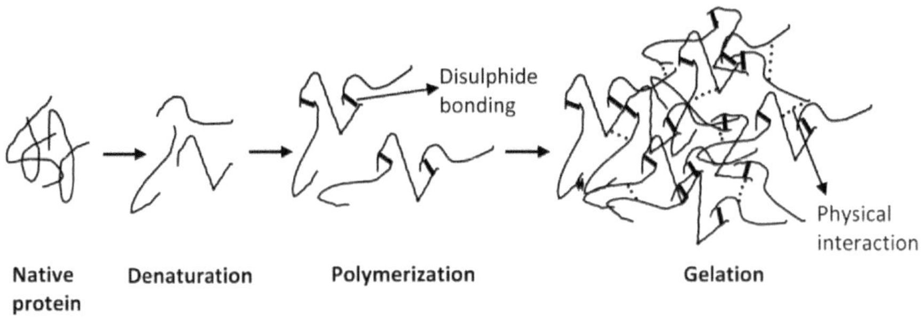

FIGURE 12.2 Sequential steps in the formation of whey gel.

Whey proteins ability to make gel at low temperature (cold set) made them an excellent gelation agent. This process follows three steps[96,99]:

- Whey protein is accustomed at higher pH than its pI to inhibit aggregation caused by electrostatic repulsion between protein–protein aggregates;
- Heating of whey proteins to denature them and to form aggregates;
- Addition of minerals or lowering the pI of whey proteins cause decrease in electrostatic repulsion and subsequently a cold-set gel forms.[96] Improvement in protein gel network was observed by addition of cellulose nanocrystals (CNC) prepared from wheat bran.

The strength of gel from heat-induced whey protein isolate was improved tremendously as CNC actively dehydrates the protein matrix; thus it encouraged the protein molecules to unfold and form links. Development of compressed and homogeneous three-dimensional network is important as protein gels are very sensitive to outer environment (salts, ions, pH, temperature). Also in foods where 3D network forming proteins are absent, PWP plays a significant role by providing gelling property to set type products like fermented foods,[4] yogurt[45] offering great texture, and lower syneresis rate.

12.5 APPLICATIONS OF WHEY PROTEINS IN FOOD AND DAIRY PRODUCTS

12.5.1 INFANT FORMULA

Ideally human milk is the best for infants. It contains all nutrients required for the proper development of a new born. Though breastfeeding has so many benefits to the new born and to the mother,[44,61] it also comes with some restrictions in cases where mother is diseased, working, nonintentional or not able to produce enough milk to feed her child. In such cases, infant formula comes to the rescue to all mothers. Typically, cow milk cannot be fed straightforward to new-borns due to its unfavorable composition. In mature human milk, the whey-to-casein ratio is 60:40,[59] which is a standard selected by the industry to manufacture infant formula.

To manufacture infant formula with standard whey/casein ratio, whey proteins like demineralized whey, whey protein concentrate or isolate are supplemented with cow milk or milk powder. Other components like fat and milk sugar are modified by incorporating vegetable oil or lactose according to the requirement. Generally, companies manufacturing infant formula tend to surplus the protein requirement in the formula so that babies can get enough amino acid and it is also important to maintain their body weight.

Hydrolyzed whey protein has unique features to be used in making infant formula as they tend to enhance the digestibility and to reduce the chances allergenicity associated with bovine milk.[37] Human milk lacks β-LG and has abundance of α-LA. In recent times, interest is to manufacture infant formula containing high amount of α-LA, whereas, complete elimination or diminishing of β-LG is favorable to get proper amino acid profiling.

Infant formula supplemented with α-LA also has high concentration of tryptophan, which possess positive behavior benefits to brain neurons.[64] Another major protein present in human milk is lactoferrin, which is an iron-binding protein. Infant formula comprises of lactoferrin and it has shown health benefits, such as anti-bacterial, antioxidant capacity, growth enhancement, high immunity, and free radical scavenging activity when compared with regular formula.[88]

12.5.2 SPORTS NUTRITION

Nutrients play a major role in the muscle metabolism including carbohydrates, fats, and proteins. Glycogens from carbohydrates are broken down and are stored in liver and muscles. Fatty acids are transported into mitochondria, where they are oxidized and are used as energy. During the exercise, processes follow the order:

protein → free amino acids → oxidation in mitochondria → energy

This free amino acid may come from muscle degradation or may get absorbed from diet. The synthesis rate is lower than degradation rate of protein, but after exercise, amino acid should be replaced in time so that protein synthesis exceeds degradation. As a result, muscle recovery takes place at faster rate to regain muscle mass.[92] Essential amino acid (EAA) include leucine, valine, and isoleucine act as a energy source as it get oxidized easily and produce energy.[2]

Currently, whey protein is widely used in sports nutrition by many industries due to its excellent EAA profile and also it gets absorbed very fast in the small intestine. Nutritional and fast acting property of whey proteins made them exceptionally great candidates in sports nutrition. Among various protein supplementation products, ready-to-mix whey protein powder is most convenient to consume. Most commonly used whey supplements in sports is powdered whey protein shake. For its preparation, artificial flavor and sweetener are used to provide maximum protein content to the consumers. WPC80 or WPI are used individually or their mixture to get ready-to-mix whey protein powder to minimize lactose (carbohydrate) in the final product.

12.5.3 PROTEIN BARS

Whey protein concentrate has been widely used in making protein bars. They are very popular instant energy supplements consumed by various athletes, adults, and even liked by children. Generally, formulation of protein bars comprises of calories from protein-30%, carbohydrates-40%, and fat-30% (30/40/30 rule). Protein goes through various reactions during making of protein bars that include:

- Aggregation—where formation of intermolecular disulfide bonds and other interactions like covalent bonding[104] takes place;

- polymerization of protein occurs due to Maillard reactions[93]; and
- migration of moisture and separation of phases occurs.[65]

In the interim, tendency to make protein bars enriched with high protein content are increasing. High protein content in bars has its own challenges like texture problem and bar hardening over the course of shelf-life. Solidification of protein ingredients occurs due to the formation of secondary structures and plasticization induced by solvent.[51] Another reasons for development of hardness is due to changes in the equilibrium content of the individual ingredients and further, they compete for available moisture.[51]

12.5.4 WHEY PROTEIN BEVERAGE

Acidified whey-protein beverage is preferred by most of the athletes because of its efficiency in quenching thirst. As whey proteins are acid stable in the pH range of 2.8–4.0 which is below its isoelectric point (4.6–5.0), they are able to survive sterilization temperature without precipitation or gelation. On the other hand, whey protein isolate (WPI) manufactured by ion-exchange method is more transparent and has low level of minerals. Pure WPI with high protein content can be used to formulate acidified beverage with high clarity and minimum turbidity.

12.6 SUMMARY

Whey proteins is abundant in heterogenous mixture of secreted proteins. It has admirable functional properties broadly used in food industry. This chapter has discussed the emerging technologies to separate whey proteins, their chemistry, functional properties that are essential to provide good texture and quality to food products. Whey proteins act as gelling agent, stabilizer or emulsifier and its foaming ability have also been discussed along with mechanism involved. Further, whey proteins utilization in infant formula and sports nutrition due to high protein content and unique functional property with regard to food application was also reviewed. However, there is a need to provide more efforts to expand certain other properties of whey proteins for food industry.

KEYWORDS

- **emulsification**
- **enzyme**
- **functional property**
- **gelation**
- **infant formula**
- **solubility**
- **sports drink**
- **whey beverage**
- **whey protein fraction**
- **whey protein processing**

REFERENCES

1. Abd El-Aziz, M. Study on Lysozyme Level, Distribution and Effect of Heat Treatment in Buffalo and Cow Milk. *Ann. Agric. Sci.* (Cairo) **2006**, *51*, E-article 439446.
2. Abumrad, N. N.; Williams, P.; Frexes-Steed, M.; Geer, R.; Flakoll, P.; Cersosimo, E.; Ghishan, F. Inter-Organ Metabolism of Amino Acids in Vivo. *Diabetes/Metab. Rev.* **1989,** *5*(3), 213–226.
3. Agyare, K. K.; Damodaran, S. The pH-stability and Thermal Properties of Microbial Transglutaminase-treated Whey Protein Isolate. *J.Agric. Food Chem.* **2010,** *58*(3), 1946–1953.
4. Alting, A. C.; Vander Meulena, E. T.; Hugenholtz, J.; Visschers, R. W. Control of Texture of Cold-Set Gels Through Programmed Bacterial Acidification. *Int. Dairy J.* **2004,** *14*(4), 323–329.
5. Anandharamakrishnan, C.; Rielly, C. D.; Stapley, A. G. Loss of Solubility of α-lactalbumin and β-lactoglobulin during the Spray Drying of Whey Proteins. *LWT Food Sci. Technol.* **2008,** *41*(2), 270–277.
6. Anonymous. *Reference Manual for US Whey Products*. US Dairy Export Council Editors: Arlington, VA, 2001; p 226.
7. Bafort, F.; Parisi, O.; Perraudin, J. P.; Jijakli, M. H. Mode of Action of Lactoperoxidase as Related to Its Antimicrobial Activity: A Review. *Enzyme Res.* **2014,** *2014,* E-article 517164.
8. Baldasso, C.; Barros, T. C.; Tessaro, I. C. Concentration and Purification of Whey Proteins by Ultrafiltration. *Desalination* **2011,** *278*, 381–386.
9. Boland, M. Whey Proteins. In *Handbook of Food Proteins*; Phillips, G. O., Williams, P. A., Eds.; Woodhead Publishing: Cambridge, UK, 2011; chapter 3, pp 30–51.

10. Bolder, S. G.; Vasbinder, A. J.; Sagis, L. M.; Vander Linden, E. Heat-Induced Whey Protein Isolate Fibrils: Conversion, Hydrolysis, and Disulphide Bond Formation. *Int. Dairy J.* **2007,** *17*(7), 846–853.
11. Boye, J. I.; Alli, I. Thermal Denaturation of Mixtures of A-Lactalbumin and B-Lactoglobulin: A Differential Scanning Calorimetric Study. *Food Res. Int.* **2000,** *33*(8), 673–682.
12. Brew, K. Milk Proteins: α-Lactalbumin. In *Encyclopedia of Dairy Sciences* (4147 pages), 2nd ed.; Fuquay, J. W., Fox, P. F., McSweeney, P. L. H., Eds.; Academic Press: San Diego, CA, 2011 (14 March); vol 3, pp. 780–786. https://www.sciencedirect.com/referencework/9780123744074/encyclopedia-of-dairy-sciences (accessed Aug 31, 2021).
13. Brew, K. α-Lactalbumin. In *Advanced Dairy Chemistry-Proteins: Basic Aspects,* 4th ed.; McSweeney, P. L. H., and Fox, P. F., Eds.; Springer: Boston, MA, 2013; vol 1A, pp 261–273.
14. Burrington, K. J. Technical Report: Whey Protein Heat Stability. Dairy Research Institute: Madison, Wisconsin, 2012; pp 1–8; Unpublished report.
15. Buys, E. M.; Donkin, F. Significance of the Lactoperoxidase System in the Dairy Industry, and its Potential Applications: A Review. *Trends Food Sci. Technol.* **2005,** *16*(4), 137–154.
16. Bylund, G. *Dairy Processing Handbook.* Tetra Pak Processing Systems AB: Lund, Swedon, 2003; p 480.
17. Caffin, J. P.; Poutrel, B.; Rainard, P. Physiological and Pathological Factors Influencing Bovine Alpha-Lactalbumin and Beta-Lactoglobulin Concentrations in Milk. *J. Dairy Sci.* **1985,** *68*, 1087–1094.
18. Carter, D. C.; Ho, J. X. Structure of Serum Albumin. *Adv. Protein Chem.* **1994,** *45*, 153–203.
19. Carvalho, F.; Prazeres, A. R.; Rivas, J. Cheese Whey Wastewater: Characterization and Treatment. *Sci. Total Environ.* **2013,** *445*, 385–396.
20. Chaneton, L.; Sáez, J. P.; Bussmann, L. E. Antimicrobial Activity of Bovine β-lactoglobulin Against Mastitis-causing Bacteria. *J. Dairy Sci.* **2011,** *94*(1), 138–145.
21. *Chemical Book.* Lysozyme; 2016 [Online]. https://www.chemicalbook.com/ChemicalProductProperty_EN_CB2709257.htm (accessed Mar 01, 2021).
22. Chevalier, F.; Chobert, J. M.; Genot, C.; Haertlé, T. Scavenging of Free Radicals, Antimicrobial, and Cytotoxic Activities of the Maillard Reaction Products of β-lactoglobulin Glycated With Several Sugars. *J. Agric. Food Chem.* **2001,** *49*(10), 5031–5038.
23. Claeys, W. L.; Verraes, C.; Cardoen, S.; De Block, J.; Huyghebaert, A.; Raes, K.; Herman, L. Consumption of Raw or Heated Milk From Different Species: An Evaluation of the Nutritional and Potential Health Benefits. *Food Control* **2014,** *42*, 188–201.
24. Cony, G.; Pedrolm, B.; Marildet, B.L. Effect of Thermal Treatment on Whey Protein Polymerization by Transglutaminase: Implications for Functionality in Processed Dairy Foods. *LWT Food Sci. Technol.* **2010,** *43*(2), 214–219.
25. Creamer, L. K.; Loveday, S. M.; Sawyer, L. Milk Proteins: β-Lactoglobulin A2. In *Encyclopedia of Dairy Sciences*, 2nd ed.; Fuquay, J. W., Fox, P. F., McSweeney, P. L. H., Eds.; Academic Press: San Diego, CA, 2011; vol 2, pp 787–794.
26. Cryan, R. Whey: Ready for Takeoff? *U.S. Dairy Mark - Outlook* **2001,** *7*(3), 1–4.
27. Damodaran, S. Protein Stabilization of Emulsions and Foams. *J. Food Sci.* **2005,** *70*(3), R54–R66.

28. De Wit, J. N. Structure and Functional Behaviour of Whey Proteins. *Neth. Milk Dairy J.* **1981,** *35,* 47–53.
29. De Wit, J. N. Thermal Stability and Functionality of Whey Proteins. *J. Dairy Sci.* **1990,** *73*(12), 3602–3612.
30. DeBoer, R. Information Sheets: Milk, Lipids, Cheese, Whey. In *From Milk By-Products to Milk Ingredients: Upgrading the Cycle*; DeBoer, R., Eds.; John Wiley & Sons Ltd.: Chichester, UK, 2014; pp 209–261.
31. Delaney, R. A. M. Composition, Properties and Uses of Whey Protein Concentrates. *Int. J. Dairy Technol.* **1976,** *29*(2), 91–101.
32. Desrumaux, A.; Marcand, J. Formation of Sunflower Oil Emulsions Stabilized by Whey Proteins With High-pressure Homogenization (up to 350 MPa): Effect of Pressure on Emulsion Characteristics. *Int. J. Food Sci. Technol.* **2002,** *37*(3), 263–269.
33. Dionysius, D. A.; Herse, J. B.; Grieve, P. A. Extraction of Lactoperoxidase and Lactoferrin From Whey Using Batch Ion-Exchange Techniques. *Aust. J. Dairy Technol.* **1991,** *46*, 72–76.
34. Dionysius, D. A.; Milne, J. M. Antibacterial Peptides of Bovine Lactoferrin: Purification and Characterization. *J. Dairy Sci.* **1997,** *80*, 667–674.
35. Divsalar, A.; Saboury, A. A.; Ahmad, F.; Moosavi-Movahedi, A. A. Effects of Temperature and Chromium(III) Ion on the Structure of Bovine β-Lactoglobulin-A. *J. Braz. Chem. Soc.* **2009,** *20*(10), 1782–1789.
36. Eallen, F.; Jackp, D. Food Protein Functionality: A Comprehensive Approach. *Food Hydrocoll.* **2011,** *25*(8), 185–1864.
37. Exl, B. Review of Recent Developments in the Use of Moderately Hydrolysed Whey Formulae in Infant Nutrition. *Nutr. Res.* **2001,** *21*(1–2), 355–379.
38. FAO/WHO. Benefits and Potential Risks of the Lactoperoxidase System of Raw Milk Preservation. In *Report of FAO/WHO Technical Meeting*; FAO Headquarters: Rome, Italy, **2006**.
39. Farkye, N. Y.; Bansal, N. Enzymes Indigenous to Milk: Other Enzymes. In *Encyclopaedia of Dairy Sciences,* 2nd ed.; Fuquay, *W.*, Fox, P. F., McSweeney, P. L. H., Eds.; Academic Press: San Diego, CA, 2011; vol 2; pp 327–334.
40. Farrell, H. M.; Jimenez-Flores, R.; Bleck, G. T.; Brown, E. M.; Butler, J. E.; Creamer, L. K.; Swaisgood, H. E. Nomenclature of the Proteins of Cows' Milk. *J. Dairy Sci.* **2004,** *87*, 1641–1674.
41. Gasparini, A.; Van Gool, M. P.; Bultsma, M.; Cutroneo, S.; Sforza, S.; Tedeschi, T. Modifications Induced by Controlled Storage Conditions On Whey Protein Concentrates: Effects On Whey Protein Lactosylation and Solubility. *Int. Dairy J.* **2020,** *109*, E-article 104765.
42. Gupta, V. K. Overview of Processing and Utilization of Dairy By-products. *Indian Dairyman, U.S. Dairy Export Council* (USDEC); **2000,** *52*, 55–59 [Online]. http://usdec.files.cmsplus.com/PDFs/2008ReferenceManuals/Whey_Lactose_Reference_Manual_Complete2_Optimized.pdf (accessed Feb 27, 2021).
43. Hendricks, G. M.; Guo, M. Component Interactions and Processing Damage During the Manufacture of Infant Formula. In *Human Milk Biochemistry and Infant Formula Manufacturing Technology*; Woodhead Publishing: London, 2014; pp 233–245.
44. Henning, D. R.; Baer, R. J.; Hassan, A. N.; Dave, R. Major Advances in Concentrated and Dry Milk Products, Cheese, and Milk Fat-Based Spreads. *J. Dairy Sci.* **2006,** *89*, 1179–1188.

45. Henriques, M., Gomes, D., Rodrigues, D.; Pereira, C.; Gil, M. Performance of Bovine and Ovine Liquid Whey Protein Concentrate on Functional Properties of Set Yoghurts. *Procedia Food Sci.* **2011,** *1*, 2007–2014.
46. Hillier, R. M.; Lyster, R. L.; Cheeseman, G. C. Thermal Denaturation of α-lactalbumin and β-Lactoglobulin in Cheese Whey: Effect of Total Solids Concentration and pH. *J. Dairy Res.* **1979,** *46*(1), 103–111.
47. Hiraoka, Y.; Segawa, T.; Kuwajima, K.; Sugai, S.; Murai, N. α-Lactalbumin: A Calcium Metalloprotein. *Biochem. Biophys. Res. Commun.* **1980,** *95*(3), 1098–1104.
48. Hirayama, K.; Akashi, S.; Furuya, M.; Fukuhara, K. I. Rapid Confirmation and Revision of the Primary Structure of Bovine Serum Albumin by ESIMS and Frit-FAB LC/MS. *Biochem. Biophys. Res. Commun.* **1990,** *173*(2), 639–646.
49. Hoffmann, M. A. M.; Van Miltenburg, J. C.; Van Mil, P. J. J. M. The Suitability of Scanning Calorimetry to Investigate Slow Irreversible Protein Denaturation. *Thermochim. Acta* **1997,** *306*(1–2), 45–49.
50. Hoffmann, M. A.; Van Mil, P. J. Heat-induced Aggregation of β-lactoglobulin: Role of the Free Thiol Group and Disulfide Bonds. *J. Agric. Food Chem.* **1997,** *45*(8), 2942–2948.
51. Hogan, S. A.; Chaurin, V.; O'Kennedy, B. T.; Kelly, P. M. Influence of Dairy Proteins on Textural Change in High-Protein Bars. *Int. Dairy J.* **2012,** *26*, E-article 5865.
52. Hurley, W. L.; Theil, P. K. Immunoglobulins in Mammary Secretions. In *Advanced Dairy Chemistry-Proteins*: *Basic Aspects*; McSweeney, P. L. H., Fox, P. F., Eds.; Springer: New York, NY, 2013; vol 1A; pp 275–294.
53. Jafary, F.; Kashanian, S.; Sharieat, S. Z. S. Purification, Immobilization, and Characterization of Bovine Lactoperoxidase. *Int. J. Food Prop.* **2013,** *16*, 905–916.
54. Jaros, D.; Partschefeld, C.; Henle, T.; Rohm, H. Transglutaminase in Dairy Products: Chemistry, Physics, Applications. *J. Texture Stud.* **2006,** *37*(2), 113–155.
55. Kilian, J.; Aparecida Fernandes, I.; Luize Lupatini Menegotto, A.; Steffens, C.; Abirached, C.; Steffens, J.; Valduga, E. Interfacial and Emulsifying Properties of Whey Protein Concentrate by Ultrafiltration. *Food Sci. Technol. Int.* **2020,** *26*(8), 657–665.
56. Klajnert, B.; Bryszewska, M. Fluorescence Studies on PAMAM Dendrimers Interactions With Bovine Serum Albumin. *Bioelectrochemistry* **2002,** *55*(1), 33–35.
57. Kensinger, R. S.; Magliaro-Macrina, A. L. Lactation: Induced Lactation. In *Encyclopaedia of Dairy Sciences;* Fuquay, J. W., Fox, P. F., McSweeney, P. L. H., Eds.; Academic Press: San Diego, CA, 2011; vol 2, pp. 20–25 [Online]. https://www.sciencedirect.com/referencework/9780123744074/encyclopedia-of-dairy-sciences (accessed Aug 31, 2021).
58. Kováčová, R.; Synytsya, A.; Štětina, J. Characterisation of Whey Proteins-Pectin Interaction in Relation to Emulsifying Properties of Whey Proteins. *Czech J. Food Sci.* **2009,** *27*, S4–S8.
59. Kunz, C.; Lönnerdal, B. Re-evaluation of the Whey Protein/Casein Ratio of Human Milk. *Acta Paediatr.* **1992,** *81*(2), 107–112.
60. Kussendrager, K. D.; van Hooijdonk, A. C. Lactoperoxidase: Physicochemical Properties, Occurrence, Mechanism of Action and Applications. *Br. J. Nutr.* **2000,** *84*(S1), 19–25.
61. Lajnaf, R.; Zouari, A.; Trigui, I.; Attia, H.; Ayadi, M. A. Effect of Different Heating Temperatures on Foaming Properties of Camel Milk Proteins: A Comparison With Bovine Milk Proteins. *Int. Dairy J.* **2020,** *104*, E-article 104643.
62. Lawrence, R. A. Breastfeeding: Benefits, Risks and Alternatives. *Curr. Opin. Obstet. Gynecol.* **2000,** *12*(6), 519–524.

63. Lien, E. L. Infant Formulas With Increased Concentrations of α-lactalbumin. *Am. J. Clin. Nutr.* **2003,** *77*(6), 1555S–1558S.
64. Liu, J.; Tian, J.; He, W.; Xie, J.; Hu, Z.; Chen, X. Spectrofluorimetric Study of the Binding of Daphnetin to Bovine Serum Albumin. *J. Pharm. Biomed. Anal.* **2004,** *35*(3), 671–677.
65. Loveday, S. M.; Hindmarsh, J. P.; Creamer, L. K.; Singh, H. Physicochemical Changes in Model Protein Bar During Storage. *Food Res. Int.* **2009,** *42*, 798–806.
66. Majorek, K. A.; Porebski, P. J.; Dayal, A.; Zimmerman, M. D.; Jablonska, K.; Stewart, A. J.; Minor, W. Structural and Immunologic Characterization of Bovine, Horse, And Rabbit Serum Albumins. *Mol. Immunol.* **2012,** *52*(3–4), 174–182.
67. Mantovani, R. A.; Pinheiro, A. C.; Vicente, A. A; Cunha, R. In vitro Digestion of Oil-in-Water Emulsions Stabilized by Whey Protein Nanofibrils. *Food Res. Int.* **2017,** *99*, 790–798.
68. Maqsood, S.; Al-Dowaila, A.; Mudgil, P.; Kamal, H.; Jobe, B.; Hassan, H. M. Comparative Characterization of Protein and Lipid Fractions From Camel and Cow Milk: Their Functionality, Antioxidant and Antihypertensive Properties Upon Simulated Gastrointestinal Digestion. *Food Chem.* **2019,** *279*, 328–338.
69. Marnila, P.; Korhonen, H. Immunoglobulins. In *Encyclopedia of Dairy Sciences*; Fuquay, J. W., Fox, P. F., McSweeney, P. L. H., Eds.; Academic Press: San Diego, CA, 2011; vol 2, pp 807–815.
70. McKenzie, H. A.; Shaw, D. C.; Ralston, G. B. Location of Sulfhydryl and Disulfide Groups in Bovine β-lactoglobulins and Effects of Urea. *Biochemistry* **1972,** *11*, 4539–4547.
71. Mehra, R.; Marnila, P.; Korhonen, H. Milk Immunoglobulins for Health Promotion. *Int. Dairy J.* **2006,** *16*(11), 1262–1271.
72. Merin, U.; Bernstein, S.; Bloch-Damti, A.; Yagil, R.; van Creveld, C.; Lindner, P.; Gollop, N. Comparative Study of Milk Serum Proteins in Camel (*Camelus dromedarius*) and Bovine Colostrum. *Livest. Prod. Sci.* **2001,** *67*, 297–301.
73. Meza, B. E.; Verdini, R. A.; Rubiolo, A. C. Effect of Freezing on the Viscoelastic Behaviour of Whey Protein Concentrate Suspensions. *Food Hydrocoll.* **2010,** *24*(4), 414–423.
74. Morr, C. V.; Josephson, R.V. Effects of Calcium, N-Ethylmaleimide and Casein Upon Heat Induced Whey Protein Aggregation. *J. Dairy Sci.* **1968,** *58*, 977–985.
75. Mutilangi, W. R. M.; Kilara, A. Functional Properties of Heat-Denatured Whey Protein. I Solubility. Milchwissenschaft **1985,** *40*(6), 338–340.
76. Nielsen, M. A.; Coulter, S. T.; Morr, C. V.; Rosenau, J. R. Four Factor Response Surface Experimental Design for Evaluating the Role of Processing Variables Upon Protein Denaturation in Heated Whey Systems. *J Dairy Sci.* **1973,** *56*(1), 76–83.
77. O'Mahony, J. A.; Fox, P. F.; Kelly, A. L. Indigenous Enzymes of Milk. In *Advanced Dairy Chemistry Proteins: Basic Aspects*; McSweeney, P. L. H., Fox, P. F., Eds.; Springer: New York, NY, 2013; chapter 12, vol 1A, pp 337–385.
78. Oevermann, A.; Engels, M.; Thomas, U.; Pellegrini, A. Antiviral Activity of Naturally Occurring Proteins and Their Peptide Fragments After Chemical Modification. *Antivir. Res.* **2003,** *59*, 23–33.
79. Oftedal, O. T. Origin and Evolution of the Major Constituents of Milk. In *Advanced Dairy Chemistry Proteins: Basic Aspects,* 4th ed.; McSweeney, P. L. H., Fox, P. F., Eds.; Springer: Boston, MA, 2013; chapter 1, vol 1A, pp 142–148.
80. Padrao, J.; Gonçalves, S.; Silva, J. P.; Sencadas, V.; Lanceros-Méndez, S.; Pinheiro, A. C.; Dourado, F. Bacterial Cellulose-Lactoferrin as an Antimicrobial Edible Packaging. *Food Hydrocoll.* **2016,** *58*, 126–140.

81. Pan, Y.; Shiell, B.; Wan, J.; Coventry, M. J.; Michalski, W. P.; Lee, A.; Roginski, H. The Molecular Characterisation and Antimicrobial Properties of Amidated Bovine β-lactoglobulin. *Int. Dairy J.* **2007,** *17*(12), 1450–1459.
82. Papiz, M. Z.; Sawyer, L.; Eliopoulos, E. E.; North, A. C. T.; Findlay, J. B. C.; Sivaprasadarao, R.; Kraulis, P. J. The Structure of β-lactoglobulin and its Similarity to Plasma Retinol Binding Protein. *Naturè* **1986,** *324*(6095), 383–385.
83. Pelegrine, D. H. G.; Gasparetto, C. A. Whey Proteins Solubility as Function of Temperature and pH. *LWT Food Sci. Technol.* **2005,** *38*(1), 77–80.
84. Pellegrini, A. Antimicrobial Peptides From Food Proteins. *Curr. Pharm. Des.* **2003,** *9*(16), E-article 1225.
85. Peters, T. Serum Albumin. In *Advances in Protein Chemistry*; Anfinsen, C. B., Edsall, J. T., Richards F. M., Eds.; Academic Press: San Diego, CA, 1985; chapter 3, vol 37, pp 161–245.
86. Ramos O. L.; Pereira R. N.; Rodrigues R. M.; Teixeira J. A.; Vicente A. A.; Malcata F. X. Whey and Whey Powders: Production and Uses. In *Encyclopedia of Food and Health*; Academic Press: New York, USA, 2016; pp 498–505.
87. Richert, S. H. Physicochemical Properties of Whey Protein Foams. *J. Agric. Food Chem.* **1979,** *27*, 665–670.
88. Satué-Gracia, M. T.; Frankel, E. N.; Rangavajhyala, N.; German, J. B. Lactoferrin in Infant Formulas: Effect on Oxidation. *J. Agric. Food Chem.* **2000,** *48*(10), 4984–4990.
89. Schmidt, R. H.; Packard, V. S.; Morris, H. A. Effect of Processing on Whey Protein Functionality. *J. Dairy Sci.* **1984,** *67*(11), 2723–2733.
90. Size, G. M. *Share and Trends Analysis Report by Application (Electronics, Composites, Energy), By-product (Graphene Nanoplatelets, Graphene Oxide), by Region, and Segment Forecasts, 2020–2027*; 2020 [Online]. https://www. grandviewresearch. com/ industry-analysis/graphene-industry (assessed Mar 4, 2021).
91. Soliman, T. N., Farrag, A. F., Shendy, A.; El-Sayed, M. M. Denaturation and Viscosity of Whey Proteins Solutions as Affected by Frozen Storage. *J. Am. Sci.* **2010,** *6*, 49–62.
92. Tang, J. E.; Manolakos, J. J.; Kujbida, G. W.; Lysecki, P. J.; Moore, D. R.; Phillips, S. M. Minimal Whey Protein With Carbohydrate Stimulates Muscle Protein Synthesis Following Resistance Exercise in Trained Young Men. *Appl. Physiol. Nutr. Metab.* **2007,** *32*(6), 1132–1138.
93. Tran, H.; Datta, N.; Lewis, M. J.; Deeth, H. C. Predictions of Some Product Parameters Based on the Processing Conditions of Ultra-High-Temperature Milk Plants. *Int. Dairy J.* **2008,** *18*(9), 939–944.
94. Wagner, J. *Membrane Filtration Handbook: Practical Tips and Hints*, 2nd ed.; Osmonics: Minnetonka, Cambrige, 2001; p 129.
95. Wang, W.; Zhong, Q.; Hu, Z. Nanoscale Understanding Of Thermal Aggregation of Whey Protein Pre-treated by Transglutaminase. *J. Agric. Food Chem.* **2010,** *61*(2), 435–446.
96. Wang, W. B.; Bao, Y. H.; Hendricks, G. M.; Guo, M. Consistency, Microstructure and Probiotic Survivability of Goats' Milk Yoghurt Using Polymerized Whey Protein as a Co-Thickening Agent. *Int. Dairy J.* **2012,** *24*(2),113–119.
97. Warny, M.; Fatimi, A.; Bostwick, E. F.; Laine, D. C.; Lebel, F.; LaMont, J. T.; Kelly, C. P. Bovine Immunoglobulin Concentrate-Clostridium Difficile Retains C-Difficile Toxin Neutralizing Activity After Passage Through the Human Stomach and Small Intestine. *Gut* **1999,** *44*(2), 212–217.

98. Wynn, P. C.; Sheehy, P. A. Minor Proteins, Including Growth Factors. In *Advanced Dairy Chemistry- Proteins: Basic Aspects*; McSweeney, P. L. H., Fox, P. F., Eds.; Springer: New York, NY, 2013; chapter 11, vol 1A, pp 317–335.
99. Xiao, Y.; Liu, Y.; Wang, Y.; Jin, Y.; Guo, X.; Liu, Y.; Xu, H. Heat-Induced Whey Protein Isolate Gels Improved by Cellulose Nanocrystals: Gelling Properties and Microstructure. *Carbohydr. Polym.* **2020**, *231*, E-article 115749.
100. Zadow, J. G. Utilization of Milk Components: Whey. In *Modern Dairy Technology, Advances in Milk Processing*, 2nd ed.; Robinson, R. K., Ed.; Chapman & Hall: London, UK; 1994; vol 1, pp 313–373.
101. Zhang, S.; Vardhanabhuti, B. Effect of Initial Protein Concentration and pH on in Vitrogastric Digestion of Heated Whey Proteins. *Food Chem.* **2014,** *145*, 473–480.
102. Zhong, Q.; Wang, W.; Hu, Z.; Ikeda, S. Sequential Preheating and Transglutaminase Pre-treatments Improve Stability of Whey Protein Isolate at pH 7 During Thermal Sterilization. *Food Hydrocoll.* **2013,** *31*(2), 306–316.
103. Zhou, P.; Xiaoming, L. X.; Labuza, T. P. Moisture-Induced Aggregation of Whey Proteins in a Protein/Buffer Model System. *J. Agric. Food Chem.* **2008,** *56*, 2048–2054.

PART IV

PROCESSING AND CHARACTERIZATION OF MILK AND MILK PRODUCTS

CHAPTER 13

PROTEOLYSIS IN ULTRA-HIGH-TEMPERATURE (UHT) MILK: CAUSES, ASSESSMENT, AND REMEDIES

SUVARTAN G. RANVIR, MANJU SINGH, PRANALI NIKAM, SONIYA RANVEER, HARISH KUMAR, and THEJUS JACOB

ABSTRACT

The ultra-high-temperature (UHT) milk is stable for a long storage period; however, several types of physicochemical changes may occur that can degrade the shelf-life of milk during the storage by developing bitterness and gelation, which leads to consumer rejection of products. Proteolysis is the major spoilage due to principal enzymes (plasmin: indigeneous milk proteinase) and microbial produced heat-resistant proteinases especially by *psychotropic* bacterial. Several types of methods are used for assessment of proteolysis, such as colorimetric, fluorimetric, chromatographic, RP-HPLC, ATR-FTIR, and SDS-PAGE method. To overcome these problems and to improve shelf-life of UHT milk, dfferent methods can be used, such as enzymes inactivation by heating treatments, inactivation of enzymes by using the low-temperature treatment, innovative steam injection heating, and the addition of additives.

13.1 INTRODUCTION

Milk contains abundant amount essential constituents, such as proteins, carbohydrates, vitamins, and minerals (major and trace). Its high nutrient content and low acidity provide suitable conditions for the growth of spoilage-causing

The Chemistry of Milk and Milk Products: Physicochemical Properties, Therapeutic Characteristics, and Processing Methods. Megh R. Goyal, Suvartan Ranvir, & Junaid Ahmad Malik (Eds.)

microorganisms,[54] which can be eliminated by the use of ultra-high-temperature (UHT), pasteurization, sterilization, and thermization.

The UHT treatment is a potential treatment involving 135–150°C for 1–10 s, which effectively makes milk sterile so that aseptically packaged milk can be stored for several months at ambient temperature.[54,63] For the first time in 1893, indirect heating system (125°C for 6 min) with continuous flow was used. Later in 1912, the direct heating system (steam was directly mixed with milk) to reach 130–140°C was patented.[9,13] The major aim of UHT treatment is to inactivate enzymes and killing the pathogenic and nonpathogenic microorganisms. The holding time in UHT processing for milk is very low at elevated temperature, which is advantageous and it showed very less conceivable undesired alterations in organoleptic and physicochemical parameters of milk.[9,14]

This chapter presents review on the challenging issues in UHT milk and its causes of spoilage and remedies, and recent developments in assessment of UHT milk.

13.2 UHT MILK PROCESSING METHODS

13.2.1 UHT MILK PROCESSING: DIRECT HEATING METHOD

Before subjecting milk to the direct heating method, the milk is preheated at temperature range from 60 to 85°C for 2–6 s. The combination of time temperature for preheating treatment varies from 76 to 80°C for 40–70 s for maximum positive effects and for eliminating undesirable effects. Then, the preheated milk is directly mixed (no barrier between milk and heating medium) with superheated steam under pressure to achieve desired processing temperature.[15,52] The optimum preheating temperature is advantageous for minimal protein destabilization to prevent sediment formation during storage, minimum undesirable flavors, color development, and reasonable inactivation of native plasmin.[16]

Direct heating systems is again classified into infusion and injection on the basis of contact with the heating medium during milk processing.

- **Infusion:** In this system, the fluid milk is pumped into a big chamber of high-pressure steam through a distribution nozzle. This system is comprised of a small product volume (milk) and large steam volume, circulation in large surface area of the product. The adequate temperature of milk is maintained by steam pressure. By using a plate or tubular heat exchangers, the additional holding time can be attained; afterward, the product is flash cooled in the vacuum chamber. The infusion system is suitable for low- and high-viscous products; it allows instant heating and quick cooling, and no localized burn-on is observed.[6,8,15]

- **Injection:** In preheated milk, high-pressure steam is injected through a steam injector leading to a quick rise in the temperature of milk. Followed by holding, product is flash-cooled in a vacuum to remove water, which is replaced by an equal quantity of condensed steam. The injection system is suitable for low- and high-viscos products; it allows rapid heating and cooling and also helps in volatile removal. However, this method is limited due to hot equipment contact with the product which leads to increase in the chances of flavor damage.[9,15,36]

13.2.2 UHT MILK PROCESSING: INDIRECT HEATING SYSTEMS

In this system, milk is not in direct contact with heating medium (barrier between milk and heating medium). The raw milk is preheated at 80 to 95°C for 15 s to several minutes; and later on is subjected to UHT processing.[42] Normally, tubular- or plate-heat exchanger is used in this heating system. The plate-heat exchanger is usually used in the pasteurization process. It can be inspected very quickly and it requires less floor space as compared to other indirect exchangers. The major benefit of plate-heat exchangers is the saving more energy because of the regeneration process.[18]

The tubular type is a simple and low-cost heat exchanger. It does not have gasket limitations or plate fatigue because it consists of fewer seals. The tubular-heat exchanger is difficult to inspect and occupies larger floor space. Heating may be more uniform in this exchanger, but there may be chances of burn-on and browning due to the broad area needed to hit the target.[33] This type of exchanger is low cost and is more suitable in small-scale operations because of high heat transfer per unit length.

Another type of heat exchanger in the indirect system is scraped surface exchanger. This exchanger consists of a cylindrical tube with a moving shaft fixed at the center, and the scraped blades are located on the middle of moving shaft so that as the shaft is rotated, they run within the cylindrical wall. The product material is driven into annular space (amid the heated wall and the rotator shaft).[33,55,59] The product is more uniformly heated and there are very few chances of burn-on or browning of the product. This exchanger is useful for highly viscous products.[26,59] Advantages and difficulties of UHT milk processing are presented in Table 13.1.

TABLE 13.1 Advantages and Difficulties of UHT Milk Processing.

Advantages	Difficulties
Ability to transport it long distances without deviating the quality	Particle size limitation because chances of surface overcooking and its required to transport material
Condition of manufacturing are independent on the filling size of the container, this ensures that large containers can be filled or sold to food suppliers for food service	Heat resistant native and bacterial produce proteases and lipases may cause abnormal flavor, gelation, and increased viscosity during UHT milk storage
Extensive graphics are possible due to the use of laminated packaging	Due to high heat temperature in UHT processing proteins molecules gets denaturation and they are deposited on the surfaces of heat transfer by causing a fouling layer which is responsible for reducing heat transfer. Also cooked flavored may develop during UHT processing
It is time saving and energy saving process and also reduce labor numbers	Due to expensive packaging and a more specific requirement, the cost of the UHT process is higher
UHT processed milk suitable to drink up to several months	Nutritional quality of milk can affect due to high-temperature heating.

Source: Gedam et al.[26]

13.3 THE UHT MILK PROCESSING: ISSUES AND CHALLENGES

Mostly two types of issues occur in UHT milk processing, such as due to heating and enzymatic. Enzyme-induced issues occurring in UHT milk are due to proteolysis and the enzymes responsible for proteolysis in indigenous milk are plasmin or heat-resistance extracellular proteinases produced by a certain type of bacterial especially psychotropic.[9,44,53] During heating, maximum number of psychrotrophic bacteria get killed except *Bacillus* spp., these bacteria produce very heat stable extracellular enzymes. Many studies have been reported that these bacterially produced proteases have been shown to survive even after UHT treatment.[56,70] Proteolysis is the root cause of the development of gelation and bitter taste during the storage of UHT milk, which affects its shelf-life and customer acceptance.[9,56]

13.3.1 INDIGENOUS PROTEINASE

Milk is a complex fluid and it consists of several types of indigenous enzymes.[22,23] The plasmin is the naturally occurring principal proteinase in milk, in the plasmin system. The system contains five elements, such as, plasmin, plasminogen, plasminogen activators, and inhibitors of plasminogen activators.

Plasminogen is a precursor of plasmin, which is initially present in the inactive form in milk, and it is activated by plasminogen activator tissue-type (tPA) or urokinase (uPA). Plasminogen is transformed into the plasmin by splitting the Arg_{557}–Ile_{558} bond through the action of plasminogen activators.[6,20,56] The plasmin inhibitors and plasminogen activator inhibitors regulate the activity of plasmin and plasminogen activators.[2,35,43] Plasmin and plasminogen activators are very heat stable, while the plasminogen inhibitors are heat liable (Figure 13.1).

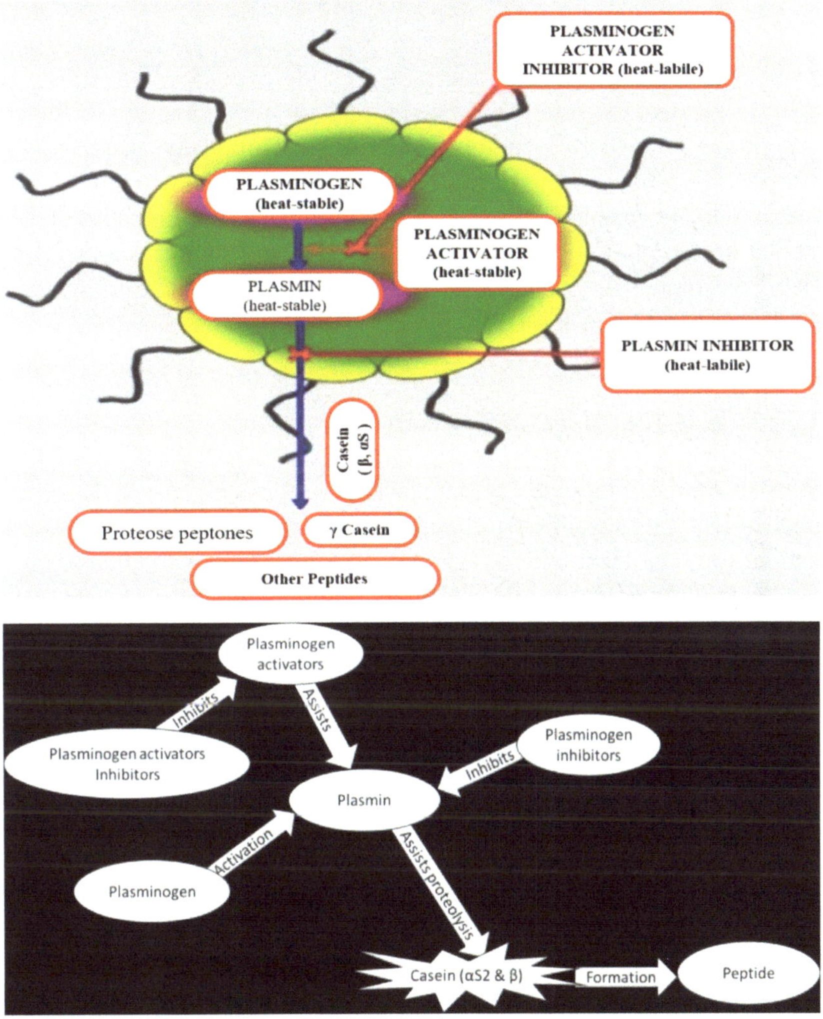

FIGURE 13.1 Bovine milk plasmin system; modified from anema.[3]

The plasmin activators are accompanied with casein (CN) and serum fraction of milk and these also exist in somatic cells,[22,66] whereas plasmin inhibitors and plasmin activator inhibitors are present in serum phase of milk.[6,28] In raw milk, level of plasmin and plasminogen is about 0.7–2.4 μg/mL^{-1} and 0.07–0.15 μg/mL^{-1}, respectively.[57] In mastitis milk, there is higher amount of plasmin and plasminogen because of high load of somatic cells.[58] In milk, plasmin specifically split the peptide bonds consisting arginine and lysine amino acid at N terminal side. Their specificity on hydrolysis of the β and α-CN has been identified. However, it has no activity or very low activity on hydrolysis of κ-CN and whey protein.

The plasmin specifically acts on β-CN and producing γ^1, γ^2, and γ^3 protein fractions and other protein fraction as peptide peptone (PP5, $PP8_{slow}$, and $PP8_{fast}$).[12,61] Also, plasmin hydrolyzed α_{s2}-CN at 8 sites and producing about 14 peptides, out which 3 peptides found might be bitter, that is, αs2- CN fraction of $f_{198-207}$, αs2- CN fraction of $f_{182-207}$, and α_{s2} CN fraction of $f_{189-207}$.[40,67] The activity of plasmin on αs1-CN is significant, although some research studies have reported that it was on αs1-CN and produces several splits sites.[40,48] Among all principal four casein, the κ-CN is the most resistant casein to the action of plasmin.[28,29]

13.3.2 BACTERIAL PROTEINASES

Because of facile culture conditions and simple cell manipulation, their biochemical and physiologic characteristics, microorganisms are the most popular sources of commercial enzymes.[23] Contamination of microorganisms in raw milk majorly occurs due to three sources: teats of cows, udder infection, and utensils of milking and storage. The proteinase formation by psychrotrophic microorganism is typically high during late stationary growth phases.[3,35] Most of the psychrotrophic bacterial produced proteinase is preferable. The most of proteinases produced from psychrotrophs bacteria can generally attack on casein molecules rather than whey protein, in that β-CN and k-CN are more liable as compared to αs-CN molecules.[19,25]

13.4 PROTEOLYSIS IN UHT MILK

Proteolysis involves the hydrolysis of protein or peptides bonds leading to formation of differenct amino acids by the action of specific enzymes. The native alkaline proteinase in milk (such as plasmin and extra-cellular heat-stable proteinases produced by psychotrophs) are principal enzymes responsible for hydrolysis of protein molecules in the UHT milk. Proteolysis is responsible for the production of bitter taste, leading to increased viscosity, with probable gel-formation during storage in UHT milk, thus restricting the shelf-life of milk.[3,9]

The age gelation describes the physical instability of a sterilized milk. There have been several studies for monitoring the reason of development of gelation throughout UHT milk storage and their relation with plasmin activity.[3,15] The plasmin-induced hydrolysis of protein is the root cause of formation of gelation; but in same milk gel formation is retarded, when the plasmin system is entirely deactivated by suppressed by plasmin inhibitors or by heat.[38,45]

Rauh et al.[56] revealed that when raw milk is preheated either at 72 or 95°C and followed by UHT processing, it showed more proteolysis in milk sample preheated at 72°C than the sample preheated at 95°C. They also found that bitter flavor appeared after storage of 6 weeks because of hydrolysis of more than 60% of whole α_S-CN and β-CN; and storage after 10 weeks of same milk, development of gelation was observed with a rise in viscosity because of complete proteolysis of α_S-CN and β-CN. However, there was no clear connection between gelation and plasmin activities according to other researchers.[4,37,60] They reported the aggregation of casein micelles because of physicochemical changes during storage of UHT milk resulting in the developement of 3D structure leading to gelation.

UHT milk subjected to an indirect heating system showed less gelation during storage as compared with the direct method.[30] The formation of gelation varies in milk treated by direct and indirect heating systems mainly due to two reasons[56]: (1) during the indirect heating system, most of the portion of enzymes get inactivated; and (2) more denaturation of whey protein specially β-lactoglobulin in the indirect heating system resulting in proteolytic activity of enzymes inhibits.

Selected research studies have reported that the formation of gelation is dependent on storage temperature of UHT milk. Datta and Deeth[16] stored UHT milk samples at 0 to 40°C, and they observed the higher degree of proteolysis in UHT milk sample stored at 40°C, but the rate of formation of gel was less than a sample stored at 30°C. The argument is that sample stored at 40°C indicated more proteolysis; however, the nature of that proteolysis is such that impaired is the free peptides formed. Datta et al.[15] clarified that milk sample stored at higher temperatures showed lack of gelation due to higher rate of protein decomposition leading to the formation of significant degraded proteins, which could not form a steady gel matrix. Initially, the gel was developed at the base side of the pack before being distributed all over in the pack as time elongates.

There are two types of mechanisms for the development of gelation throughout storage of UHT milk: (1) based on enzyme action and (2) based on physicochemical changes also known as nonenzymatic age gelation.

13.4.1 ENZYMATIC MECHANISM OF GELATION

McMahon[47] proposed that proteinases do not act directly on βκ-complex, however on the complex brackdown so that the κ-CN will break from casein micelle thus enabling the release of βκ-complex (Figure 13.2).

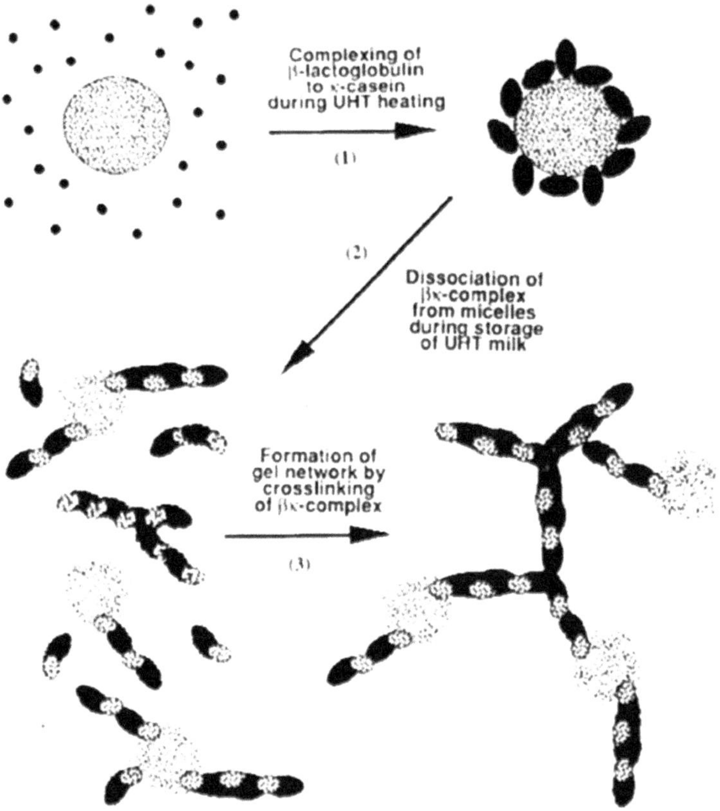

FIGURE 13.2 Enzymatic mechanism of gelation.

Source: Reprinted from Ref. [71] as reprinted in Ref. [13].

These βκ-complexes dissociation from casein micelles is known as the first step of age geleation mechanism. The second step consists of succeeding accumulation of the βκ-complexes and the development of a three-dimensional network of cross-linked proteins.

13.4.2 NONENZYMATIC MECHANISM OF GELATION

Storage promotes the polymerization of whey proteins and casein molecules resulting in the formation of gel in the UHT milk.[1] Contradictorily to this phenomenon, gel formation is retarded in UHT milk when it is stored above 35°C. Samel[60] stated that blockage of ε-NH_2 groups of lysine amino acid, prevention of casein micelles interaction, and changes in charge of casein micelles may result in retardation of age gelation in UHT milk.

Alernate theory suggested that the formation of gelation in UHT milk is due to changes in free energy of casein micelles. The potential energy differences encourage aggregation of casein micelles. The rate of aggregation depends on the number of casein micelles with low potential and possibility of their interaction. The aggregation of micelle results in a rise in the viscosity of the UHT milk.

13.5 DETECTION OF PROTEOLYSIS IN UHT MILK

The method available for assessment of proteolysis in UHT milk is based on measuring free amino acids, which liberate by the action of indigenous plasmin or bacterial proteinase, detecting protease producing organism, and directly measuring plasmin or protease activity. The quantification of a large number of amino acids, or its group and conjugate is extremely difficult[68]; also the analysis is costly and requires more efforts.[31] However, several methods have been developed and these have been broadly studied[2,10,56] in this section.

13.5.1 RP-HPLC ANALYSIS

RP-HPLC is widely used for purification, separation, and assessment of biochemical compounds in dairy and food sectors. Reverse-phase chromatography (RP-C) may separate molecules with certain hydrophobic properties, such as proteins, peptides, and nucleic acid.[39] The plasmin and bacterial proteinases act differently on milk proteins to form several types of peptides. The RP-HPLC approach has been used to distinguish these peptide products. The clarified sample extract for RP-HPLC analysis is prepared by two ways: (1) peptides soluble at pH 4.6; and (2) peptides soluble at 12% TCA. The pH 4.6 soluble clarified extract resulted in peptide peaks formed by proteolysis of milk either or both plasmin and bacterial proteinases; whereas TCA clarified extract resulted in peptide peaks formed by proteolysis of milk only by bacterial proteinase contamination.[14,54]

The ideal TCA concentration is about 4% that is used for peptide analysis. Higher and lower concentration than this value have been reported to be not suitable. The nonprecipitated protein blocked the HPLC by TCA concentration below 4% and higher concentrations contributed to yield of a lower quantity of peptides.[41] The detection limits of proteinase activity analyzed by RP-HPLC methods are nearby 600 times lower as compared to the Fluorescamine method.[41] This method is advanced, reliable, reproducible, and sensitive, while the instrument is expensive due to its restricted use in routine testing.[10]

13.5.2 2,4,6-TNBS ASSAY

This assay was originally proposed by Okuyama and Satake[51] for determining the peptides and amino acids and the reaction was more specific and the tyrosine or histidine side chains were not detected. The TNBS method is based on the principle that measurement derivative of yellow color is due to reaction of free amino acids formed by proteolysis with TNBS solution at 420 nm by spectrophotometry. Chove et al.[10] measured the proteolysis in the UHT milk exposed to different time–temperature combinations by TNBS method; and they found that the proteolysis was positively correlated with the development of yellow color; raw milk and pasteurized milk showed high activity because of survival of protease enzymes; and activity was decreased with the increase in heat treatment.

Ranvir et al.[54] observed the activity proteolysis during storage of UHT milk by spectrophotometric method at 420 nm in the clarified filtrated achieved precipitation pH 4.6 and TCA precipitation. The authors observed that pH-TCA precipitation extract provides precise readings as compared to pH 4.6 precipitation extract because it had been 10 times less diluted. This protocol is a valuable technique for measuring the low level of proteolysis. This method is rapid, and inexpensive compared with other available methods, and it is used for routine laboratory analysis.[46,59,65]

13.5.3 FLUORESCAMINE ASSAY

This method was developed by Castel et al.[7] for monitoring the protease activity. The principle of this method is based on: when primary amines in proteins, like as terminal amino groups react with fluorescamine (nonfluorescent component), then highly fluorescent moieties are formed.[24] Many researchers used the fluorescamine method for the assessment of low level of proteolysis in UHT milk by preparing the sample extract either by precipitation at pH 4.6 or precipitation by TCA.[13] The fluorescamine assay is simple, rapid, and sensitive; however this method is impractical for routine laboratory analysis mainly because of the requirement of expensive equipment and its low upper detection limit.[10]

13.5.4 GEL ELECTROPHORESIS

It is a qualitative analysis method to monitor proteolysis in the sterilized milk caused by protease. Chove et al.[10] studied proteolysis at different concentrations of added plasmin in UHT milk after incubation for 7 days at 37°C. They found no significant casein breakdown occurring in plasmin treated milk on day zero, which was similar to the control; whereas progressive breakdown of caseins was observed on third day which lead to the formation of new γ-CN bands. On day seventh, similar trend of hydrolysis was observed: both the α-CN and β-CN bands were disappearing thus representing the extreme status of proteolysis. The milk treated with higher concentration of plasmin resulted in quick vanishment of α- and β-CN bands. The major drawbacks of this method are the usage of hazardous chemicals, time-consuming process, weak resolution of lower molecular weight peptides.[65]

13.5.5 ATR-FTIR ANALYSIS

FTIR spectroscopy provides molecular changes in various compounds and the overall chemical composition of the analyzed sample.[8,62] Ranvir et al.[54] assessed proteolysis in UHT milk by applying concentrated isoelectric and TCA clarified extracts on ATR-FTIR for over mid-infra-red range of 4000 to 400 cm^{-1}. They reported that there was an increase in absorbance of amide I, II, and III regions during the storage because of progression of proteolysis.

13.5.6 FITC-CASEIN ASSAY

This assay was developed by Twining[64] for the quantitative measurement of protease activity. The FITC-labeled casein substrate could mimic to natural substrate (milk protein) to measure the protease activity. As the protease present in the sample will act on this substrate resulting in splitting of peptide bonds, it leads to formation of fluorescent dye-labeled small peptides.

13.5.7 O-PHTHALDIALDEHYDE (OPA) METHOD

This method can be used for monitoring the proteolysis reported by many researchers. The principle of this assay is based on the fact that OPA reagent reacts with released hydrolyzed α-amino groups to form an adduct, which can be spectrophotometrically measured at 340 nm.[11]

13.5.8 HULL METHOD

This is the oldest method for assessment of proteolysis in UHT milk and milk products. In this method, sodium carbonate is mixed with TCA-soluble peptides that react with a phenol reagent. The tyrosine and tryptophan, aromatic hydroxyl groups reduce the phenol reagent and produce the blue color that is observed spectrophotometrically at 650 nm. Microbial proteases release only low amounts of aromatic amino acids from casein, thus making the Hull method not very responsive for monitoring of milk proteolysis. This system does not detect TCA-soluble peptides or peptides with no aromatic amino acids.[11,34]

13.6 REMEDIES FOR CONTROLLING PROTEOLYSIS IN THE UHT MILK

Proteolyis is one of the main defects during storage of UHT milk. To control or prevent this problem, there are several techniques in use. These techniques are based on principle of reducing or inactivating bacterially produced proteinase or indigenous milk proteinase, which is primarily responsible for gelation in UHT milk. The techniques used in practices employ good quality raw milk, modification of processing or storage conditions, and application of additives.[2,9]

13.6.1 RAW MILK OF GOOD QUALITY

The production of good quality UHT milk product always require good quality raw milk. When the quality of raw milk is degraded, it cannot retain its original quality during the processing, and defects are often more pronounced. Bacterial proteinase is the primarily responsible for gelation in UHT milk and it is produced from psychrotrophic bacteria.[54]

The longer holding of raw milk at lower temperature before the sterilization process facilitates the environment for rise in the count of psychotropic organisms. These pychrotophs produce heat-stable enzymes, such as proteinases in milk. Many researchers have reported that the count of psychotropic bacteria can be reduced by storage of raw milk below at 4°C for short period (<48 h), and it can also be reduce the bacterial proteinases in milk prior heat-treatment.[17] Proteolysis occurs in UHT milk when it is prepared from high bacterial raw milk with about 1 million cfu/mL. If pH is in the normal range but the bacterial count is more than 5 million cfu/mL, then there is a high risk of an increase in the count of bacterial heat-resistant proteinase. This high bacterial proteinase in milk is not suitable for the manufacturing of UHT milk.[2,9]

13.6.2 LOW-TEMPERATURE INACTIVATION (LTI) OF PROTEINASES

Low-temperature treatment is an effective technique for inactivation or reducing the amount of heat resistance to extracellular bacterial proteinase.[9] Barach et al.[5] reported that low-temperature treatment of milk at 55°C for 30 to 60 min was suitable for inactivation of heat-resistant enzymes. Proteinase inactivation is a two-step procedure in which : (1) the first stage results in a conformational transition in protease that modifies the structure of the enzyme. This was demonstrated by the reversible loss of catalytic activity and protease susceptibile to autolysis; and (2) the second stage: the thermally altered proteases were aggregated with casein micelles resulting in the formation complex of enzyme casein. The conformational transition resulted in small irreversible loss of enzyme activity, but complex formation inactivated the enzyme and thus reduced the amount of active protease in milk.

The efficiency of LTI treatment does not depend on the concentration of proteinase. This treatment is only effective at 55°C up to 60 min; afterward, the protease inactivation rate is similarto when UHT treatment is combined with LTI treatment and UHT treatment alone.[9,65] Griffiths et al.,[27] Mitchell et al.[49] carried out research studies on a laboratory scale for assessment of inactivation of bacterial protease produced by alone UHT treatment and combined UHT treatment with LTI treatment (at 55°C for 1 h). LTI treatment given to milk followed by UHT treatment was better than the UHT treated milk alone. This method can be useful prior or poststerilization process. It was reported that this process is most effective one day after the UHT processing of milk.[9,15]

13.6.3 PREHEATING AND STERILIZATION TEMPERATURES

Preheating or forewarning is carried out to improve heat stability of UHT milk. Heat stability of milk was increased because of formation complex of β-lg and k-CN, but the formation of such complexes required adequate heating temperature.[21] As such preheating or sterilization temperature leads to inactivation of the protease activity. For similar bactericidal results, indirect UHT process milk is more stable to the formation of gelation during storage as compared with milk processing by direct heating method. It can be used for preventing gelation in UHT milk; on the other hand there is an unfortunate development of cooked flavor that is not acceptable by consumers leading to reduction in its market demand. The manufacturer must choose processing conditions which produce milk with suitable stability to gelation without excessive flavor damage.[2,9]

13.6.4 ADDITION OF SODIUM HEX METAPHOSPHATE

The sodium hex metaphosphate (SHMP) is widely used in the preservation of children's drinks.[50] Kocak and Zadow[37] reported an addition of SHMP and calcium chloride at a concentration of 0.1 and 0.05%; sodium citrate and ethylenediaminetetraacetic acid at a concentration of 0.1 and 0.3% in raw milk and followed by storage for 5–7 days at 2°C prior to UHT process. The authors observed that adding calcium chloride @ 0.05% or SHMP @ 0.1% in milk before UHT treatments resulted in significant enhancement of storage stability and gelation was not observed after storage of 500 days at 25°C. Their study concluded that SHMP can inhibit the second stage of gelation involving protein coagulation.

13.6.5 ADDITION OF SULFHYDRYL GROUP–BLOCKING AGENT

The addition of *N*-ethylmaleimide (NEM) into the milk prior to heating leads to prevention of denaturation of whey proteins and their interaction with casein molecules. Hong et al.[32] carried out a comparative study of direct and indirect UHT methods with the addition 0.5 g/L NEM. They observed that direct heating processed milk is more stable than indirect heating processed milk for the development of gelation. In the concentrated milk, there was very little effect of NEM on gelation time; however disulfide-reducing agents (such as cysteine and mercaptoethanol) accelerated significantly the formation of gelation. Lysine inhibits the activation of plasminogen for the lysine binding site in the plasmin/ plasminogen kingles.[9,14]

13.6.6 MEMBRANE MILK PROCESSING

Milk can be processed using novel membrane technology so that high-quality concentrate milk can be manufactured without additives with long-term stability through UHT heating. The order of increase in heat stability was significant for UF>NF>RO concentrate due to reduced ash content UF<NF<RO.[9]

13.7 SUMMARY

The UHT processing carried out by direct heating and indirect heating systems followed by aseptically packed the milk could result in elongatation of shelf-life of milk for 6 months at room temperature. The major advantages of UHT processing include: shelf-life extended, it saves time/labor/energy, allow to it to transport over long distances without deviating the quality. There are many factors that affect the marketable success of UHT milk, such as physicochemical changes, heat-induced changes (development of cooked flavor), post-process contamination. Among all these changes, proteolysis is a major spoilage during the storage of milk and it degrades shelf-life by developing bitterness and gelation. Gelation can be enlightened by enzymatic and nonenzymatic mechanisms. The plasmin and bacterial-producing proteinase are principal enzymes to cause the proteolysis in UHT milk. To overcome these problems and improvement of the shelf-life, UHT milk can be obtained by using different strategies, such as enzymes inactivation by heating treatment, inactivation of enzymes by using the low-temperature treatment, innovative steam injection heating, and the addition of additives, etc.

KEYWORDS

- **additives**
- **enzymes inactivation**
- **gelation**
- **physicochemical changes**
- **steam injection heating**
- **UHT**

REFERENCES

1. Akkerman, M.; Johansen, L. B.; Rauh, V.; Poulsen, N.A.; Larsen, L.B. Contribution of Casein Micelle Size and Proteolysis on Protein Distribution and Sediment Formation in UHT Milk during Storage. *Int. Dairy J.* **2021,** *117*, E-aricle 104980 [Online]. https://doi.org/10.1016/j.idairyj.2021.104980 (accessed Oct 21, 2021).
2. Andrews, A. T.; Cheeseman, G. C. Properties of Aseptically Packed Ultra-High-Temperature Milk: II. Molecular Weight Changes of the Casein Components During Storage. *J. Dairy Res.* **1972,** *39*, 395–408.
3. Anema, S. G. Age Gelation, Sedimentation, and Creaming in UHT Milk: A Review. *Compr. Rev. Food Sci. Food Saf.* **2019,** *18*, 140–166.
4. Auldist, M. J.; Coats, S. J.; Sutherland, B. J.; Hardham, J. F.; McDowell, G. H.; Rogers, G. L. Effect of Somatic Cell Count and Stage of Lactation on the Quality and Storage Life of Ultra High Temperature Milk. *J. Dairy Res.* **1996,** *63*, 377–386.
5. Barach, J. T.; Adams, D. M.; Speck, M. L. Low Temperature Inactivation in Milk of Heat-Resistant Proteases From Psychrotrophic Bacteria. *J. Dairy Sci.* **1976,** *59*, 391–395.
6. Bastian, E. D.; Brown, R. J. Plasmin in Milk and Dairy Products: An Update. *Int. Dairy J.* **1996,** *6*, 435–457.
7. Castell, J. V.; Cervera, M.; Marco, R. Convenient Micromethod for the Assay of Primary Amines and Proteins with Fluorescamine: Reexamination of the Conditions of Reaction. *Anal. Biochem.* **1979,** *99*, 379–391.
8. Cebi, N.; Yilmaz, M. T.; Sagdic, O. Rapid ATR-FTIR Spectroscopic Method for Detection of Sibutramine Adulteration in Tea and Coffee Based on Hierarchical Cluster and Principal Component Analyses. *Food Chem.* **2017,** *229*, 517–526.

9. Chavan, R. S.; Chavan, S. R.; Khedkar, C. D.; Jana, A. H. The UHT Milk Processing and Effect of Plasmin Activity on Shelf-life: A Review. *Compr. Rev. Food Sci. Food Saf.* **2011,** *10*, 251–268.
10. Chove, L. M.; Grandison, A. S.; Lewis, M. J. Comparison of Methods for Analysis of Proteolysis by Plasmin in Milk. *J. Dairy Res.* **2011,** *78*, 184–190.
11. Church, F. C.; Swaisgood, H. E.; Porter, D. H.; Catignani, G. L. Spectrophotometric Assay Using O-Phthaldialdehyde for Determination of Proteolysis in Milk and Isolated Milk Proteins. *J. Dairy Sci.* **1983,** *66*, 1219–1227.
12. Crudden, A.; Afoufa-Bastien, D.; Fox, P. F.; Brisson, G.; Kelly, A. L. Effect of Hydrolysis of Casein by Plasmin on the Heat Stability of Milk. *Int. Dairy J.* **2005,** *15*, 1017–1025.
13. Datta, N.; Deeth, H. C. Age Gelation of UHT Milk: Review. *Food Bioprod. Process.* **2001,** *79*, 197–210.
14. Datta, N.; Deeth, H. C. Diagnosing the Cause of Proteolysis in UHT Milk. *LWT Food Sci. Technol.* **2003,** *36*, 173–182.
15. Datta, N.; Elliott, A.; Perkins, M.; Deeth, H. Ultra-High-Temperature (UHT) Treatment of Milk: Comparison of Direct and Indirect Modes of Heating. *Aust. J. Dairy Technol.* **2002,** *57*(3), 211–227.
16. Deeth, H. C.; Datta, N. Heat Treatment of Milk: Ultra-High Temperature Treatment (UHT): Heating Systems. In *Encyclopedia of Dairy Sciences*; Elsevier: New York, 2011; pp 699–707.
17. Deeth, H. C.; Lewis, M. J. Eds.; *High Temperature Processing of Milk and Milk Products*. John Wiley & Sons: New York, 2017; p 556.
18. Emond, S. P. Continuous Heat Processing. In *Thermal Technologies in Food Processing*; Richardson, P., Ed.; Woodhead Publishing: London, 2001; chapter 3, pp 29–48.
19. Fairbairn, D. J.; Law, B. A. Proteinases of Psychrotrophic Bacteria: Their Production, Properties, Effects and Control. *J. Dairy Res.* **1986,** *53*, 139–177.
20. Fang, W.; Sandholm, M. Inhibition of the Proteinase Activity in Mastitic Milk. *J. Dairy Res.* **1995,** *62*, 61–68.
21. Fox, P. F. Heat-Induced Changes in Milk Preceding Coagulation. *J. Dairy Sci.* **1981,** *64*, 2127–2137.
22. Fox, P. F.; Kelly, A. L. Indigenous Enzymes in Milk: Overview and Historical Aspects, Part 2. *Int. Dairy J.* **2006,** *16*, 517–532.
23. Fox, P. F.; Uniacke-Lowe, T.; McSweeney, P. L. H.; O'Mahony, J. A. Enzymology of Milk and Milk Products. In *Dairy Chemistry and Biochemistry,* 2nd ed.; Springer: New York, 2015; pp 377–414.
24. Funk, G. M.; Hunt, C. E.; Epps, D. E.; Brown, P. K. Use of a Rapid and Highly Sensitive Fluorescamine-Based Procedure for the Assay of Plasma Lipoproteins. *J. Lipid Res.* **1988,** *27*, 792–795.
25. Gebre-Egziabher, A.; Humbert, E. S.; Blankenagel, G. Hydrolysis of Milk Proteins by Microbial Enzymes. *J. Food Prot.* **1980,** *43*, 709–712.
26. Gedam, K.; Prasad, R.; Vijay, V. K.; Chanda, G. C. The Study on UHT Processing of Milk: Versatile Option for Rural Sector. *World J. Dairy Food Sci.* **2007,** *2*, 49–53.
27. Griffiths, M. W.; Phillips, J. D.; Muir, D. D. Thermostability of Proteases and Lipases From a Number of Species of Psychrotrophic Bacteria of Dairy Origin. *J. Appl. Bacteriol.* **1981,** *50*, 289–303.
28. Grufferty, M. B.; Fox, P. F. Milk Alkaline Proteinase. *J. Dairy Res.* **1988,** *55*, 609–630.

29. Guinot-Thomas, P.; Al Ammoury, M.; Le Roux, Y.; Laurent, F. Study of Proteolysis during Storage of Raw Milk at 4°C: Effect of Plasmin and Microbial Proteinases. *Int. Dairy J.* **1995,** *5*, 685–697.
30. Harwalkar, V. R. Age Gelation of Sterilized Milks. *Develop. Dairy Chem.* **1982,** *1*, 229–269.
31. Hesse, A.; Weller, M. G. Protein Quantification by Derivatization-Free High-Performance Liquid Chromatography of Aromatic Amino Acids. *J. Amino Acids* **2016,** *2016,* E-article 7374316.
32. Hong, Y. H.; Guthy, K.; Klostermeyer, H. On the Influence of SH-Groups in UHT Milk During Storage. *Milchwissenschaft* **1984,** *39*, 285–287.
33. Hui, Y. H. Ed.; *Handbook of Food Science, Technology, and Engineering*. Boca CRC Press: Raton, FL, 2006; vol 4, p 3632.
34. Hull, M. E. Studies on Milk Proteins. II. Colorimetric Determination of the Partial Hydrolysis of the Proteins in Milk. *J. Dairy Sci.* **1947,** *30*, 881–884. DOI: 10.3168/jds. S0022-0302(47)92412-0.
35. Ismail, B.; Nielsen, S. S. Review: Plasmin Protease in Milk: Current Knowledge and Relevance to Dairy Industry. *J. Dairy Sci.* **2010,** *93*, 4999–5009.
36. Jelen, P. Experience With Direct and Indirect UHT Processing of Milk-a Canadian Viewpoint. *J. Food Prot.* **1982,** *45*, 878–883.
37. Kocak, H. R.; Zadow, J. G. Age Gelation of UHT Whole Milk as Influenced by Storage Temperature. *Aust. J. Dairy Technol.* **1985,** *40*, 14.
38. Kohlmann, K. L.; Nielsen, S. S.; Steenson, L. R.; Ladisch, M. R. Production of Proteases by Psychrotrophic Microorganisms. *J. Dairy Sci.* **1991**, *74*, 3275–3283.
39. Kumar, S. D.; Kumar, D. H. Importance of RP-HPLC in Analytical Method Development: A Review. *Int. J. Pharm. Sci. Res.* **2012,** *3*, 4626–4636.
40. Le Bars, D.; Gripon, J. C. Specificity of Plasmin Towards Bovine As2-Casein. *J. Dairy Res.* **1989,** *56*, 817–821.
41. Le, T. X.; Datta, N.; Deeth, H. C. A Sensitive HPLC Method for Measuring Bacterial Proteolysis and Proteinase Activity in UHT Milk. *Food Res. Int.* **2006,** *39*, 823–830.
42. Lewis, M. J.; Deeth, H. C. Heat Treatment of Milk. In *Milk Processing and Quality Management*; Tamime, A. Y., Ed.; Wiley-Blackwell Publishing Co.: London, 2009; pp 168–204.
43. Lu, D. D.; Nielsen, S. S. Isolation and Characterization of Native Bovine Milk Plasminogen Activators. *J. Dairy Sci.* **1993,** *76*, 3369–3383.
44. Malmgren, B. Evaluation of UHT Milk Processed by Direct Steam Injection and Steam Infusion Technology. PhD Thesis, Department of Food Technology Engineering and Nutrition at Lund University, 2007; p 218.
45. Manji, B.; Kakuda, Y. Effect of Storage Temperature on Age Gelation of UHT-Milk Processed by Direct and Indirect Heating Systems. *Can. Inst. Food Sci. Technol. J.* **1986,** *19*, 2994–3001.
46. McKellar, R. C. Development of Off-Flavors in Ultra-High Temperature and Pasteurized Milk as a Function of Proteolysis. *J. Dairy Sci.* **1981,** *64*, 2138–2145.
47. McMahon, D. J. Age-Gelation of UHT Milk: Changes that Occur during Storage, Their Effect on Shelf-life and the Mechanism by which Age-Gelation Occurs. In Proceedings of the Heat treatments and alternative methods. Unpublished Paper at *International Dairy Federation Symposium, Vienna (Austria)*, Sept 6–8, 1995, p 9.

48. McSweeney, P. L. H.; Fox, P. F. Chemical Methods for the Characterization of Proteolysis in Cheese During Ripening. *Le lait* **1997,** *77*, 41–76.
49. Mitchell, G. E; Ewings, K. N; Bartley, J. P. Physicochemical Properties of Proteinases. Physicochemical Properties of Proteinases From Selected Psychrotrophic Bacteria. *J. Dairy Res.* **1986,** *53*(1), 97–115.
50. Murphy, J.; Beighton, D.; Clark, D.; Bartlett, D. The Potential for Sodium Hexametaphosphate (SHMP) Found in Common Children Drinks to Limit Acid Production in the Oral Biofilm. *J. Dent.* **2007,** *35*, 214–217.
51. Okuyama, T.; Satake, K. Preparation and Properties of 2,4,6-Trinitrophenyl-Amino Acids and Peptides. *J. Biochem. (Tokyo)* **1960,** *47*, 454–466.
52. Ranken, M. D.; Kill, R. C.; Baker, C., Eds.; *Food Industries Manual*, 24th ed.; Blackie Academic & Profesional: London, 1997; p 650.
53. Ranvir, S.; Sharma, R.; Gandhi, K.; Mann, B. Assessment of Physicochemical Changes in UHT Milk During Storage at Different Temperatures. *J. Dairy Res.* **2020,** *87*, 243–247.
54. Ranvir, S.; Sharma, R.; Gandhi, K.; Upadhyay, N.; Mann, B. Assessment of Proteolysis in Ultra-High Temperature Milk Using Attenuated Total Reflectance Fourier Transform Infrared Spectroscopy. *Int. J. Dairy Technol.* **2020,** *73*, 366–375.
55. Rao, C. S.; Hartel, R. W. Scraped Surface Heat Exchangers. *Crit. Rev. Food Sci. Nutr.* **2006,** *46*, 207–219.
56. Rauh, V. M.; Sundgren, A.; Bakman, M.; Ipsen, R.; Paulsson, M.; Larsen, L. B.; Hammershøj, M. Plasmin Activity as Possible Cause for Age Gelation in UHT Milk Produced by Direct Steam Infusion. *Int. Dairy J.* **2014,** *38*, 199–207.
57. Rollema, H. S.; Poll, J. K. The Alkaline Milk Proteinase System: Kinetics and Mechanism of Heat-Inactivation. *Milchwissenschaft* (Dairy Sci.) **1986,** *41*, 536–540.
58. Saeman, A. I.; Verdi, R. J.; Galton, D. M.; Barbano, D. M. Effect of Mastitis on Proteolytic Activity in Bovine Milk. *J. Dairy Sci.* **1988,** *71*, 505–512.
59. Sakaridis, I.; Lewis, M. J. Detecting a Rapid Method for Measuring the Proteolytic Activity of Raw and UHT Milk. *SDRP J. Food Sci. Technol.* **2016,** *1*, 68–77.
60. Samel, R.; Weaver, R. W. V.; Gammack, D. B. Changes on Storage in Milk Processed by Ultra-High-Temperature Sterilization. *J. Dairy Res.* **1971,** *38*, 323–332.
61. Schaar, J. Plasmin Activity and Proteose-Peptone Content of Individual Milks. *J. Dairy Res.* **1985,** *52*, 369–378.
62. Sivakesava, S.; Irudayaraj, J. Prediction of Inverted Cane Sugar Adulteration of Honey by Fourier Transform Infrared Spectroscopy. *J. Food Sci.* **2001,** *66*, 972–978.
63. Sunds, A. V.; Rauh, V. M.; Sørensen, J.; Larsen, L. B. Maillard Reaction Progress in UHT Milk During Storage at Different Temperature Levels and Cycles. *Int. Dairy J.* **2018,** *77*, 56–64.
64. Twining, S. S. Fluorescein Isothiocyanate-Labeled Casein Assay for Proteolytic Enzymes. *Anal. Biochem.* **1984,** *143*, 30–34.
65. Vaghela, K. D.; Chaudhary, B. N.; Mehta, B. M. A Review on Proteolysis Rate in UHT Milk: Its Mechanism, Pattern, Assessment and Enzymatic Changes during Storage: Research & Reviews. *J. Dairy Sci. Technol.* **2018,** *6*, 1–16.
66. Verdi, R. J.; Barbano, D. M. Effect of Coagulants, Somatic Cell Enzymes, and Extracellular Bacterial Enzymes on Plasminogen Activation. *J. Dairy Sci.* **1991,** *74*, 772–782.

67. Visser, S.; Noorman, H. J.; Slangen, C. J.; Rollema, H. S. Action of Plasmin on Bovine β-Casein in a Membrane Reactor. *J. Dairy Res.* **1989,** *56*, 323–333.
68. Walker, V.; Mills, G. A. Quantitative Methods for Amino Acid Analysis in Biological Fluids. *Ann. Clin. Biochem.* **1995,** *32*, 28–57.
69. Yataghene, M.; Legrand, J. 3D-CFD Model Thermal Analysis Within a Scraped Surface Heat Exchanger. *Comput. Fluids* **2013,** *71*, 380–399.
70. Zhang, C.; Bijl, E.; Hettinga, K. Destabilization of UHT Milk by Protease Apr-X From *Pseudomonas Fluorescens* and Plasmin. *Food Chem.* **2018,** *263*, 127–134.
71. McMahon, D. J., 1996, Age-gelation of UHT milk: Changesthat occur during storage on shelf-life and the mechanism by which age-gelation occurs, Heat Treatments and Alternative Methods, (International Dairy Federation—Ref. S.I. 9602, Vienna, Austria), pp 315–325.

CHAPTER 14

HEAT AND ACID COAGULATED MILK PRODUCTS: PHYSICOCHEMICAL CHANGES DURING PROCESSING AND STORAGE

RONIT MANDAL, PAYAL KARMAKAR, KUNTAL ROY, and REKHA RANI

ABSTRACT

The heat-acid coagulated dairy products are produced by acidification of hot milk, which yields a coagulated mass of curd that is separated from whey. *Chhana* and *paneer* are examples of traditional Indian milk products. *Chhana* is used in the manufacture of various milk-based sweets. On the other hand, some soft cheeses like *Queso Blanco*, quark, ricotta are also included in such classification. The raw materials used for production of such products undergo various physicochemical changes during processing, which affect their properties, shelf life, and consumption.

14.1 INTRODUCTION

Milk products can be classified into fluid milk products, concentrated milk products, dried products, heat-desiccated milk products, fermented milk products, and heat-acid coagulated milk products. Heat-acid coagulated milk products are produced by direct acidification of the heat-treated milk. The heat-treated milk when acidified results in the coagulation of milk proteins and aggregation. Such a class of milk products is usually specific to regional culinary and delicacies. For example, the products such as, chhana or paneer

The Chemistry of Milk and Milk Products: Physicochemical Properties, Therapeutic Characteristics, and Processing Methods. Megh R. Goyal, Suvartan Ranvir, & Junaid Ahmad Malik (Eds.)

are indigenous to South Asian countries, that is, India, Pakistan, etc., where they are used in the local cuisines and dishes. Some other examples include the soft cheeses, such as *Queso Blanco*, quark, and ricotta. However, in recent years, these milk products are available for consumption globally due to globalization and migration of people around the world.

The processing and storage of heat-acid coagulate milk products lead to several physicochemical changes in these products. Therefore, this chapter elucidates to the readers about the physicochemical changes that take place during the processing and storage of heat-acid coagulated milk products.

14.2 PROCESSING METHODS

14.2.1 CHHANA

Chhana (also called *chhena*) is a heat-acid coagulated milk product, widely popular in India, and it is used as a base product for preparation of various traditional sweetmeats. While the traditional sweets are of particular interest in the small-scale and non-organized sectors, the commercial interest in chhana and other traditional products is rising in the recent decades.[3] Conventional preparation of chhana starts with the heating of milk to sub-boiling temperature followed by adding coagulants to coagulate the milk proteins and then whey is drained off. A typical process diagram of chhana production is given in Figure 14.1. Due to commercial interest, continuous chhana making line has been developed in the recent decades. This involves metering heated milk and citric acid solution into a mixing tank and separating coagulated milk solids from whey by the assistance of a vacuum strainer. Chhana coagulum can be further utilized for the production of paneer.

14.2.2 PANEER

Paneer (called Indian cottage cheese) is an indigenous heat-acid coagulated milk product in India. It is used in several culinary dishes throughout India, and few other South Asian countries. Paneer has been widely commercialized product which makes it readily available not only in India, but also in other parts of the world. Paneer production in small-scale industries is carried out by heating milk in a pan up to sub-boiling temperatures and then adding the coagulant to coagulate the milk. The aggregated mass of milk solid is separated from the whey using a straining cloth and pressed to form a block to facilitate further expulsion of whey. The process flow diagram of paneer making is given in the Figure 14.2. The National Dairy Development Board (NDDB) has standardized the production of paneer using preferably buffalo milk by heating up to 90°C without holding, followed by cooling to 76°C and adding 1% citric acid solution.[4] The whey is separated from the curd by transferring into a cloth-lined hoop and pressing for about 15 min.

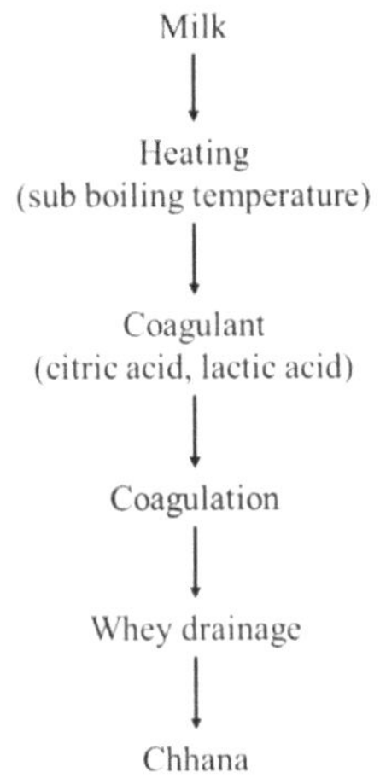

FIGURE 14.1 Chhana making process.

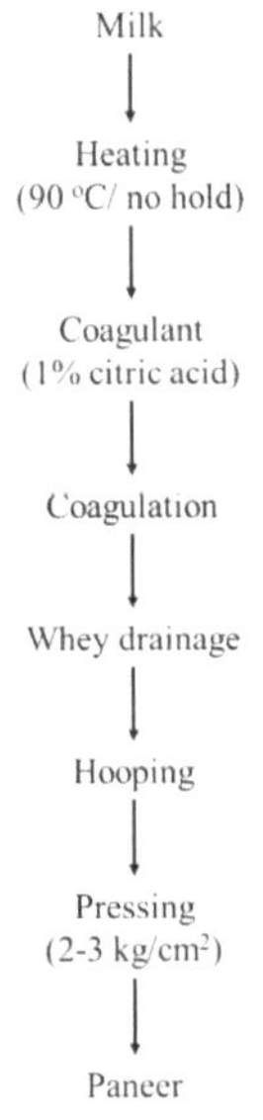

FIGURE 14.2 Production process of paneer.

14.2.3 QUESO BLANCO CHEESE

The *Queso Blanco* is a Latin American variant of the Indian paneer. Typically, it is prepared from cow milk by heating and adding acidifying agents (such as vinegar or lime juice) and stirring until a soft curd is obtained. Then whey is drained off using a straining cloth or cheesecloth. The curd is pressed up to 5 h to obtain a firmer curd. The preparation steps of *Queso Blanco* are given in Figure 14.3.

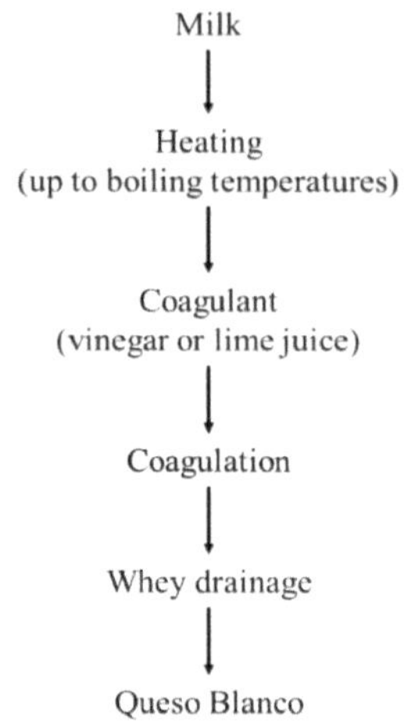

FIGURE 14.3 Queso Blanco making process.

14.2.4 QUARK CHEESE

Quark or quarg cheese is another type of heat-acid coagulated product, which is analogous to *Chhakka*. It is used in various traditional dishes in Germany, Dutch-speaking, Slavic and Scandinavian countries. It is also called acid—set cheese, whereby it is prepared by heating of soured milk. Milk is soured using mesophilic lactic cultures to a pH of 6.0 and then heated. The curd mass is then filtrated using a cheesecloth to separate the whey and quark cheese is obtained. Typical processing method of quark cheese is given in Figure 14.4.

14.2.5 RICOTTA CHEESE

Ricotta cheese is an Italian variety of whey cheese, which is obtained by the heating and acidification of the whey left over from production of other cheese varieties. Ricotta literally translates into a recooked cheese in Italian. The whey is typically obtained from goat or sheep milk, although buffalo milk whey is also utilized. The whey for the manufacture of cheese is collected in pans and acidified further at room temperature. Then, the acidified whey is heated to denature whey proteins (typically globulin and albumin), which form into a white curd mass. The Ricotta cheese so obtained is creamy white and lumpy in appearance with a sweet taste. The process flow diagram of Ricotta cheese is given in Figure 14.5.

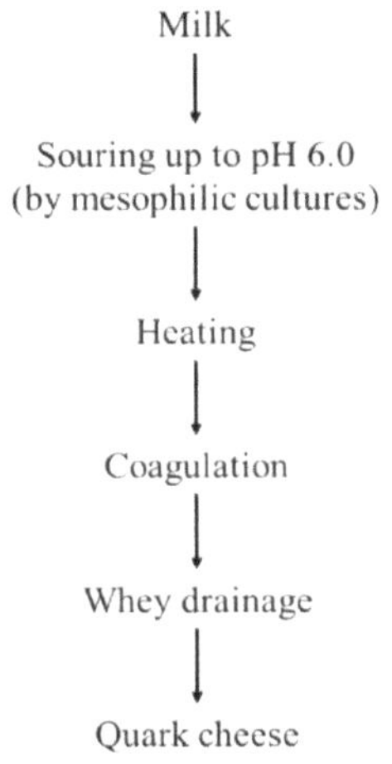

FIGURE 14.4 Quark cheese making process.

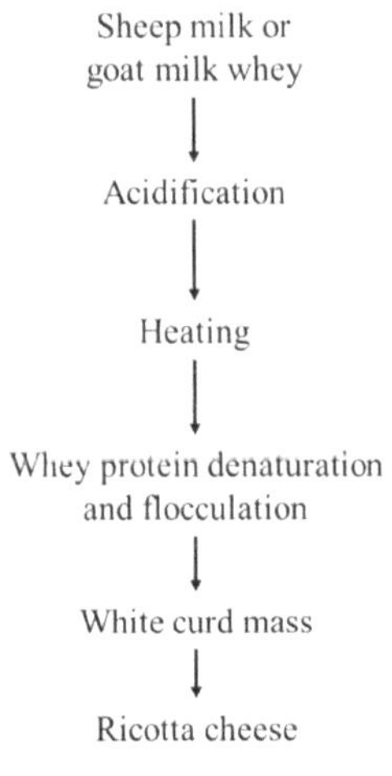

FIGURE 14.5 Ricotta cheese making process.

14.3 PHYSICOCHEMICAL CHANGES DURING PROCESSING AND STORAGE

14.3.1 CHHANA AND CHHANA-BASED SWEETS

Chhana comprises of coagulated mass of milk proteins (such as casein), which can entrap moisture, fat, and other components like minerals, vitamins. Best quality chhana is obtained by using cow milk, which yields a soft chhana. Using buffalo milk, it results in a firmer, chewy, rubbery chhana, which renders it unsuitable for making chhana-based sweetmeats. This is due to the presence of higher calcium content, protein, and curd tension.[7] However, the yield of chhana from cow milk is less than buffalo milk.

During the processing of chhana, the heating up to sub-boiling temperature, leads to the interaction of denatured whey protein β-lactoglobulin with k-casein fraction. The heat-treated and acidified milk proteins also have enhanced water-binding capacity, which is affected by temperature, pH, surface hydrophobicity, and polar groups.[5,7] The whey proteins also interact with the milk fat globules and coat them. Due to lowering of the pH up to 4.6, the micellar casein system destabilizes, and the casein-whey protein complexes aggregate while entrapping the fat and moisture in the coagulum.

Color of the product relies on the milk used for the processing. It is known that the coagulation of casein leads to their aggregation and clustering. This enhances the tendency to scatter light and this chhana appears white in color.[2,10] Higher the casein and fat content in the chhana, greater is the lightness or whiteness.

Textural changes also occur during chhana processing. During conversion of milk into chhana, and due to milk protein-protein interactions, a close firm network is formed.[7] Using buffalo milk for chhana making leads to further increase of firmness due to higher protein content and larger size of casein micelles. Also, higher calcium content enhances the curd tension and forms a close-knit structure.[2] This gives the chhana from buffalo milk a harder, chewy, and rubbery texture.[14] On the other hand, the cow milk chhana exhibits soft, adhesive, and sticky body.[7] Textural parameters of chhana are also affected by the type of coagulant used, which results in a firmer chhana as compared to lactic acid. Sweetmeats made from chhana (such as *Rasogulla*) also tend to be harder when made coagulated using citric acid.[3]

Another notable change during processing is the solubilization of the colloidal calcium phosphate (CCP), hence loosening of bonding of casein submicelles occurs. When the pH decreases to <5.0, the net negative change on casein surface reduces, which then reduces electrostatic repulsion between the casein micelles.[13] Due to hydrophobic interactions, the nature of milk changes into a weak viscous gel-like structure.

Chakraborty et al.[7] carried out the Fourier-Transform Infrared Spectroscopy on milk and chhana samples and observed that major difference among the samples was due to removal of water and concentration of proteins in chhana. Also, fat and water reorganization occurs during coagulation, which leads to stable protein network without any disintegration.

Chhana tends to be spoiled quickly and thus needs to be consumed or transformed into final products. Typically, the shelf-life chhana prepared from cow milk is about 3 days at 22°C.[8] Chhana was observed to undergo slight changes in total solids and protein content, while there was enhanced lipolysis and proteolysis during storage. The microbial growth is also an important factor that accompanies the storage of chhana and chhana-based products.[16]

14.3.2 PANEER

Paneer, like chhana, follows the same process of production, except an additional pressing step. While chhana is preferable made with cow milk, good quality is prepared from buffalo milk. However, paneer prepared from cow milk results in weak and softer body with an open texture, which is unsuitable for the cooking process.[9] Buffalo milk, owing to its higher calcium, fat, and protein content, and bigger casein micelles imparts firm and rubbery body to paneer. Characteristics of paneer make it a non-melting type of milk product when heated. It also possesses the property to be deep fried without any disintegration.

Since, the production process of paneer is similar to the chhana making, the physicochemical changes that take place during their processing are same. Heating of milk to 90°C denatures the whey proteins and leads to aggregation with casein micelles (interaction of beta-lactoglobulin with kappa-casein by sulphydryl-disulphide interchange reactions). The CCP insolubilizes and precipitates. The milk is then lowered to a coagulation temperature of 80°C. Temperature of coagulation influences the fat content, moisture, and solids retention in paneer, which affects the body and texture.[11] During coagulation, the tricalcium phosphate from the casein surface is removed to form monocalcium phosphate and soluble calcium salt. The calcium is also removed from the calcium hydrogen caseinate and free casein is obtained. When the pH decreases by acidification, then casein dispersion is destabilized, and it forms a coagulum, which has entrapped fat and moisture. Calcium plays an important role during paneer manufacturing because it results in a firm body and close texture in paneer.[11] During pressing of paneer, whey, with coagulant is expelled, which further makes the paneer firm. Inadequate pressing of paneer might lead to a softer body with slight acidic flavor.

Aghav et al.[1] observed that the milk fat undergoes some chemical changes during paneer manufacturing. The Reichert-Meissl, Polenske, and saponification values are decreased while milk is converted into paneer. And there was increase in the iodine value, free fatty acid content, and Butyro-Refractometer reading while conversion into paneer.

Paneer is best consumed when fresh. Typical shelf life of paneer at room temperature is not more than 1 day.[11] It can be stored for 6 days at 10°C without any change in body and texture while the freshness is lost after 3 days.[6] Due to microorganism's growth, a greenish yellow slime develops on the surface accompanied by off-flavor development. Also, during long-term storage of paneer, moisture is lost, with an increase in nonprotein nitrogen and substantial changes in pH, fat, and total nitrogen content. Also, the body and texture are deteriorated.

14.3.3 OTHER DIRECT ACIDIFIED CHEESES

Heat treatment during manufacture of some cheeses like Quark prior to acidification plays an important role in incorporation of whey protein into the fresh cheese matrix, while increasing the yield. The heat treatment of milk denatures the whey proteins, aggregates with casein and forms cross-links within the protein network during acidification. Heat treatment of milk decreases the gelation time and coagulation starts at higher pH of 5.2–5.4[12] This is mainly due to the higher isoelectric point of whey protein (pH 5.2) especially beta-lactoglobulin, which starts to show isoelectric precipitation and aggregation. The solubilization of CCP in the network loosens it and enhances syneresis. Heat treatment also increases the dynamic moduli of heat-acid milk gels. This shows greater gel strength of heat-treated milk gels as compared to non-heat-treated gels.

Ricotta cheese, which is an important whey cheese, plays an important role in utilization of the whey after manufacture of cheese. During cheese making, the whey is expelled and termed as sweet whey. Since, the milk is not extensively heat-treated before cheese making, the whey proteins remain in their native form in the expelled whey. The whey is acidified and heat-treated thereafter. The whey proteins denature and without any casein micelle, do not interact with caseins. The denatured whey proteins flocculate and then are strained to form Ricotta cheese. Ricotta cheese keeps fresh for 7 days at refrigerated temperature.[15] During storage, Ricotta cheese undergoes deterioration changes by oxidative reactions and microbial activity. There is also a development of slight reddish color during storage of Ricotta cheese.[15]

14.4 SUMMARY

Heat-acid coagulated milk products include the soft type of cheese including paneer, chhana, *Queso Blanco*, Ricotta, etc., which are essentially made by acidification of heat-treated milk sample. During processing of such products and subsequent storage, several physicochemical changes take place to affect overall eating quality and shelf life. With the availability of such type of cheese all around the world, the study of these changes is very important when designing processing lines for the production and further commercialization of these products. Also, the study of these changes will be beneficial in carrying out shelf-life studies and microbial analysis of such products.

KEYWORDS

- **milk products**
- **coagulation**
- **chhana**
- **paneer**
- **direct acidified cheese**

REFERENCES

1. Aghav, A. D.; Gandhi, K.; Upadhyay, N.; Kumar, A.; Lal, D. A Study on the Physico-Chemical Changes Occurring in the Milk Fat During Preparation of Paneer. *Indian J. Dairy Sci.* **2014,** *67*(5), 398–404.
2. Ahmad, S.; Piot, M.; Rousseau, F.; Grongnet, J. F.; Gaucheron, F. Physicochemical Changes in Casein Micelles of Buffalo and Cow Milks as a Function of Alkalinization. *Dairy Sci. Technol.* **2009,** *89*, 387–403.

3. Bandyopadhyay, M.; Chakraborty, R.; Raychaudhuri, U. The Effect of Coagulants on the Texture of Chhana (an Acid and Heat Coagulated Product Made From Milk). *Int. J. Food Sci. Technol.* **2005,** *40*(8), 799–810.
4. Bandyopadhyay, P.; Khamrui, K. Technological Advancement on Traditional Indian Desiccated and Heat-Acid Coagulated Dairy Products. *Bulletin-Int. Dairy Fed.* **2007,***415*, 4–10.
5. Bech, A. M. The Physicochemical Properties of Whey Proteins. *Dairy Ind. Int.* **1980,** *46*, 25–33.
6. Bhattacharya, D. C.; Mathur, O. N.; Srinivasan, M. R.; Samlik, O. L. Studies on the Method of Production and Shelf Life of Paneer (Cooking Type of Acid Coagulated Cottage Cheese). *J. Food Sci. Technol.* **1971,** *8*, 117–120.
7. Chakraborty, P.; Singh, T.; Shivhare, U. S.; Basu, S. Understanding the Effect of Milk Composition and Milking Season on Quality Characteristics of Chhana. *J. Texture Stud.* **2021,** *52*(1), 45–56.
8. De, S.; Bhattacharya, D.C.; Mathur, O.N.; Srinivasan, M. R. Studies on Shelf-Life of Chhana. *Indian Dairym.* **1971,** *23*, 224–230.
9. Dongare, S. A.; Syed, H. M. Physicochemical and Sensory Properties of Paneer Prepared from Buffalo Milk. *J. Pharmacogn. Phytochem.* **2018,** *7*(2), 3852–3854.
10. Johnson, M. Casein-how it Colors Cheese: Dairy Pipeline-a Technical Resource for Dairy Manufacturers. *Wis. Cent. Dairy Res.* **1999,** *11*(2), 1–12.
11. Kumar, S.; Rai, D. C.; Niranjan, K.; Bhat, Z. F. Paneer: An Indian Soft Cheese Variant: a Review. *J. Food Sci. Technol.* **2014,** *51*(5), 821–831.
12. Lucey, J. A. Formation, Structural Properties, and Rheology of Acid-Coagulated Milk Gels. In *Cheese*; McSweeney, P. L. H., Fox, P. F., Cotter, P. D.; Everett, D. W., Eds.; Academic Press: London, 2017; pp 179–197.
13. Lucey, J. A.; Singh, H. Formation and Physical Properties of Acid Milk Gels: A Review. *Food Res. Int.* **1997,** *30*(7), 529–542.
14. Sindhu, J. S. Suitability of Buffalo Milk for Products Manufacturing. *Indian Dairym.* **1996,** *48*, 41–48.
15. Souza, J. L. F.; da Silva, M. A. P.; da Silva, R. C. F.; do Carmo, R. M.; de Souza, R. G.; Célia, J. A.; de Oliveira, K. B.; Plácido, G. R.; Lage, M. E.; Nicolau, E. S. Effect of Whey Storage on Physicochemical Properties, Microstructure and Texture Profile of Ricotta Cheese. *Afr. J. Biotechnol.* **2016,** *15*(47), 2649–2658.
16. Yadav, B.; Gupta, M. Effect of Preservatives on Physicochemical Changes and Shelf-Life of Chhana. *J. Rural Agric. Res.* **2012,** *12*(1), 19–22.

CHAPTER 15

PROCESSING AND CHARACTERIZATION OF DRY MILK POWDERS

NAVEEN JOSE, MENON REKHA RAVINDRA, and
GAJANAN P. DESHMUKH

ABSTRACT

Conversion of milk into powder is a common strategy adopted by the industry to lengthen the shelf life of milk and to meet requirements during the lean season. Study of engineering properties is essential in optimization of process design, product characterization, process calculations, and to ensure the desired quality and safety of processed foods. Sorption properties help in designing the storage conditions, package selection, and prediction of stability of the product during storage.

15.1 INTRODUCTION

Milk is a balanced diet with limited shelf-life stability due to which the widening of the market is difficult. An alternative to this difficulty is the conversion into dry powder by removal of moisture resulting in an enhanced keeping quality and shelf life. Powders have complex structure, in which the components are randomly dispersed, some in amorphous while others in crystalline forms. Milk powders are highly prone to deteriorative reactions such as caking, softening, browning, compaction, or collapse even though it possesses various advantages such as ease of transportation, handling,

The Chemistry of Milk and Milk Products: Physicochemical Properties, Therapeutic Characteristics, and Processing Methods. Megh R. Goyal, Suvartan Ranvir, & Junaid Ahmad Malik (Eds.)

and convenience in processing and for new product formulations.[1] Physical properties and chemical composition are mainly considered as the quality indicators of milk powders.

Milk powders directly derived from milk includes skim milk powder (SMP), obtained by drying of defatted milk and whole milk powder (WMP) obtained by drying milk after concentration. Various other dairy-based powders have been reported which includes whey powder, cheese powder, ice cream powder, kulfi powder, and other instant dry mixes such as *kheer* mix, *phirni* mix, *gulab jamun* mix powder, dry-crystallized *palada payasam* mix.

Engineering properties and sorption behavior of both milk and dried dairy powders have been reviewed in this chapter.

15.2 ENGINEERING PROPERTIES OF DRY POWDER

Engineering properties of foods include those attributes, which interfere with the processing and material handling of the product. Barbosa et al.[5] classified these properties into physical, thermal, optical, geometrical, and structural properties. An understanding of the food properties and its transformation and evolution during processing is important for optimizing and economizing the process/product design and simulation, package selection, and to ensure the desired quality and safety of processed foods.

15.2.1 STRUCTURAL AND GEOMETRIC PROPERTIES

Structural and geometric properties such as density, shrinkage, and porosity find major application in quality control, process design, product characterization, and process calculations. Structural variations can occur in foods during processing, storage, and distribution, and is directly influenced by factors such as internal and external pressure, temperature, and relative humidity. The variation in the structural and geometric properties of a product can alter other physical properties and these properties are widely employed in the process calculations for mixing, transportation, storing, and packaging of particulate materials. Various properties such as nature and composition of feed, viscosity, temperature, TS content, and drying conditions largely affect the physical characteristics of dairy powders.[42]

15.2.1.1 DENSITY

Density is an inherent property of the product, and it affects almost all other engineering properties. Bulk also known by apparent density, the quantity of the matter contained per unit volume of the product, encompasses both occluded air (air entrapped within the structure) and interstitial air (air entrapped between the particles). Geometry and size of particle was found to have a greater influence on the bulk density of a commodity.[68] The density of grains and other food commodities are applied in the design calculations and constructions of storage bins, silos, powder handling systems, pneumatic conveying, and separation of undesirable materials from the product, cleaning, sorting, grading, and final quality evaluation of the product.[75]

Bulk density of a product, which is a predominant property in packaging, storing, and transportation of powders and can be expressed as following parameters:

- Compact density/packed density, density determined after application of compressive force to the powder bulk mass using impact, vibration, and mechanical pressure.
- Tapped density (TD), determined as the mass per unit volume of the powder that has been subjected to fixed number of tapping or vibrated under standard conditions.
- Loose bulk density, evaluated as the mass per unit volume after allowing the material to flow into a container under free flow under gravity so that the material is allowed to loosely pack.
- Particle density (PD), determined as mass of the product per unit volume of a particle, considering the closed pores for calculation and eliminating the open pores. The PD has been found to decrease linearly with increasing compressibility.[53]

TABLE 15.1 Densities of Common Dairy Powder.

Powder	Particle density (kg/m^3)	Bulk density (kg/m^3)
*MPC 70	1200	580
*MPC 85	1000	380
Instant SMP	1300	450
Instant WMP	1200	480
SMP	1300	750
WMP	1200	600

*MPC—milk protein concentrate

According to Touhy,[75] milk powders prepared as SMP, WMP, and FFMP reported bulk density values between 300 and 620 kg/m^3, whereas on packing, the values for the same commodity were in the range of 440 to 880 kg/m^3. Pugliese et al.[51] reported the average loose bulk density of SMP to be 541.36 and WMP be 449.75 kg/m^3, respectively. Various authors have listed the values of loose and tapped bulk density of SMP as follows, 480 and 640 kg/m^3,[26] 562 and 655 kg/m^3,[71] 530 and 630 kg/m^3.[44] The loose and tapped bulk density of WMP reported by various authors were 360 and 550 kg/m^3,[26] 461 and 551 kg/m^3.[71] The TD of spray dried malted milk powder was reported to be 650 kg/m^3.[58]

Density of dairy-based powders and several instant dry mixes was determined by various authors. Jha et al.[30] reported the bulk, packed, and PD of Instant *Kheer* mix to be 690, 810, and 1250 kg/m^3, respectively. Jose[32] analyzed the physicochemical properties of instant *palada payasam* mix and determined the bulk, packed, and particle densities to be 724, 764, and 1602 kg/m^3, respectively. Mitra[45] reported that the BD, TD, and PD of *basundi* mix were 622, 810, 1546 kg/m^3, respectively while the respective density values were 351, 558, and 1436 kg/m^3, respectively, for ice cream powder. Ice cream mix powders prepared on lab-scale and pilot-scale had bulk, tapped, and particle densities to be 447.47 kg/m^3, 580.64 kg/m^3, and 1215.03, 1288.43 kg/m^3, respectively.

15.2.1.2 POROSITY

Porosity is a measure of the volume of void space within the material and is determined as a ratio of the air occupied volume (voids) to that of the entire material volume. Porosity finds application in the characterization of food products and processing calculations and is reported to affect the thermophysical properties like specific heat, thermal conductivity, and diffusivity. These thermal properties were found to vary with changes in the volume of internal and external pores.[2] The size and distribution of pores within the food product are stated to influence its heat transfer and mass transfer capacities and the texture characteristics and quality attributes associated with dry and intermediate moisture foods.[31] It also affects the hydration properties of the powders. Highly porous powders contain large volume of air contained within it, which may result in reduced shelf life of the product.

TABLE 15.2 Porosity of Common Milk-based Powders.

Powder	Porosity (%)	Ref.
Basundi mix powder	47.02%	[45]
Ice cream mix powder	62.73%	[45]
Instant *kheer* mix	44.8%	[30]
Instant *palada payasam* mix	54.77%	[32]
Kulfi mix powder	47.77%	[64]
Skim milk powder	45%	[54]

15.2.2 MICROSTRUCTURAL PROPERTIES

Milk powders are mainly composed of interstitial gas molecules, water molecules, and solid matter. The study of the microstructural attributes of a product is important, since during processing, the internal structure of the product may be altered but not detected by the naked eye. Scanning electron microscopy (SEM) is extensively used to study shape of the particle, size, and morphological behavior of dry powders such as wrinkles or pores or to detect the presence of sugar crystals (lactose/sucrose) in milk powders which are not been able to be visualized by normal vision.[3,21] Structural data generated through micrographs when combined with the physical properties yield substantial information related to transport phenomena and moisture distribution within the product that could be used to optimize operations such as drying. Reports suggest that spray-dried WMP particles possess spherical shape with some have wrinkles on surface while some have extending parts.[20]

Analysis of scanning electron micrographs of milk powders by Zhong et al.[82] concluded that the processing method influenced the interaction between product components such as lactose crystals and free and encapsulated fat which in turn altered the surface morphology in terms of roughness and discrete or fuzzy boundaries between the microstructural granules as well as the porosity within the powder matrix. Kim et al.[35] compared the surface structural characteristics of SMP with WMP and reported that surface of SMP was found to have deep and shallow dents over the surface of while that of WMP particulates possessed smooth-surface, spherical in shape. Similarly, Koc et al.[39] studied that surface structure of yoghurt powder was found to possess generally spherical in shape but had crater like structures.

Jha et al.[30] presented the scanning electron micrograph of instant kheer mix with irregular shaped agglomerates (5–50μ); it was reported that the individual powder particles in the agglomerates were of varying size, spherical shape, and was having more than one surface infolds. A few particles were found to have perfectly spherical shape with no visible surface infolds and surface cracks or wrinkles were also observed in the analysis.

Pahwa[49] reported the scanning electron micrograph of fresh instant vermicelli kheer mix powder having smooth surfaces with no lactose crystals. They also reported that the individual powder particles in the agglomerates were essentially spherical in shape, varying size, and were also free of surface cracks and wrinkles. The scanning electron micrograph of stored instant vermicelli kheer mix depicted morphological changes (loose matrix with uneven surface and needle-type lactose crystals on particle surface) which revealed that lactose crystallization has a huge impact on the microstructure of milk powder particles. Spray dried ice cream powder particles analyzed using SEM indicated that the particles were of spherical shape but with wide distribution of particles.[45]

15.2.3 THERMAL PROPERTIES

Thermal properties are important physical properties that find major application in the food processing operation. They comprise of properties such as thermal conductivity, sticky point temperature, thermal diffusivity, specific heat, glass transition temperature, and enthalpy. The information about these properties is crucial in the design of heating and cooling equipment, dryer capacity evaluation, temperature change prediction, estimation of energy consumption for various processes and calculation of heating, cooling, and freezing times.[24] Thermal properties alter significantly by variations in the physical properties such as microstructure, porosity, and density of food. Yang et al.[81] reported that variations in moisture content and temperature of the product can significantly affect the thermal conductivity and specific heat since presence of water crucially affects these properties.

15.2.3.1 GLASS TRANSITION TEMPERATURE

The specific temperature at which reversible transition occurs in amorphous food from hard and brittle glassy state into viscous or rubbery state is glass transition temperature (T_g)[6] and the transition is time dependent. Glassy state of the product possesses a rigid structure while transition into amorphous state result in nonequilibrium metastable state. It is an important property considered to establish the storage behavior of a commodity, especially for low and intermediate moisture content foodstuffs and is also considered as an important indicator of stickiness in powders. At this stage, most of the physical changes such as caking, agglomeration, and structural collapse occur.[22] T_g possess an endothermal range is apparent specific heat capacity which is usually measured using DSC.[56] The phase transition usually follows a second order reaction, which is affected by variations in the temperature and amount of moisture present in the food product[8] and T_g values increases with increase in molecular weight.[76] Mitra[45] determined the glass transition temperature of basundi mix and ice cream powder using DSC and was found to be in the range of 23.5–44.79°C and 32.28–50.29°C, respectively, and it decreased inversely with product moisture content (3–9%). Ozmen and Langrish[48] determined the T_g of SMP using DSC and found a similar decreasing trend with increase in moisture content. It was reported to be 87.7°C for low moisture powder (1.65% d.b.) and 46.7°C for powder with high moisture content (4.52% d.b.) and found that T_g and sticky point temperature were same for the product. T_g of *kulfi* mix powder was stated to be in the range 43.1–46.05°C with two transition zones observed of which first in the range of 33–39°C, attributed to fat melting while the second region between 43–50°C, attributed to moisture-dependent glass transition. T_g of various milk powders prepared using spray-drying technique are reported in Table 15.3.

TABLE 15.3 Glass Transition Temperature of Powders.

Product	Glass transition temperature (°C)	Ref.
Lactose powder	97.8	[65]
Skim milk powder (SMP)	93	[65]
SMP with hydrolyzed lactose	48	[65]
WMP with partially hydrolyzed lactose	58	[15]

15.2.3.2 STICKY POINT TEMPERATURE

Sticky point temperature of a powder refers to that temperature causing the powder to either adhere to one another or stick to its contact surfaces; resulting in maximum resistance to the flow of the particles. Liquid bridging between the particles is found to be the major cause of this, which may occur from amorphous sugar, melted fat, or strong sugar solution.[16] Stickiness is most common in sugar rich powders and dairy powders. Stickiness and caking can lead to agglomeration of the powder particles finally resulting into a reduced shelf life, quality losses, risk of explosions in dryers, frequent shutdowns for dryer cleaning reduced flowability and packaging problems. Control of the particle surface temperature and moisture content within the environment are essential in controlling the stickiness in powders. SPT of ice cream powder decreased from 45°C to approximately 10°C with increase of moisture (2–10%) while that of *basundi* mix powder decreased from 40 to 10°C (an increase from 2 to 8%).[45]

15.2.3.3 THERMAL CONDUCTIVITY, DIFFUSIVITY, SPECIFIC HEAT, AND ENTHALPY

Thermal conductivity quantifies the rate of propagation of heat while thermal diffusivity is the material property of how fast the heat is propagated. These properties can be easily determined using KD2-Pro Thermal property analyzer, which typically records the change in temperature manipulated through an electrical heat source placed in a needle along with a temperature sensor, as a rapid method to determine these properties. Specific heat of the product is also recorded using a DSC.

TABLE 15.4 Thermal Properties of Dairy-Based Powders.

Product	Thermal conductivity	Thermal diffusivity	Volumetric specific heat	Ref.
Basundi mix	0.126 W/mK	0.101 m^2/s	1.23 MJ/m^3K	[45]
Ice cream mix	0.131 W/mK	0.099 m^2/s	1.37 MJ/m^3K	[45]
Instant *palada payasam* mix	0.125 W/mK	0.091 m^2/s	1.35 MJ/m^3K	[32]
Skim milk powder	0.079–0.088 W/m°C	–	–	[46]
Whole milk powder	0.064–0.071 W/m°C	–	–	[46]

15.2.4 OPTICAL PROPERTIES

Optical properties are used for quality control and continuous inspection during processing which helps in predicting the color of final product. Color is considered as a major parameter by consumers in evaluating quality and product acceptance. It is important in setting the process time and processing conditions so that the final desired color of the product and the uniformity within different batch of production can be achieved. Color can be expressed as CIELAB parameters in L*, a*, and b* values, which are then mathematically transformed to color descriptors of the product such as hue, chroma, whitening index, browning index, etc.

TABLE 15.5 Color Parameters Derived From the CIELAB Values.

Color parameter and range	Formula
Browning index (BI)	$BI = \frac{100(x - 0.31)}{0.17}$ Where, $x = \frac{a^* + 1.75L^*}{(5.645\,L^* + a^* - 0.3012b^*)}$
Chroma	$\text{Chroma} = \sqrt{(a^* + b^*)}$
Hue	$\theta = \frac{180}{\pi} * \tan^{-1}(\frac{b^*}{a^*})$
Lightness (L*) Black (L*=0) to White (L*=100)	$L^* = \frac{100 \times L}{255}$
Red–green chromaticity (a*) Range: +60 (red) to −60 (green)	$a^* = \frac{240 \times a}{255} - 120$
Whiteness index (WI)	$WI = 100 - \sqrt{(100 - L^*)^2 + a^{*2} + b^{*2}}$
Yellow–blue chromaticity (b*) Range: +60 (yellow) to −60 (blue)	$b^* = \frac{240 \times b}{255} - 120$
Yellowness index (YI)	$YI = \frac{142.86 \times b^*}{L^*}$

Previously, color of the products were measured using Hunter colorimeters which directly measured the L, a, b values. This has since been replaced with spectrophotometric techniques, that can directly measure the L*, a*, and b* values. Image analysis technique is also gaining wide popularity in color measurement; the methodology uses a flatbed scanner or an image capturing device (e.g., camera) to obtain the images that are then further processed in suitable software platforms, such as, Adobe Photoshop or Image J, to obtain L, a, and b values.

Estimation of color using image analysis using a flat-bed scanner and Adobe Photoshop has been reported for espresso drinks,[47] *gulab jamun*,[14] and ghee.[80] Cho et al.[13] examined the browning of banana peel under different packaging conditions using image analysis and colorimetry and found the values comparable.

The L* values of SMP were found to be slightly higher than that of WMP, while due to the yellowish color of WMP, b* values were found to be higher positive and less green as compared to SMP resulted in a* values less negative. Variations in the color values of the product may be due to variations in particle size, presence of particle and pigments, amount of moisture removed during dehydration and drying, brown colored pigment formation due to Maillard reactions.[79] The browning index of *basundi* mix ranged from 24.81 to 25.27, which could be classified as medium-white powder while that of ice cream powder ranged from 13.68 to 13.71 which indicates that *basundi* mix is browner in color due to the Maillard reactions occurring between sugars and proteins.[45] Browning index of Instant *palada payasam* mix was reported to be in the range of 13.98–14.02.[32]

TABLE 15.6 Average Color Values of Common Milk-based Powders.

Product	L*	a*	b*	Ref.
Instant *palada payasam* mix	63.67	9.57	22.11	[32]
SMP	93–94.9	−1.7 to −2.6	11.3–18.3	[37]
SMP	96.94	−2.32	11.12	[51]
WMP	79	0 to 1.7	7.9	[39]
WMP	96.01	−1.74	14.45	[51]

15.2.5 FLOW PROPERTIES

Flow properties, also known as frictional properties, are important in understanding the flow of the material under various conditions. These properties are found to be essential for the design of the handling equipment and gives information regarding the behavior of the product while fed into bins, hoppers, and feeders.

Flow characteristics of the product can be measured in terms of angle of repose, recorded as the angle made by the boundary of the heap formed when the tested material is dropped freely with the horizontal axis. Carr[11] classified powders based on its flowability, as defined by the angle of repose of the product as follows: <35°—free flowable, 35–45°—Moderate flowing (partly cohesive), 45–55°—Loss of free flow (Cohesive) and >55°—no flowability (highly cohesive). The flowability of milk powders was expressed as a function of its angle of repose.[77,70]

TABLE 15.7 Angle of Repose of Milk-Based Powders.

Powder	Angle of repose	Ref.
Basundi mix powder	41.49°	[45]
Cheddar cheese whey powder	32°	[28]
Ice cream mix powder	46.30°	[45]
Instant *kheer* mix	40.09°	[30]
Instant *palada payasam* mix	26.28°	[32]
Khoa powder	64.1°	[54]
Kulfi mix powder	42.65°	[64]
Malted milk powder	62.8°	[58]
Skim milk powder	45°	[54]
Whey-based mushroom soup powder	58°	[18]
Whole milk powder	65°	[54]

The angle of repose ranged from 33 to 38° for SMP to 40 to 58° for WMP.[75] Hausner ratio (HR) and Carr index (CI) are ratios used for evaluating the flowability of powders, which are based on packed and bulk density of the product. CI is used as a measure, indicating the compressibility of the powder (ability to reduce volume by tapping) and the powder will be more flowable if it is more compressible. Touhy[75] reported that compressibility of powders varied in the range of 32–40%. Powders with a CI value more than 28% are having lower flowability[27] and are classified into cohesive powders. HR is expressed as the quotient of tapped and bulk density, and establishes the cohesion/cohesive forces between the powder particles; flowability of the powders was deemed to be poorer at greater values of HR and CI. Powders, which are difficult to flow, were found to have loose density higher than the TD.[19] According to Saw et al.,[59] particle size has a strong influence on HR and there is an inverse relationship between both. Thalberg et al.[73] introduced a new term Modified Hausner Ratio (MHR) to determine the flowability in which the packed density was replaced by PD. Powders having good flowability are those having greater proportion of bigger agglomerates and lesser fine particles,[66] and as per studies of Ilari and Mekkaoui,[27] powders with larger size have lower cohesive forces between particles resulting in reduced friction and Van der Walls forces acting are weak.

Hayes[23] classified the flowability of powders based on HR as 1.0–1.1—excellent flowability, 1.1–1.25—moderate flowing, 1.25–1.4 difficult in flowing, and >1.4 very difficult to flow.

TABLE 15.8 Carr Index and Hausner Ratio for Classification of Flowability of Powders.

Flowability	Carr index (CI) %	Hausner ratio (HR)
Excellent/very free flow	0–10	1.00–1.11
Fair	16–20	1.19–1.25
Good/free flow	11–15	1.12–1.18
Passable	21–25	1.26–1.34
Poor/cohesive	26–31	1.35–1.45
Very poor/very cohesive	32–37	1.46–1.59
Very-very poor/no flow	>38	>1.60

TABLE 15.9 Carr Index and Hausner Ratio of Common Milk-based Powders.

Product	Carr index (CI)	Hausner ratio (HR)	Ref.
Basundi mix powder	15.71	1.18	[45]
Ice cream mix powder	37.0	1.58	[45]
Instant *Palada payasam* mix	4.58	1.04	[32]
Kulfi mix powder	1.36	1.46	[64]
SMP	26.71	1.37	[51]
SMP	25	1.33	[26]
SMP	14.2	1.17	[71]
SMP	15.9	1.19	[44]
WMP	28.13	1.39	[51]
WMP	34.5	1.53	[26]
WMP	16.3	1.20	[71]

Flowability of milk powders were also found to be affected by the particle size as powders having particle size larger than 200 microns have good flowable characteristics whereas powders with fine particles tends to adhere each other which results in increased resistance to flow.[36]

The frictional resistance offered to normal flow of powder particles due to its sliding along the wall of the hopper or the bin itself is quantified using wall friction angle. The *basundi* mix was reported to have a wall friction angle of 29.60 and 29.33° for ice cream mix.[45]

15.2.5.1 FLOW FUNCTION

The relation between the components of the stress tensor (normal to shear) is expressed as the flow function; accordingly, powders are designated as cohesive (2 <FF< 4) and very cohesive (1<FF<2).[29] It is usually computed by dividing the value of principal stress (major) with the unconfined yield stress.

Angle of internal friction is majorly used in the hopper design, which is used to measure the friction exerted between the particles under steady flow. More the angle of internal friction, flow of the material along the sides of the wall is more difficult. These dynamic flow properties can be measured using flow rheometers.

TABLE 15.10 Flow Properties of Dairy-Based Powders.

Product	Flow function	Angle of internal friction, deg	Ref.
Basundi mix	4.55	50.41	[45]
Ice cream mix	1.86	42.92	[45]
Kulfi mix powder (Lab scale)	1.40	49.77	[64]
Kulfi mix powder (Pilot scale)	2.51	45.71	[64]

15.2.6 HYDRATION PROPERTIES

Major hydration properties include sinkability, wettability, dispersibility, and solubility of which the sinkability is the time taken by the powder to sink, once it is wet whereas wettability is the time to wet the surface of the particle after it is added on to the solvent surface, while solubility is the actual solubility of a solid in solution or colloidal dispersion when dispersed.

15.2.6.1 SOLUBILITY

According to Chen and Patel,[12] solubility is the final consideration in reconstitution; usually achieved after the powder has undergone various steps for dissolution, such as sinkability, dispersibility, and wettability. Tamime[72] reported that the solubility of the milk powders (SMP, WMP, and partially skimmed milk powder (PSMP)) obtained by spray-drying technique was more than 99%, while tray dried cheese powder and roller dried milk powders had 91% and 85%, respectively. Therefore, indicating that the method of production significantly influences the hydration properties of the dried powders. The insolubility index of *Kulfi* mix powder, ice cream powder, and *basundi* mix was reported to be 1.42 mL,[64] 1.70 mL,[45] and 0.90 mL,[45] respectively.

15.2.6.2 WETTABILITY

The ability of a material to wet the surface is called as wettability. Schubert[61] pointed out that the particle size and structure including its porous nature, surface characteristics, and air entrapped inside the structure (occluded air content) significantly affects the wettability of powders. Conventionally, wettability is expressed in terms of the time elapsed before complete wetting. The effect of particle size on wetting time of SMP and partial SMP was correlated by Gaiani et al.[17] who reported that wetting time showed a linear decrease with increase in particle size which can be resulted from a lower wetting angle. A wetting time of 90 s for ice cream powder and 45 s for *basundi* mix powder was reported.[45] Lab scale prepared *kulfi* mix powder, the wetting time was determined as 91 s and 86 s for pilot scale prepared powder.[64] The wettability of fresh kheer mix powder was 120 s and the higher value was ascribed to the rice solids as particulates in the mix.[30] Jose[32] determined the wetting time of instant *palada payasam* mix to be 118 s and Salooja and Balachandran[58] reported the wettability of malted milk powder to be 13 s and sinkability as 2 s. WMP was found to be wetted within 30 to 60 s.[34] Sawhney et al.[60] reported that porosity and occluded air within buffalo SMP powder increased its wetting time.

15.2.6.3 DISPERSIBILITY

Dispersibility is the process by which solids disperse or diffuse from the particles. Vilder et al.[78] observed that fat content was a factor that hindered dispersibility of milk powders. Tamime[72] also concurred that dispersibility of WMP was less than that of SMP.

Dispersibility of ice cream powder and *basundi* mix powder were reported to be 84.87 and 73.67% at 3% moisture content and the dispersibility decreased with the moisture content of the powder.[45] Shivanand[64] mentioned the dispersibility of pilot-scale *kulfi* mix powder to be 71%, which was slightly higher than the value observed in lab-scale dried powder (66.2%)

which could be due to the differences in moisture and occluded air contents. Higher moisture content in powders reduces their dispersibility[55] and the presence of wetting and surface-active agents improved the wettability and dispersibility of milk powders.[43]

15.2.7 PARTICLE SIZE

Powder properties such as appearance of powder, reconstitution properties, and flowability characteristics of the product are closely related to its particle size. Major factors that influence the particle size are milk composition and characteristics, processing conditions to which product is exposed and the type of equipments used for drying. Milk powders normally have a particle size ranging from 20–60 μm while that of agglomerated powder was large up to 1 mm and irregular in shape. Roller dried milk powders are usually irregular in shape with cracks on the surface. Spray dried powder particles mostly possess spherical shape with particle size range of 10–250 μm. Research studies revealed that powders having very good dispersibility possessed particle size ranging from 150 to 200 μm.

According to Caric,[10] agglomerated milk powder particle is found to be large and more irregular in shape. Superior dispersibility was found in powders with large particle size and it was found that increase in the fines (<90 μm) in the powder reduced its dispersibility.[67] The size of particles of fine powders can be determined using SEM analysis whereas that of particulates can be determined using sieve analysis in which sieves of different mesh size were used. The mean particle size was found to be 85 μm for regular SMP and 230 to 250 μm for fat-filled milk powders.[75] Particle size distribution of *basundi* mix and ice cream mix were 91.44 and 20.98 μm, respectively. Ice cream mix and *Basundi* mix were found to have $d_{0.1}$, $d_{0.5}$, $d_{0.9}$ as 4.28, 20.98, 100.54 μm and 21.91, 91.44, 258.07 μm at 3% moisture content.[45]

15.3 SORPTION PROPERTIES

15.3.1 WATER ACTIVITY

Research studies revealed that moisture content of the food cannot be alone considered as a stability determining factor, since it depends on the water activity (a_w). Water activity which is obtained as the ratio of partial pressure of water present in food to that of the vapor pressure of pure water, gives information regarding the availability of the water for the reactions. Water present in food can be classified into monolayer water, multilayer water, and capillary water of which the monolayer water is attached by chemical bonds which are difficult to be removed and unavailable for chemical reactions. Products must be dried to its monolayer moisture content to obtain the maximum extended shelf life possible. Products like milk powders possess a shelf life in the range of 6–12 months due to its water activity range of 0.3–0.35. Schuck et al.[62] found that the water activity present in WMP at different spray drying inlet air temperatures (140, 190, and 210°C) as 0.32, 0.20, and 0.16, respectively. The water activity of *instant palada payasam* mix was found to be 0.41–0.42 (Jose, 2019) while that of *gulab jamun* mix was 0.53–0.55.[52]

15.3.2 MOISTURE SORPTION ISOTHERMS (MSI)

The graphical plot of the product moisture (dry basis) against its corresponding water activity or ERH is known as moisture sorption isotherms (MSI). Practically, these plots have been applied in the (1) prediction of water activity at particular moisture of the product, (2) estimation of thermodynamic properties associated with sorption like Gibbs free energy and isosteric heat of sorption, (3) estimation of monolayer (bound), multilayer, and free water. Brunauer classified isotherms into five types of which Type II and Type III isotherms are most commonly found in foods. Type II isotherms are also known as sigmoidal-shaped isotherms which shows asymptotic trend toward water activity 1.0. Type III isotherm is mostly found in high sugar-containing foods where water gaining ability is low at the initial stages.

MSI of milk powders commonly followed a sigmoidal shape with a discontinuity which was attributed to lactose crystallization and could be divided into three specific zones.[7] The initial zone is because of the gradual moisture adsorption at bulk phase interaction sites, while the middle region is constituted of adsorption by the lactose glass. The ultimate region of sorption (corresponding to a_w between 0.6 and 1.0) in the isotherms of milk powders was attributed to the water diffusion into the particles, swelling of proteins, and dissolution of sugars. The studies by them also revealed that the whey powders shown lower moisture sorption characteristics at the initial sorption range in comparison to that of milk powders. At levels greater than critical moisture content (CMC), the deteriorative reactions such as nonenzymatic browning and other hydrolytic reactions may be initiated. Milk powder was found to have higher rate of moisture in the region of moderate to high

RH, which may cause an increase in the rate of the reactions. Joupilla and Roos[33] mentioned that dried products usually report monolayer moisture of 2.1–3.6g/100g dry solids.

TABLE 15.11 Monolayer Moisture Content of Various Products.

Product	BET monolayer moisture content (g/100g) (% d.b.)	Ref.
Basundi mix (10°C)	3.30	[45]
Basundi mix (5°C)	3.58	[63]
Casein powder	5.06 - 6.16	[57]
Casein powder (50°C)	4.91	[4]
Chhana powder (50°C)	1.95	[4]
Gulab jamun mix	3.32	[52]
High heat NFDM*	11.34	[24]
Ice cream powder (10°C)	3.16	[45]
Instant *Palada payasam* mix (25°C)	2.71	[32]
Lactose powder (50°C)	2.20	[4]
Low heat NFDM	9.30	[24]
Medium heat NFDM	10.00	[24]

* NFDM—nonfat dry milk

According to Boquet et al.,[9] Hasley equation is found to be ideal for evaluating the sorption characteristics of milk products followed by Mizrahi equation.

Thanuja and Ravindra[74] determined the sorption properties for cheese-puri mix between 25 and 45°C by the static method. Various mathematical models were evaluated for describing the experimental data and GAB model was found to provide an excellent fit. Decrease of sorption ability with increase in temperature was found for SMP by Stencl[69] who stated that the Modified Oswin equation showed better fit for sorption data of SMP recorded using both cycles of sorption (i.e., adsorption/desorption). Ko et al.[38] characterized the sorption isotherms for milk powders at 15, 25, and 35°C, and found that higher monolayer moisture content was found at 15°C in comparison to those at 25 and 35°C. The adsorption enthalpy of product decreased as EMC increased from 2.23 to 18.66 g/100g. Sharma et al.[66] conducted studies on MSI of *basundi* mix (ready-to-use) by isopiestic method and found that curves were sigmoidal in shape with EMC decreasing with increase in temperature from 5° to 45°. Excellent agreement between experimental and predicted values was showed by GAB model and sorption mechanism was enthalpy driven.

Pushpadass et al.[52] determined the EMC values of *gulab jamun* mix powder at temperatures <40°C, observing a type II isotherm (best described using the GAB model) with reduced sorption at higher temperature. The increase in moisture content also resulted in decrease of isosteric heat of sorption of the product. Shivanand[64] evaluated the moisture sorption characteristics of *kulfi* mix powder prepared lab scale at 10, 25, and 40°C and reported that the isotherm followed Type II sigmoidal shaped curve with GAB model found to be most suitable for the sorption data at all three temperatures.

Mitra[45] studied the sorption behavior of *Basundi* mix and found that the adsorption isotherm of sigmoidal shape (Type II) with two bending regions at 0.02–0.08 and 0.71–0.85 a_w. GAB monolayer value decreased with increase in the sorption temperature. GAB monolayer value at 10°C of 2.291 g/100 g dry solids was reported. His studies on ice cream powder revealed that isotherm obtained were sigmoidal (Type II) curve, having two bending regions and decreasing nature of monolayer value on increasing temperature and the GAB monolayer value at 10°C was 2.27 g/100 g dry solids, Caurie's model fitted well with the data of both *Basundi* mix and ice cream powder. Moisture sorption behavior of yoghurt powder prepared by spray-drying technique was determined by Koc et al.[39] Three temperatures (10, 25, and 40°C) were selected for study and GAB model having better fit for the data followed by Oswin and Modified Henderson model. BET and GAB models were used for the determination of monolayer moisture content and were found to be 0.070–0.092 and 0.136 kg/kg dry solids, respectively.

Kumar[40] conducted moisture sorption studies of *toukir* powder and determined that the isotherm followed a sigmoidal shape classified as Braunner Demming Demming Teller (BDDT) type II isotherm. Almost all dairy products were having GAB model as the ideal model for fitting the sorption data due to its wide range applicability.

15.4 SUMMARY

Dairy-based powders are gaining popularity nowadays due to its extended shelf life and convenience of storage and transport. Analysis of engineering properties associated with these products will help in process engineering calculations, machine design, evaluation of product quality during intermediate and final stages, formulation of new products. Sorption study helps in predicting the shelf stability of a product under given conditions of temperature and RH, this in turn can help in designing the suitable packaging toward optimal retailing and marketing of a product.

KEYWORDS

- **engineering properties**
- **glass transition temperature**
- **microstructure**
- **optical properties**
- **sorption**
- **thermal properties**
- **water activity**

REFERENCES

1. Al-Mahdi, R.; Nasirpour, A.; Banon, S.; Scher, J.; Desobry, S. Morphological and Mechanical Properties of Dried Skimmed Milk and Wheat Flour Mixtures During Storage. *Powder Technol.* **2006,** *163*(3), 145–151.
2. ASHRAE. Refrigeration. *American Society of Heating, Refrigerating and Air-Conditioning Engineers*; Atlanta, GA; 2006; p 236.
3. Baechler, R.; Clerc, M. F.; Ulrich, S; Benet, S. Physical Changes in Heat-Treated Whole Milk Powder. *Le Lait* **2005,** *85*(4–5), 305–314.
4. Bandyopadhyay, P.; Das, H.; Sharma, G. P. Moisture Adsorption Characteristics of Casein, Lactose, Skim Milk and *Chhana* Powder. *J. Food Sci. Technol.* **1987,** *24*(1), 6–11.
5. Barbosa-Canovas, G. V.; Juliano, P; Peleg, M. Engineering Properties of Foods. Chapter 2; In *Food Engineering: Encyclopedia of Life Support Systems*; Barbosa-Canovas, G. V., Ed.; EOLSS Publishers: Oxford, UK; 2009; Vol. 1; pp 39–70.
6. Becker, H.; Locascio, L. E. Polymer Microfluidic Devices. *Talanta* **2002,** *56*(2), 267–287.
7. Berlin, E.; Anderson, B. A.; Pallanch, M. J. Water Vapor Sorption Properties of Various Dried Milks and Whey. *J. Dairy Sci.* **1968,** *51*(9), 1339–1344.
8. Bhandari, B. R.; Howes, T. Implication of Glass Transition for the Drying and Stability of Dried Foods. *J. Food Eng.* **1999**, *40*(1–2), 71–79.
9. Boquet, R.; Chirife, J.; Iglesias, H. A. Equations for Fitting Water Sorption Isotherms of Foods: II. Evaluation of Various Two Parameter Models. *Int. J. Food Sci. Technol.* **1978,** *13*(4), 319–327.
10. Caric, M. Milk Powders. Chapter 4; In *Concentrated and Dried Dairy Products*; Caric, M., Ed.; VCH Publishers: New York, USA; 1994; pp 156–184.

11. Carr, R. L. Powder and Granule Properties and Mechanics. Chapter 2; In *Gas–Solids Handling in the Processing Industries*; Marchello, J. M., Gomezplata, A., Eds.; Marcel Dekker: New York, USA; 1976; pp 23–34.
12. Chen, X. D.; Patel, K. C. Manufacturing Better Quality Food Powders from Spray Drying and Subsequent Treatments. *Dry. Technol.* **2008,** *26*(11), 1313–1318.
13. Cho, J. S.; Lee, H. J.; Park, J. H.; Sung, J. H.; Choi, J. Y.; Moon, K. D. Image Analysis to Evaluate the Browning Degree of Banana (Musa spp.) Peel. *Food Chem.* **2016,** *194*(1), 1028–1033.
14. Emerald, F. M. E.; Ravindra, M. R.; Amaladhas, P. H.; Rao, K. J.; Nath, B. S. Image Analysis and Kinetics of Color Changes During Frying of *Gulabjamun.* Poster Abstract. In *XX Indian Convention of Food Scientists and Technologists*; AFSTI: Bengaluru; 2009; pp 1–9.
15. Fernandez, E.; Schebor, C.; Chirife, J. Glass Transition Temperature of Regular and Lactose Hydrolyzed Milk Powders. *Lebensm. Wiss. Technol.* **2003,** *36*(5), 547–551.
16. Foster, K. D.; Bronlund, J. E.; Paterson, A. H. J. Glass Transition Related Cohesion of Amorphous Sugar Powders. *J. Food Eng.* **2006,** *77*(4), 997–1006.
17. Gaiani, C.; Ehrhardt, J.; Scher, J.; Hardy, J.; Desobry, S.; Banon, S. Surface Composition of Dairy Powders Observed by X-Ray Photoelectron Spectroscopy and Effects on Their Rehydration Properties. *Coll. Surf. B. Biointerfaces* **2006,** *49*(1), 71–78.
18. Ghosh, S; Singh, S. Utilization of Whey for the Manufacture of Ready-To-Serve Mushroom Soup. *Mushroom Res.* **1995,** *4*(1), 34–36.
19. Guerin, E.; Tchoreloff, P.; Leclerc, B.; Tanguy, D.; Deleuil, M.; Couarraze, G. Rheological Characterization of Pharmaceutical Powders Using Tap Testing, Shear Cell and Mercury Porosimeter. *Int. J. Pharm.* **1999,** *189*(1), 91–103.
20. Hall, C. W.; Hendrick, T. I., Eds. In *Drying Milk and Milk Products*, 1st ed.; A.V.I. Publishing Company: Westport-Connecticut, USA, 1965; p 115.
21. Hardy, J.; Scher, J.; Banon, S. Water Activity and Hydration of Dairy Powders. *Le Lait* **2002,** *82*(4), 441–452.
22. Hashemi, N.; Milani, E.; Mortezavi, S. A.; Yazdi, F. T. Sticky Point Temperature as a Suitable Method in Evaluation of Shelf-Life of Food Powders. *Bulletin de la Société Royale des Sciences de Liège* **2017,** *86*(1), 7–12.
23. Hayes, G. D., Ed. *Food Engineering Data Handbook*, 1st ed.; Longman Scientific & Technical: UK; 1987; pp 58–99.
24. Heldman, D. R.; Hall, C. W.; Hedrick, T. I. Vapor Equilibrium Relationships of Dry Milk. *J. Dairy Sci.* **1965,** *48*(7), 845–852.
25. Heldman, D. R.; Lund, D. B.; Sabliov, C., Eds.; *Handbook of Food Engineering*; 3rd ed.; CRC Press: New York; 2018; p 480.
26. Ilari, J. L. Flow Properties of Industrial Dairy Powders. *Le Lait* **2002,** *82*(4), 383–399.
27. Ilari, J. L.; Mekkaoui, L. Physical Properties of Constitutive Size Classes of Spray-Dried Skim Milk Powder and Their Mixtures. *Le Lait* **2005,** *85*(4–5), 279–294.
28. Jayaprakasha, H. M. Membrane Processing Applications for Production of Whey Powder and Whey Protein Concentrate. Master's Degree Thesis, National Dairy Research Institute, Karnal, India, 1992; pp 48–85.
29. Jenike, A. W. Storage and Flow of Solids. *Bulletin of the Utah Engineering Experiment Station.* No. 123. *Bull. Univ. Utah* **1964,** *53*(26), 1–194.
30. Jha, A.; Patel, A. A., Singh, R. R. B. Physicochemical Properties of Instant *Kheer* Mix. *Le Lait* **2002,** *82*(4), 501–513.

31. Joardder, M. U.; Kumar, C.; Karim, M. A. Prediction of Porosity of Food Materials During Drying: Current Challenges and Directions. *Crit. Rev. Food Sci. Nutr.* **2018,** *58*(17), 2896–2907.
32. Jose, N. Engineering Properties and Sorption Behavior Analysis of Dry-Crystallized Palada Payasam Mix. Master's Degree Thesis, National Dairy Research Institute, Karnal, India, 2019; pp 51–88.
33. Jouppila, K.; Roos, Y. H. Glass Transitions and Crystallization in Milk Powders. *J. Dairy Sci.* **1994,** *77*(10), 2907–2915.
34. Kelly, A. L., O'Connell, J. E., Fox, P. F. Manufacture and Properties of Milk Powders. Chapter 2; In *Advanced Dairy Chemistry: Proteins, Parts A&B*; Fox, P. F., McSweeney, P. L., Eds.; Springer: Boston, MA, 2003; Vol. 1; pp 1027–1061.
35. Kim, E. H. J.; Chen, X. D.; Pearce, D. Surface Characterization of Four Industrial Spray-Dried Dairy Powders in Relation to Chemical Composition, Structure and Wetting Property. *Coll. Surf. B. Biointerfaces* **2002,** *26*(3), 197–212.
36. Kim, E. H. J.; Chen, X. D.; Pearce, D. Effect of Surface Composition on the Flowability of Industrial Spray-Dried Dairy Powders. *Coll. Surf. B. Biointerfaces* **2005,** *46*(3), 182–187.
37. Kneifel, W.; Ulberth, F.; Schaffer, E. Tri-Stimulus Color Reflectance Measurement of Milk and Dairy Products. *Le Lait* **1992,** *72*(4), 383–391.
38. Ko, S.; Hong, G.; Min, S.; Choi, M. Moisture Sorption Isotherms of Powders and Their Calculated Heat of Sorption. *Drying Technol.* **2008,** *26*(11), 1396–1400.
39. Koc, B.; Sakin-Yilmazer, M.; Kaymak-Ertekin, F; Balkir, P. Physical Properties of Yoghurt Powder Produced by Spray Drying. *J. Food Sci. Technol.* **2014,** *51*(7), 1377–1383.
40. Kumar, B. Determination of Physical and Engineering Properties of Toukir and its Application in the Preparation of Gulabjamun. Master's Degree Thesis, National Dairy Research Institute, Karnal, India, 2014; pp 48–77.
41. Lebrun, P.; Krier, F.; Mantanus, J.; Grohganz, H.; Yang, M.; Rozet, E.; Hubert, P. Design Space Approach in the Optimization of the Spray-Drying Process. *Eur. J. Pharm. Biopharm.* **2012,** *80*(1), 226–234.
42. Master, K., Ed. *Spray Drying Handbook*, 4th ed.; Longman Scientific and Technical: New York, 1991; p 465.
43. McKenna, A. B.; Lloyd, R. J.; Munro, P. A.; Singh, H. Microstructure of Whole Milk Powder and of Insoluble Detected by Powder Functional Testing. *Scanning* **1999,** *21*(5), 305–315.
44. Mistry, V. V.; Pulgar, J. B. Physical and Storage Properties of High Milk Protein Powder. *Int. Dairy J.* **1996,** *6*(2), 195–203.
45. Mitra, H. Physico Chemical Microstructural Thermal and Flow Properties of Select Dairy Powders. Master's Degree Thesis, National Dairy Research Institute, Karnal, India, 2016; pp 51–73.
46. Muramatsu, Y.; Tagawa, A.; Kasai, T. Effective Thermal Conductivity of Rice Flour and Whole and Skim Milk Powder. *J. Food Sci.* **2005,** *70*(4), 279–287.
47. Nawale, P. K.; Vyawahare, A. S.; Aravindakshan, P.; Rao, K. J. Dairy Based Espresso Drinks Preparation Using Whey as a Base Ingredient. In *Poster Paper Presented at XX Indian Convention of Food Scientists and Technologists*, 2009; pp 21–23.
48. Ozmen, L.; Langrish, T. A. G. Comparison of Glass Transition Temperature and Sticky Point Temperature for Skim Milk Powder. *Drying Technol.* **2002,** *20*(6), 1177–1192.

49. Pahwa, A. Technology of Instant Kheer Mix from Oat-Semolina Based Vermicelli. Doctoral Thesis, National Dairy Research Institute, Karnal, India, 2018, pp 39–88.
50. Pathare, P. B.; Opara, U. L.; Al-Said, F. A. Color Measurement and Analysis in Fresh and Processed Foods: A Review. *Food Bioprocess Technol.* **2013,** *6*(1), 36–60.
51. Pugliese, A.; Cabassi, G.; Chiavaro, E.; Paciulli, M.; Carini, E.; Mucchetti, G. Physical Characterization of Whole and Skim Dried Milk Powders. *J. Food Sci. Technol.* **2017,** *54*(11), 3433–3442.
52. Pushpadass, H. A.; Emerald, F. M. E.; Chaturvedi, B.; Rao, K. J. Moisture Sorption Behavior and Thermodynamic Properties of *Gulabjamun* Mix. *J. Food Process. Preservation* **2014,** *38*(6), 2192–2200.
53. Rahman, M. S., Ed. *Food Properties Handbook*, 2nd ed.; CRC Press: New York, USA, 2009; pp 23–49.
54. Ranganadham, M. Process Innovations in the Manufacture of Khoa Powder. Master's Degree Thesis, National Dairy Research Institute, Karnal, India, 1988; pp 53–87.
55. Reddy, R. S.; Ramachandra, C. T.; Hiregoudar, S.; Nidoni, U.; Ram, J.; Kammar, M. Influence of Processing Conditions on Functional and Reconstitution Properties of Milk Powder Made from Goat Milk by Spray Drying. *Small Rumin. Res.* **2014,** *119*(1–3), 130–137.
56. Roos, Y.; Karel, M. Differential Scanning Calorimetry Study of Phase Transitions Affecting the Quality of Dehydrated Materials. *Biotechnol. Prog.* **1990,** *6*(2), 159–163.
57. Ruegg, M. Water in Dairy Products Related to Quality, with Special Reference to Cheese. Chapter 4; In *Properties of Water in Foods*. Wolf, W., Spiess, W. E. L., Jung, G. E., Simato, D., Multon, J. L., Eds.; Springer: Dordrecht, Netherlands, 1985; pp 603–625.
58. Salooja, M. K.; Balachandran, R. Physical Properties of Spray-Dried Malted Milk Powder. *Indian J. Dairy Sci.* **1988,** *41*(4), 456–461.
59. Saw, H. Y.; Davies, C. E.; Paterson, A. H.; Jones, J. R. The Influence of Particle Size Distribution and Tapping on the Bulk Density of Milled Lactose Powders. *Chemeca Chall. Tomorrow*, **2013,** *15* (1), 299–304.
60. Sawhney, I. K.; Sarkar, B. C.; Patil, G. R. Engineering Properties of Buffalo Skim Milk Powder. *Indian J. Dairy Sci.* **2010,** *63*(3), 163–172.
61. Schubert, H. Food Particle Technology. Part I: Properties of Particles and Particulate Food Systems. *J. Food Eng.* **1987**, *6*(1), 1–32.
62. Schuck, P.; Blanchard, E.; Dolivet, A.; Mejean, S.; Onillon, E.; Jeantet, R. Water Activity and Glass Transition in Dairy Ingredients. *Le Lait* **2005,** *85*(4–5), 295–304.
63. Sharma, P.; Singh, R. R. B.; Singh, A. K.; Patel, A. A.; Patil, G. R. Sorption Isotherms and Thermodynamics of Water Sorption of Ready-To-Use *Basundi* Mix. *LWT-Food Sci. Technol.* **2009,** *42*(1), 441–445.
64. Shivanand. Effect of Spray Drying Conditions and Additives on Physicochemical, Microstructural and Engineering Properties of *Kulfi* Mix Powder. Master's Degree Thesis, National Dairy Research Institute, Karnal, India, 2017; pp 52–89.
65. Shrestha, A. K.; Howes, T.; Adhikari, B.P.; Bhandari, B. R. Water Sorption and Glass Transition Properties of Spray Dried Lactose Hydrolyzed Skim Milk Powder. *LWT – Food Sci. Technol.* **2007,** *40*(9), 1593–1600.
66. Sharma, A.; Jana, A. H.; Chavan, R. S. Functionality of Milk Powders and Milk-Based Powders for End Use Applications-A Review. *Compr. Rev. Food Sci. Food Saf.* **2012,** *11*(5), 518–528.

67. Singh, H.; Newstead, D. F. Aspects of Proteins in Milk Powder Manufacture. In *Advanced Dairy Chemistry*; Fox, P. F., Ed.; Elsevier Applied Science Publications: New York, USA; 1992; Vol. 1; pp 735–765.
68. Singh, R. P.; Heldman, D. R., Eds. *Introduction to Food Engineering*, 3rd ed.; Academic Press: London, UK, 2001; p 556.
69. Stencl, J. Water Activity of Skimmed Milk Powder in the Temperature Range of 20–45°C. *Acta Veterinaria Brno* **1999,** *68*(3), 209–215.
70. Svarovsky, L. Angle of Repose and Other Handling Angles. In *Powder Testing Guide: Methods of Measuring the Physical Properties of Bulk Powders*; Svarovsky, L., Ed.; Elsevier Science Publishers: London, UK; 1987, pp 71–78.
71. Szulc, K.; Nazarko, J.; Ostrowska-Ligęza, E.; Lenart, A. Effect of Fat Replacement on Flow and Thermal Properties of Dairy Powders. *LWT-Food Sci. Technol.* **2016,** *68*(1), 653–658.
72. Tamime, A. Y. Dried Milk Products. Chapter 7; In *Dairy Powders and Concentrated Milk Products*; Tamime, A. Y., Ed.; Blackwell Publishers Ltd.: Oxford, UK, 2009; pp 231–245.
73. Thalberg, K.; Lindholm, D.; Axelsson, A. Comparison of Different Flowability Tests for Powders for Inhalation. *Powder Technol.* **2004,** *146*(3), 206–213.
74. Thanuja, D.; Ravindra, M. R. Thermodynamic Analysis of Moisture Sorption Characteristics of *Cheese-Puri* Mix. *J. Food Process. Preserv.* **2014,** *38*(1), 420–429.
75. Touhy, J. J. Some Physical Properties of Milk Powders. *Irish J. Food Sci. Technol.* **1989,** *13*(1), 141–152.
76. Truong, V.; Bhandari, B. R.; Howes, T. Optimization of Co-Current Spray Drying Process of Sugar-Rich Foods. Part I – Moisture and Glass Transition Temperature Profile During Drying. *J. Food Eng.* **2005,** *71*(1), 55–65.
77. Vega, C.; Kim, E. H. J.; Chen, X. D.; Roos, Y. H. Solid State Characterization of Spray-Dried Ice-Cream Mixes. *Coll. Surf. B Bio Interfaces* **2005,** *45*(2), 66–75.
78. Vilder, J. D.; Moermans, R.; Martens, R. The Free Fat Contents and Other Physical Characteristics of Whole Milk Powder. *Milchwissenschaft* **1977,** *32*(6), 347–350.
79. Walstra, P.; Wouters, J. T. M.; Geurts, T. J., Eds.; *Dairy Science and Technology*, 3rd ed.; CRC Press: Boca Raton, 2006; p 544.
80. Wasnik, P. G.; Menon, R. R.; Surendra Nath, B.; Balasubramanian, B. V.; Manjunatha, M; Sivaram, M. Application of Particle Analysis and Color Parameters for Detection of Adulteration of Cow Ghee with Vanaspati Derived from Image Analysis. *Indian J. Dairy Sci.* **2017,** *70*(2), 34–38.
81. Yang, W.; Sokhansanj, S.; Tang, J.; Winter, P. Determination of Thermal Conductivity, Specific Heat and Thermal Diffusivity of Borage Seeds. *Biosyst. Eng.* **2002,** *82*(2), 169–176.
82. Zhong, Y.; Wu, Y.; Zheng, Y.; Zhu, H.; Liu, Z.; Jiao, S. Assessment of Radio Frequency Heating on Composition, Microstructure, Flowability and Rehydration Characteristics of Milk Powder. *Food Sci. Technol.* **2017,** *37*(4), 544–551.

CHAPTER 16

ASSESSMENT OF PESTICIDE RESIDUES IN MILK AND MILK PRODUCTS

JYOTIKA DHANKHAR, RITA MEHLA, and PREETI KUNDU

ABSTRACT

The presence of pesticide residues in milk and dairy products has raised concerns among consumers and is responsible for incurring economic losses to dairy industry. Pesticides represent one of the key elements required to increase the productivity of agricultural crops. However, due to their persistent nature, pesticides enter the food chain via the feed and fodder given to the animals. The presence of pesticides in milk is attributed to their absorption into bloodstream, subsequent storage into fat and bones followed by partial excretion into milk. Among the main classes of pesticides used for agriculture purposes, namely, organochlorides, organophosphates, carbamates, pyrethroids, triazines, organochlorine pesticides are considered the most harmful as they are highly toxic and are persistent organic pollutants with deleterious health and environmental effects compared to the others. To protect people from the risks associated with the contaminants, the *Codex Alimentarius*, the World Health Organization, and the Food Agriculture Organization have proposed the standards for acceptable daily intake and maximum residue limit (MRLs) of pesticides in milk and dairy products. Since there is rampant use of pesticide on crops for increased productivity, pesticide residues are very frequently identified in milk and dairy products at levels above the MRLs. Standard analytical procedures generally used for detection of pesticides involve procedures like sample cleanup, separation of pesticides from matrix, and quantification of pesticides using gas chromatography or high-performance liquid chromatography coupled

The Chemistry of Milk and Milk Products: Physicochemical Properties, Therapeutic Characteristics, and Processing Methods. Megh R. Goyal, Suvartan Ranvir, & Junaid Ahmad Malik (Eds.)

with mass selective detectors. The present chapter attempts to compile the information about different strategies and techniques that have been used to analyze the occurrence of pesticides in milk and dairy products.

16.1 INTRODUCTION

Milk is a balanced and indispensable diet for almost everyone, including youngsters, adults, and elderly people. Milk is considered an ideal candidate for healthy and balanced diet plans and is a significant source of lipids, proteins, vitamins, and minerals (calcium, phosphate, sodium, and potassium).[44,51] The consumption of milk is related to lower the chances of various severe disorders like cancer, inflammation, obesity, and coronary heart diseases.[70] However, the contamination of milk with residues of pesticides, antibiotics, and heavy metals is a critical issue for the safety of consumers.

The intensive and indiscriminate use of pesticides for increasing the agricultural productivity of crops, protection of crops from agricultural pests as well as parasite control in domestic animals has generated danger to human health and environment.[30,85] Because of lipophilic nature of pesticides, high-fat animal-based diets are the main contributing factors for these contaminants.[4] In addition, pesticides are mostly unaffected by chemical changes, and because of their low polarity, they are likely to deposit in the fat of animals and animal products.[24,26] The extent of pesticide residues in food commodities is largely determined by nature of food material, environmental conditions, and the nature of pesticide.[10]

India ranks seventh position in the world for the application of pesticides on agricultural feed and fodder.[89] Pesticides gradually start to disperse once being sprayed. The pesticides used on crops require some waiting time prior to harvesting that varies among different pesticides as well as crops. Food is considered safe for eating only after completion of the waiting period. If a crop is harvested before ending this period, the crop is likely to contain pesticide residues in larger quantity which are harmful to health. Despite the prohibitions and regulations on the application of pesticide, one-third cases of pesticide poisoning around the world are reported in India.[10]

The presence of pesticide residues in excessive quantity can adversely affect the human health by causing chronic toxicity resulting in loss of sight, cancer, and ailments of liver and nervous system. Continuous intake of pesticide residues for a long time may lead to a rise in cholesterol levels and infant mortality rates in addition to numerous metabolic and hereditary disorders due to their biomagnification in the human system.[90]

Thus, to lessen the pesticide residue level in food products, some simple and cost-effective strategies are needed to ensure food safety both at domestic and industry levels. Various handling and processing operations are being carried out to address the present scenario of unsafe food. Several studies have shown the high efficiency of processing techniques such as cooking, pasteurization, drying, sterilization, fermentation, storage processes, and microwave oven cooking to reduce the pesticide concentration. Nevertheless, several processes are likely to increase the pesticide level because of the concentration effect and tendency of adherence to fat molecules.[10,54,83]

Consequently, it is becoming increasingly necessary to generate knowledge from regulatory and public perspectives on the impact of various processing practices on the residual pesticide level for food products.

This chapter aims to cover the different extraction and analytical techniques to monitor the pesticide content, toxicity of pesticides, regulations for their residual levels as well as effect of processing techniques on fate of residual pesticides in dairy commodities.

16.2 TOXICITY OF PESTICIDES

Despite being a highly nutritious diet, milk is a potential source of some toxic compounds such as pesticides which are persistent organic contaminants. The possible sources of milk contamination are the use of insecticides on animal fodder during the growing season, feeding on contaminated water, use of pesticides on animals for prevention and control of external parasites, and application of hygienic treatments for pest control in milk-processing industries.[79] These compounds may get into animal body through contaminated feed and subsequently pass into milk that contains a significant amount of fat and is able to dissolve other environmental pollutants. These toxicants are likely to be excreted through simple diffusion and thus can be transferred to milk and other fat-rich dairy products after several paths owing to their lipophilic nature.

Pesticides are highly potent because of their persistence and stability and may adversely affect the cattle body as well as human body by causing several toxicological implications. Most of the pesticides are either banned or their application is strictly restricted due to their toxic nature.[15] The concept of organic farming is gaining popularity around the globe; however, the utilization of pesticides on large agricultural area is also being carried out in some parts of the world. Moreover, despite being constrained to use in agriculture, these compounds are still allowed to use in limited amount to control vector-borne diseases in many developing countries.[49,86]

The persistent organic contaminants include organochlorine pesticides (OCPs), organophosphorus pesticides, polychlorinated biphenyls (PCBs), polyaromatic hydrocarbons, synthetic pyrethroid and carbmate.[64,82] As per the World Health Organization standards, pesticides are classified into acute toxic, very toxic, toxic, and slightly toxic pesticides. These compounds comprise a broad range of octanol–water partitioning coefficient ($K_{ow} = -0.8$ and log $P = -6.6$) values. This induces a series of storage as well as transport patterns in human system.[84] These pesticides may shift from their storage sites to the other parts of the human body by means of partitioning. Log P value measures the lipophilicity of a pesticide molecule and the pesticide with high K_{ow} and log P values can be accumulated in the fat-rich tissues and released in milk discharge, thus may cause serious health problems to infants and mother. A suitable way out to tackle these issues is the execution of control measures and frequent analysis of residual pesticides in food samples.

Organophosphate pesticide residues can cause several adverse changes to the placenta cells including incitation of enzymes linked to cell proliferation as well as tissue repair.[20] Organochlorine pesticides are more potent and lipid-soluble by nature. The residues of this group of pesticides cause damage to environment and are also fatal to human and other animals. Because of their long half-lives and highly toxic nature, OCPs have been banned but still can be detected in traces in food products.[37]

16.3 REGULATIONS FOR PESTICIDE RESIDUES

The *Codex Alimentarius* is a group of globally recognized standards, guidelines, and recommendations that are best-established measures for protecting consumers' health from hazards emerging from toxicants in foodstuffs. It covers all the processed, semiprocessed, and raw foods and is referred to as an international reference point for consumers to ensure food safety as well as for "food producers and processors, national food control agencies and the international food trade" to ensure fair practices in trade. *Codex Committee on Pesticide Residues* and the Joint Meeting on Pesticide Residues for the pesticide residues monitoring are charged with management of pesticide pollutants in foods. Several organizations, such as Food and Agriculture Organization, US Food and Drug Administration, and European Union (EU), have established the threshold values, that is, the maximum residue limits (MRLs) and acceptable daily intake values for pesticides in dairy products.

MRL (mg/kg) is the maximum amount of a pesticide residue legally allowed in foodstuffs or animal feed when used correctly as per the guidelines. Acceptable daily intake is defined as the maximum quantity of a pesticide residue that can be consumed daily without a substantial health risk. It is expressed as milligram per kilogram of body weight per day.[21] EU has laid down strict regulations for pesticides in infant formulae according to which the residual level of individual pesticides in infant foods must not be above 10 μm/kg. This limit is the minimum detectable level by means of the officially accepted detection methods.[35]

16.4 IMPACT OF DIFFERENT PROCESSING TECHNIQUES ON RETENTION OF PESTICIDE RESIDUES IN DAIRY COMMODITIES

Due to high fat content, milk, and other dairy products get easily polluted with pesticides from contaminated water and animal feed. Thus, the pesticide residues can possibly be present in higher quantities in dairy products (such as butter, cheese, and ghee) than milk, as these organic compounds have high affinity for fats.[60] The absorbance, penetration, and degradation of pesticide molecules in different food products depend on the composition and properties of pesticide itself as well as surrounding environmental conditions.

Pesticide residues are persistent in nature and their potential to resist degradation in different circumstances is estimated as half-life of pesticide, which refers to time needed for half of the molecule to disintegrate or evaporate. It may vary from several hours or days, to years for highly persistent pesticides relying on environmental conditions.[46] The concentration of pesticide residues enduring after half-life relies on concentration of pesticide initially used.[65] There are numerous extrinsic and intrinsic factors that favor the degradation of pesticides, for example, temperature, photolysis, pH, hydrolysis, metabolism, oxidation, and reduction.[10]

Various food processing techniques are employed to increase the shelf-stability, palatability, nutrient accessibility, add variety, and lessen the residue levels. The extent of pesticide reduction by processing is affected by the physicochemical properties of food, site, and nature of pesticide molecule and processing steps used.[55] The different mechanisms resulting in reduction in pesticide concentration during processing include hydrolysis, thermal degradation, photolysis, dissolution, volatilization, and microbial degradation leading to enzymatic alteration, oxidation, and reduction.[34] The impact of these processing technologies on pesticide residual level is discussed in this section.

16.4.1 PASTEURIZATION AND STERILIZATION

Different thermal processing techniques such as sterilization and pasteurization are utilized to enhance the shelf life of food commodities. Due to heating effect, these treatments induce changes in pesticide components and cause degradation of pesticides in different ways that leads to reduction in pesticide residual level.

Singh and Nelapati[90] studied the impact of pasteurization, boiling, and sterilization on hexachlorohexane (HCH) content in normal and spiked milk samples containing α, β, γ, and δ-hexachlorohexane at 1 ppm. It was observed that the total hexachlorohexane content in normal milk was reduced by 10.19%, 37.55%, and 48.69% in pasteurized, boiled, and sterilized milk samples, respectively, whereas in case of spiked milk, the initial HCH content was reduced by 17.88% in pasteurized milk, 66.85% in boiled milk, and 65.16% in sterilized milk. A significant difference was also observed for residual levels within treatments for both normal and spiked milk samples. Likewise in another experiment, scientists evaluated effect of pasteurization on dichlorodiphenyltrichloroethane (DDT) content in milk and observed a decrease in the residues of DDT and its metabolites by 15.6–58.5%.[52]

In the study conducted by Abd-Rabo et al.,[1] processing of milk by different treatments, such as sterilization and pasteurization, resulted in considerable reduction in the content of dimethoate and malathion. It was noted that on pasteurization, malathion was completely degraded while dimethoate was reduced by 73.42%. Sterilization of milk resulted in a decrease of 75.72% and 95.99% in the levels of dimethoate and malathion, respectively.

16.4.2 DRYING AND DEHYDRATION

Drying and dehydration are simple processing methods of food preservation performed at both domestic and commercial levels. These processes have been found to decrease the residual pesticides significantly due to thermal degradation. Liska[61] investigated effect of spray drying on lindane level of milk and reported a reduction of the lindane residues by 80% in raw milk. Heptachlor epoxide was found to be more difficult to destroy than heptachlor. Among different insecticides examined, methoxychlor was found to be unaffected by heat treatments used in milk processing. In another study, it was found that the intensity of heating process affected degradation of pesticides (dicofol, β-endosulfan, and α-endosulfan) in milk products and highest concentrations of pesticide residues were found in butter than khoa.[75]

16.4.3 FERMENTATION

During fermentation, the proteins get hydrolyzed into amino acids and peptides due to enzymatic action as well as starch breaks down to simple sugars that are fermented mainly to lactic acid, alcohol, and carbon dioxide.[68] Fermentation is applied for the manufacture of several foods such as pastry, dairy, wine, and beer and found to be effectual in decreasing the pesticide residues in many of these food products.

Abou-Arab[3] reported a steady decrease in HCH content (1.4–8.9%) in yogurt made from spiked milk on 3 days of refrigerated storage. The study also showed that at the end of ripening period of 6 months, HCH, and its metabolites were reduced by 36.7% in Ras-cheese prepared from spiked milk. The acid–enzyme coagulation step was found more effective in reducing the HCH level by 3.6% during the manufacture of Domiati cheese than an enzyme coagulation step (1.4%). In another study, Ali et al.[5] evaluated the impact of processing of milk into yogurt on the pesticide level initially present in the milk and noted a reduction in its natural pesticide residue content.

16.4.4 FREEZING

Freezing is one of the most convenient methods of preserving foods that enables the conservation of sensorial as well as nutritional attributes of food products better than any other method. Freezing is ranked as a thermal process as it involves the application of extreme temperature, although very low ones. Extremely low temperatures slow down the decay and most chemical reactions and can also reduce the pesticide residues by causing reduction in concentrations. The impact of freeze-drying on the reduction of pesticide level was evaluated.[72] It was observed that freeze-drying partially decreased the pesticide residues while pesticide residues present in low concentrations exhibited no considerable impact on the processing of milk into cheese which might be attributed to degradation of pesticide residues during cheese ripening by microflora used in cheese manufacture.

16.4.5 MANUFACTURE OF DAIRY PRODUCTS

It has also been considered useful to process milk into other products to lower the amount of residues of pesticides in final products. In the production of some milk products, initial concentrations of HCH (3 ppm), lindane

(1.0 ppm), and DDT (2.0 ppm) in milk were decreased by 77.3%, 2.0%, and 9.0% in yogurt; 80.4%, 92.7%, and 77.4% in samna; 20.5%, 23.3%, and 24.6% in butter; 3.5%, 3.8%, and 2.8% in cheese; and 0%, 5.3%, and 0.1% in cream, respectively.[106]

Dabiza et al.[25] monitored the production of Domiati cheese, ghee, and butter from bovine milks with and without addition of malathion (1.75 ppm). It was observed that butter contained around 92% of malathion incorporated to bovine milk, but retention in ghee was only 12.5%. The retention of malathion in fresh cheese was 72.2% which further decreased during storage. In an experiment, the reduction of "pesticide residues" in milk by several kitchen practices was examined. It was reported that heat-processing methods such as boiling, cream separation, and ghee formation reduced the hexachlorohexane residues by 11.54–26.78% because of boiling only and 35.86–50.88% because of boiling followed by cream separation. A decrease in DDT residual level was also observed.[62]

16.5 EXTRACTION AND CLEANUP OF PESTICIDES

To evaluate the pesticides in milk, systematic and comprehensive approaches are required since milk presents a complicated matrix with a low concentration of the desired analyte. Therefore, efficient sample extraction is of paramount importance in determination of pesticide residues in milk and milk products. Different techniques like liquid–liquid extraction (LLE), accelerated solvent extraction, solid-phase extraction (SPE), supercritical fluid extraction, Soxhlet extraction, solid-phase microextraction (SPME), and so on are used for sample extraction.

Conventional methods of extraction for food samples, LLE, and Soxhlet extraction generally take longer time and need high quantity of organic solvents. In case of LLE and SPE techniques of extraction for pesticides, liquid food utilizes the tendency of the analyte to partition between two immiscible solvents or onto sorbent (SPE) based on its differential solubility. The following section provides a brief description of some important "sample preparation" procedures for pesticide determination.

16.5.1 LIQUID–LIQUID EXTRACTION

In LLE, the pesticide residue in liquid samples is extracted with an immiscible solvent by partitioning. The technique involves different degrees of solubility of a compound in two liquid phases of different miscibility depending on its partition coefficient. The compound under interest can be separated based on its tendency to selectively polarize between the two liquid phases. The LLE method is specifically for compounds dissolved in aqueous phases. In general, separation of the desired compound is achieved by blending the two liquid phases in a separating funnel and then allowing the separation to take place. To prevent emulsion formation, salt addition is often done along with the centrifugation step, or the matrix solid-phase dispersion method may be used. The two layers are pooled separately, and multiple extractions are commonly followed for achieving complete extraction of an analyte. For analyzing multiresidues in milk and dairy products, the choice of extraction solvent should be based upon its ability to function suitably in the broad range of polarity in diverse matrices consisting of different constituents, like water, fat, carbohydrates, and other associated components, in different proportions.

Extraction of compound under interest is generally done by solvent-aided breakdown of the matrix in a colloid mill or homogenizer. Therefore, even the most used original standard methods for instance, AOAC method, has been modified subject to the conditions.[91] Many modifications have been done with respect to use of solvents and salts for extraction of solvents. The most used solvents are ethanol, methanol, acetone, ethyl acetate, toluene, and hexane. Some methods use electrolytes or salts in the mixture, which dissolve into the water phase as solvent and help extract the analytes into the water-soluble organic phase by salting out mechanism. Many studies executed with GC and HPLC have shown that acetone can be used successfully in extraction of nonpolar and polar pesticides. Moreover, it is more volatile, nontoxic, and cheaper compared to other solvents used for extraction.[91]

Raslan et al.[79] performed multiresidue extraction in milk samples from different animals in Egypt with hexane and acetonitrile. Extracted samples were subjected to cleanup step using activated florisil and were analyzed using GC on HP–5MS capillary column. In another study conducted in Pakistan for milk, Arif et al.[8] performed LLE technique to analyze chlorinated pesticides in samples by extraction with petroleum ether as solvent. A florisil column was employed for cleanup of extract and hexane was used as elution solvent. Milk was analyzed using gas chromatography equipped with electron capture detectors (ECDs) for OCPs.

Gill et al.[39] conducted determination of pesticide residue in bovine milk collected from five separate points in India using dichloromethane, and subsequently its elution using a combination of dichloromethane–acetone. While performing extraction process, the mixture of sample–solvent should be shaken or reversed multiple times to ensure close contact of the two liquid phases with each other. Vigorous shaking can result in emulsion formation, which can be addressed by centrifugation. For achieving the extraction of maximum amount of analyte from the matrix, extraction should be done at least for 2–3 times. Some procedures entail the back extraction of compound of interest using a slight volume of acid in methanol or other similar solvent for removal of compounds that may hamper with the chromatographic analysis subsequently.

16.5.2 SOLID-PHASE EXTRACTION

SPE is an optional method to LLE for specific extraction, purification, and concentration of compound under interest in a matrix.[98] SPE may be used straight away for carrying out extraction in liquid samples, or as a cleanup approach in cases when extracts are solubilized in solvents. The approach encompasses passing of sample containing multiresidue over the stationary phase (solid phase) for separation of residues corresponding to the extent to which they have been partitioned or adsorbed by stationary phase.

The analytes may display affinity for adsorption at the solid phase or they may be retained in the liquid phase. In case the analytes are adsorbed onto stationary phase, then a solvent with stronger elution tendency may be used to selectively desorb the analytes. While the analytes that display affinity for the liquid phase may be pooled and prepared for further analysis, and the interfering compounds are adsorbed by the solid-phase sorbent. Efficient separation by SPE may be accomplished by choice of appropriate solid-phase sorbent as well as suitable solvents for elution. SPE demonstrates versatility for its application in all the phases like solids, liquids, and gases subject to the proper choice of sorbent and solvent; however, it is commonly used for selective extraction and enrichment of liquid samples.

The third step involves passing suitable solvent, generally with low elution power, through the solid sorbent to remove the sample components adsorbed by the sorbent while retaining the analytes. In case of aqueous matrix like milk, drying is recommended to remove traces of water from sorbent to prevent interference in the concentration of extract. The last step involves elution of analyte from sorbent using proper solvent. It is necessary to use appropriate volume of solvent to achieve minimum dilution of analyte.

The SPE technique is characterized by two important attributes, standardization, and good reproducibility. As a result, SPE can be employed on a broad spectrum of applications, like normal phase, reversed phase (RP), and ion-exchange materials, where it allows for its use on the aqueous solution with further trapping mechanisms. Commonly used sorbents include cartridges, syringe barrels, microcolumns, and discs.

In general, cartridges are meant for one-time use and are a better option than conventional solvent extraction methods due to the low costs incurred and the lesser time involved. The introduction of flat disks represents the additional range of SPE material, and they have stationary phase held on mesh that assists in rapid extraction of large volume. Moreover, recently alterations have been made in technique to improve the flow rates with the setting up of "turbulent flow extractions" in cartridges. Many studies have suggested an arrangement of more than one SPE for executing cleanup in normal phase. Weak anion-exchange sorbents like primary secondary amine (PSA), aminopropyl, or diethylaminopropyl-modified silica, strong anion-exchange sorbents are mostly recommended for cleanup of food samples.[87] Additional preconcentration strategies involve the grouping of GCB (graphitized carbon black) and PSA columns.[2]

The execution of SPE technique is carried out when samples have already been subjected to different treatments like homogenization, filtration, sonication, centrifugation, and so on. Ho and Yu[47] used C18 SPE cartridges for estimation of OCPs residues in milk. Before subjecting to SPE treatment, samples were passed on Kieselguhr columns and a solvent mixture of hexane–acetone–ethyl acetate mixed in optimum proportion was used for elution of analytes.

In another study, scientists executed SPE technique on a florisil SPE cartridge for estimation of multiple chlorinated pesticides (OCP) in milk and dairy product samples.[105] Samples (milk or milk powder) were prepared with *n*-hexane and acetone mixture in equal ratio by centrifugation at 8875*g* for 10 min at 8°C. It was followed by separation of organic layer in a volumetric flask and again reextraction of residue with solvent mixture. Organic layers were pooled and vacuum dried at 40°C. Reconstitution of the dried extract was done using cyclohexane and ethyl acetate mixture in equal ratio. The extract was then subjected to SPE treatment. The SPE cartridge was conditioned with *n*-hexane followed by sample elution due to gravity. The sample was passed on to the conditioned cartridge and elution was allowed to take place. Eluted sample was collected in a test tube containing mixture of *n*-hexane and dichloromethane mixture in optimum ratio; eluent was vacuum dried. Dried extract was diluted with minimum amount of hexane before subjecting it to GC–MS/MS analysis. Scientists have demonstrated benefit of using a dual-layer SPE, with an arrangement of PSA and GCB for sample cleanup in simultaneous detection of many pesticide residues in food and related products.[88]

The adhesion of fatty acids by sorbent-like PSA was evaluated and the effects of the eluting solvent on fatty acids adsorbed on SPE cartridge were assessed. The choice of solvents with higher eluting strength to elute some pesticides from graphitized carbon was reported to hinder with ability of PSA to retain fatty acids properly. The tendency of the PSA to bind fatty acids was related to the conditioning and elution methods. SPE technique has significant advantages over the conventional liquid extraction techniques as follows:

- SPE is a suitable, easier, and simpler technique[2]
- involves less amount of solvent that allows flexibility of use[3]
- the SPE cartridges can be used to retain and hold the compounds for shorter duration[4]
- efficiency can be improved by passing water and eluting solvents from cartridges

16.5.2.1 LIMITATIONS

Efficient functioning of SPE cartridges requires continuous smooth packing of adsorbent in prepacked columns. Although SPE technique is efficient, the functioning is found to be negatively affected by presence of interfering compounds like solid and oily components, which prevent flow of the solvent by blocking the sorbent pores, and hence resulting in overloading of samples. This leads to lack of reproducibility for certain samples. Also, conventional sorbents sometimes prove inadequate with respect to adsorption of polar compounds.

16.5.3 SOLID-PHASE MICROEXTRACTION

Arthur and Pawlyszin[9] developed SPME to address limitations related with "solid-phase extraction" and LLE. SPME is a novel sample preparation procedure where a fused-silica fiber is covered on the external surface with polysiloxane stationary phases having different polarities. SPME has been established as a potent and valuable technique for analyte extraction and preconcentration. It is emerging as a promising technique which omits the use of solvents for carrying out extraction and thereby preventing occurrence of poisonous residues. SPME entails sampling, extraction, preconcentration, and introduction of the sample into the analytical system in a single continuous process to prevent interference from the matrix. It is essentially an open-bed, diffusion rate-controlled extraction technique that contributes to effective analyte enrichment from varied sample matrices by means of suitable flexible designs corresponding to the application.[19,74]

There are presently few SPME modes available for use. For functioning properly, there is a requirement of fused-silica fibers or GC capillary columns. Fiber extraction mode is generally represented by headspace (HS) and direct immersion (DI) SPME, whereas liquid chromatography or high-performance liquid chromatography instrument operates with the in-tube SPME mode. Depending upon the sample and analytes of interest, fiber may be used for either DI or HS extraction. In HS SPME, the fiber is brought in contact with the air above a liquid sample, which is present in equilibrium with the liquid phase. To ensure effectiveness of this approach, the analytes should be able to partition positively into the vapor phase. In this technique, a fused-silica fiber coated with thin layer of sorbent or adsorbent material is positioned above food material. Migration of compounds from vapor phase of food sample to silica fiber where they undergo absorption or adsorption. They are subsequently desorbed by insertion of silica fiber in the GC through its injector. SPME technique can be employed on extensive range of compounds without the risk of artifact formation as in other techniques.

In general, DI-SPME which is the popular mode of analysis involves injection of the fiber directly into the sample medium and compounds of interest are transferred straight to liquid extraction medium. In case of the samples of aqueous nature, more effective mixing methodologies, like quick sample flow, fast fiber or vial movement, agitation or ultrasound-assisted mixing are carried out. Such procedures are followed as they overcome the influence of depletion zone that originates due to the tendency of fiber to interfere with fluid flow and contribute to the decreased flow rate of analytes.

There is usually a coating of thin polymeric film upon the fiber in SPME which aids in concentration of analyte from sample by either absorption or adsorption. Depending upon the characteristics of fiber, analyte is extracted by aforesaid phenomena. In case the fiber has liquid phase, extraction is carried out by absorption while if coating of porous particle combination is used, extraction is done by adsorption. During absorption, analyte solubilizes in the majority of liquid phase in a competitive manner, whereas in adsorption, binding of analyte to the solid surface takes place through competitive process.

During extraction through adsorption, availability of binding sites for analyte is scarce and fiber gets saturated on occupation of all the sites. As adsorption occurs in competitive manner, analytes with more affinity desorb the ones with lower affinity.

There is a wide range of "fiber coatings based on solid sorbents" present commercially along with commonly used poly(dimethylsiloxane) (PDMS) and poly(acrilate)-coated fibers, namely PDMS/divinylbenzene (DVB), Carbowax/DVB, Carbowax/template resin, and polyethylene glycol. Among these porous polymers, SPME fibers with provision of different polarity attributes may prove beneficial for the rapid and simultaneous analysis of many pesticides, expanding range of SPME applications.[23] However, limitations associated with commonly used SPME fibers involve their brittleness. Different characteristics of coating material like polarity, thickness, and surface can be manipulated for extraction of wide variety of pesticides.

SPME method is a sophisticated and rapid technique compared to existing methodologies. Complete automation of technique may be realized by connecting an automated sampling device with SPME equipment. Moreover, SPME exhibits good precision and selectivity, which can be performed without sample pretreatment and use of solvent. This facilitates the improved sample purity by decreasing the chances of sample variation and contamination. The key drawbacks entail poor reproducibility for fiber-based transfer and may suffer from lack of accuracy. The procedure is restricted to rather semivolatile or volatile compounds, and matrix effects are also evident in the complicated food matrix. Since the technique may require costly consumables, trained staff, and standardization of processing parameters, SPME is not very easy techniques.

16.5.4 ACCELERATED SOLVENT EXTRACTION

This technique is also called as pressurized fluid extraction. ASE is a solid–liquid extraction method that combines high temperatures and elevated pressures with organic solvents and is performed for short time periods usually 5–10 min. Elevated temperatures (80–200°C) are used to enhance the extraction efficiency of the analyte and elevated pressures (10–20 MPa) are employed to keep the organic solvent in its liquid state. This results in increased salvation power and low viscosities of the solvent, which, in turn, permits elevated diffusion rates for the analytes.[95] Therefore, extraction efficiency enhances, reducing quantity of solvent required and accelerating extraction process. The extraction efficiency is largely influenced by temperature conditions, while the sample mass has no effect on the time needed for extraction. Thus, ASE is an interesting method as it is rapid and time-saving; requires a smaller amount of solvent with no need of filtration after extraction.

The ASE instrumentation can accommodate different sample sizes and allows extraction in unattended operation. However, this extraction technique has limited selectivity due to its necessity for additional cleanup of analyte obtained and high initial cost than some other advanced techniques like microwave-assisted extraction. Moreover, the recovery of fat-soluble and thermally labile components is difficult while extracting the wet samples with hot water. In some cases, a further concentration or extraction step is needed before analysis which can be performed using LLE or SPE.[92] To avoid further cleanup, a trapping agent (such as SPE disc) can be placed in the extraction container, which is then afterward extracted. The final separation and determination can be accomplished by using gas chromatography, high-performance liquid chromatography, liquid chromatography, and mass spectrometry (MS). In an experiment, ASE–LC–MS/MS method was examined to find out the six mostly used insecticides in cattle milk by using hot water extraction and reported a 70–92% recovery of the analytes.[18]

16.5.5 MICROWAVE-ASSISTED EXTRACTION

Microwave-assisted extraction (MAE) is a conventional method that employs microwave energy (300 MHz–300 GHz) to heat the solvents containing samples.[40] Due to dipole rotation and ionic conductance phenomena, microwaves rapidly heat the sample–solvent mixture leading to very fast extraction of analytes and are also applicable for thermosensitive components. The microwave heating generated in the sample based on dielectric constant of extraction solvent mixture as higher the dielectric constant, the more heat energy is released resulting in fast heating at a given frequency. Therefore, the characteristics of solvent and sample matrix largely influence effect of microwave radiation.[36]

Generally, dielectric constant of extraction solvent is high to facilitate strong absorption of microwave energy. But, in case of thermally unstable substances, the solid matrix may absorb microwave radiation followed by heating of the matrix and discharge of the solutes into the cold solvent. The efficiency of microwave-assisted extraction is largely influenced by the nature of solvent, applied power, the extraction temperature, and time. The solvent should well solubilize desired compounds in sample by absorbing the microwave energy without resulting into degradation of analytes due to strong heating. Microwave-assisted extraction directly heat sample so restricts contamination from vessel. The major advantages of microwave-assisted extraction include shorter time, low solvent consumption, low-temperature requirement, higher yield, totally automated, feasibility of simultaneous extraction of several various analytes at a time with minimum interference and lower cost in comparison to traditional extraction techniques.[13,32] However, it has also some limitations such as requirement of polar solvents, necessity for filtration, and cleanup of extract after extraction.

16.5.6 SOXHLET EXTRACTION

Soxhlet extraction is a simple and extensively used extraction process. Despite being a traditional method, it is a very useful practice for preparative purposes in which the compound of interest with partial solubility in a solvent is concentrated in the presence of insoluble impurity. In the first step, a porous thimble loaded with a solid sample is positioned inside the Soxhlet apparatus. Continuous extraction takes place as the extraction solvent is continuously refluxed through the thimble using a condenser and a siphon side arm. This technique usually requires further cleanup and concentration steps.

This extraction technique allows unattended extraction and just one batch of solvent is recycled until complete extraction is anticipated. Thus, it saves time and solvent as well as enhances extraction efficiency in comparison to other traditional solvent extraction techniques. However, it is related to high heat consumption and is only suitable for "extraction" of thermally stable compounds as prolonged heating may cause degradation of thermosensitive components. Soxhlet extraction has also been used in combination with other advanced extraction processes such as UAE and MAE to increase the extraction efficiency. Soxhlet extraction has been applied for the determination of pesticide pollutants, for instance, PCBs and polycyclic aromatic hydrocarbons.[94,102]

16.6 COMMON ANALYTICAL TECHNIQUES

The common procedures for investigation of pesticide residue in milk and milk products involve use of GC and LC-based methods. The initial methods for evaluation of pesticides residue analysis involved use of detectors: ECD, nitrogen and phosphorus detector (NPD), and flame photometric detector (FPD).[81]

The detectors display high sensitivity for analyte of choice. ECD is a selective detector, which is extremely particular for electronegative compounds like halogens and hence OCPs. The NPD is specific to the

TABLE 16.1 Extraction and Identification of Pesticide Residues in Dairy Products.

Pesticide residues	Matrix	Extraction technique	Extractants	Analytic method	Reference
Dioxins	Cheese	Liquid-phase extraction	Hexane/Methylene chloride	Quadrupole ion storage collision-induced dissociation mass spectrometry/mass spectrometry	[45]
Dioxins	Yogurt, cheese, milk	Soxhlet extraction	Toluene	Gas chromatography–mass spectrometry	[16]
Dioxins	Powdered milk	Soxhlet extraction	Acetonehexane (1:1)	Gas chromatography	[76]
Dioxins	Cheese, butter	Pressurized liquid extraction	Hexane (10 MPa)		[38]
Herbicides	Milk	Solid-phase extraction	CH_2Cl_2/MeOH (8:2)		[17]
Organochlorides	Milk	Solid-phase extraction	Hexane		[71]
Organochlorines	Cheese, butter	Liquid-phase extraction	Chloroform		[12]
Organochlorines, organophosphates	Cheese	Supercritical fluid extraction	Carbon dioxide (80°C, 69 MPa)		[48]
Organochlorines, polychlorinated biphenyls	Yogurt	Soxhlet extraction	Cyclohexane/Acetone (1:1)		[104]
Organophosphates	Milk	Solid-phase microextraction	Uncoated fiber	Gas chromatography; nitrogen phosphorous detector	[29]
Polychlorinated biphenyls	Milk	Liquid–liquid extraction	Hexane	Gas chromatography–mass spectrometry	[97]
Polychlorinated biphenyls	Powdered milk	Supercritical fluid extraction	Carbon dioxide (50°C, 13–23 MPa)	Gas chromatography	[77]

TABLE 16.1 *(Continued)*

Pesticide residues	Matrix	Extraction technique	Extractants	Analytic method	Reference
Polychlorinated biphenyls	Powdered milk	Pressurized liquid extraction	Hexane (100°C, 10 MPa)	Gas chromatography–mass spectrometry	[63]
Polychlorinated biphenyls	Milk	Solid-phase microextraction	Uncoated fiber	Gas chromatography–electron capture detector	[56]
Polychlorinated biphenyls, organochlorines	Butter	Liquid-phase extraction	Hexane	Gas chromatography	[53]
Polycyclic aromatic hydrocarbons	Cheese	Liquid-phase extraction	Cyclohexane	High-performance liquid chromatography–fluorescence	[6]
Pyrethroids	Milk	Solid-phase extraction	Light petroleum saturated with acetonitrile and ethanol	Gas chromatography	[31]
Triazines	Milk	Solid-phase extraction	CH_2Cl_2/CH_3CN (6:2)	Gas chromatography	[57]
Triazines	Milk	Solid-phase microextraction	Polydimethylsiloxane/ Divinylbenzene-coated fiber	Gas chromatography–mass spectrometry	[14]

presence of nitrogen and/or phosphorus atom and, hence, selectively detects organophosphorus and nitrogenated pesticides. Likewise, FPD responds particularly to sulfur and phosphorous pesticides.[58]

The detectors are, therefore, mostly used in detection of specific pesticides compared to multiresidues during analysis. UV or diode array detectors were the primary detectors used in LC-based methods. However, in recent years, most laboratories opt for MS since it is more selective and sensitive to the entire range of pesticides.[69,96,103] Furthermore, tandem mass spectrometry (MS/MS) and high-resolution mass spectrometry (HRMS) are presently popular for rapid and accurate analysis. Different kinds of MS analyzers frequently used for detection of pesticide residues include single quadrupole (Q), triple quadrupole (QqQ), ion trap (IT), hybrid quadrupole ion trap (QTrap), time of flight (TOF) and Orbitrap.

16.6.1 GC–MS-BASED ANALYSIS

The Q has been the most frequently used gas chromatography–MS detector for analysis of pesticide residue in milk. In general, sample is separated on a fused-silica dimethylpolysiloxane-coated capillary column after being injected through split/splitless injection and fragmentation is accomplished via electron ionization (EI, 70 eV) mode.[59,67,78] In comparison to the standard EI, chemical ionization (CI) is not popular for pesticide determination.[7,33]

Analytes can be analyzed using either (CI) or EI mode. In case the negative (CI) is used, analytes require addition of a group with acidic nature or halogen atoms with electronegativity for stabilization of negative charge. Therefore, negative CI mode is more effective for analysis of pesticides due to presence of electronegative groups. Selected ion monitoring (SIM) mode provides more precise information compared to full scan mode. Full scan mode operates in a range of *m/z* 50–600, while SIM mode analyzes through selection of 2–3 ions in SIM mode. Data acquisition through the IT enables the ion monitoring after the run and mass spectrum is generated by ejection of ions of increasing *m/z* ratio.[41] In recent years, synchronized arrangement of MS analyzers contributes to the improved efficiency of analysis, for instance, the use of QqQ helps in achieving the better performance compared to Q.

With the use of tandem MS, accuracy of determination method improves because of its ability to accomplish selective analysis through SIM. It assists in overcoming matrix interferences because of its potential to eliminate background noise in chromatogram and thereby realizing lower limits of detection in analysis.[11,73] In general, the most followed QqQ methods involve selection of at least 2 SIM transitions per analyte. However, selection of two appropriate transitions is mostly complicated owing to poor signal intensity of ions of the analyte. The choice of correct mass fragments for identification of analytes is typically complicated for chlorine as it is not easy to differentiate between diverse OCPs which elute quite close to each other. This analysis in SIM mode often leads to difficulty in accuracy of results in terms of identification and the quantification.

16.6.2 LC–MS-BASED ANALYSIS

Chromatographic elution of pesticides is usually accomplished with RP chromatography depending upon polar nature of analytes and C18 is the most used column. In most cases, assessment of pesticides is done with mobile phase comprising mixture of water–methanol and water–acetonitrile. Studies have shown that addition of different reagents or salts, like ammonium acetate and formic acid (mostly concentration level of 0.1%), to mobile phase improves its ionization ability during analysis.[50,66,69,96,99] For carrying out analysis, the conditions generally set are as follows: the volume for injection ranges from 1 to 30 μL and flow rate varies from 200 to 300 μL/min. For efficient working of ultrahigh-pressure liquid chromatography, there is a marked reduction in particle size of adsorbent phase up to 3–5 μm.[80]

The operating condition with different modifications helps improve the resolution with the reduction in the total time required for analysis. Ionization in liquid chromatography–mass spectrometry is often accomplished by atmospheric pressure ionization (API) sources. API source is typically used for ionization of pesticides in "milk and milk products." The ionization can facilitate formation of whole protonated molecules and electrospray in positive mode is frequently used for analysis.[22]

Atmospheric pressure chemical ionization may be an ideal option for analytes without acid or a basic center that can be ionized in abundance with the aid of the mode. Because of distinctive soft ionization technique of source API, different tandem MS analyzers like QqQ or QTrap may be used for analysis. Therefore, QqQ and QTrap MS analyzers are the most frequently used in LC-based detection methods for pesticide residues in milk. With the aid of SIM mode, characteristic SIM ions are chosen for estimation of pesticide residue which includes ions for confirmatory and quantification purposes. QTrap assists in increasing the sensitivity because of the provision of scan facility in the second analyzer.

16.6.3 HIGH-RESOLUTION MS-BASED METHODS

Nowadays, the application of high-resolution MS method has been adopted as the latest technique for quantitative analysis of pesticides in milk and milk-based products. Q TOF and Q Orbitrap are the most used HRMS analyzers for analysis of pesticide residues.[27,28] HRMS has emerged as a popular technique for analysis due to its potential of precise mass measurement and because of the ability to consider characteristic parent and daughter ion mass spectra in operation for improved specificity and selectivity. For carrying out analysis, the characteristic ion is chosen for quantification and particular software for the deconvolution is employed to achieve pure mass spectra for accurate resolution in case of cooccurrence of analytes.[93] Studies have shown the employing of Q Orbitrap for the investigation of pesticides in different samples facilitates rapid and accurate determination.[42,43,100,102]

Data acquisition is carried out generally in positive mode either through MS–SIM or information is obtained in MS/MS mode. The presence of collision cell allows for efficient mass selection; when the detection of the pesticide is carried out in the ion abundance threshold, then the product-ion spectra may be done in a particular *m/z* mass window with specified mass error level and the further accuracy achieved by sending the selected mass fragment to "HCD collision cell" of the Q Orbitrap mass spectrometer.[41]

Therefore, the correct mass measurement is realized through identification and confirmation of analyte by accurate selection of precursor and daughter ion. The use of HRMS analyzers would most likely surge in future for quantitative purposes. Because of the sophisticated software and user-friendly operation, HRMS would most probably be more in demand. The instrument facilitates simultaneous screening and determination of analytes with high sensitivity. HRMS displays extremely good selectivity during data acquisition which enables accurate analysis of high number of pesticides.

16.7 SUMMARY

Different studies for analytical procedures over the past few years have shown that an extensive variety of techniques are being implemented to analyze the different pesticide residues available for milk and dairy products. Extraction solvents the most widely used are ethyl acetate and acetonitrile. SPE or dSPE sample cleanup is preceded by extraction and QuEChERS technique has become popular for assessment of pesticide residue in milk and dairy products. LC and GC are mostly associated with MS for identification and quantification of residues. In recent years, different kinds of advancements in the analytical protocols to determine the "pesticides" in "milk and dairy products" have been realized. These new developments are:

- the lower sample requirement and reduction in the number of reagents or other chemicals for entailing a miniaturized extraction process.
- the rapid and automated methods for sample preparation that ensure accurate analytical procedure with enhanced reproducibility and repeatability.
- efficient separation methods for an improved resolution and decrease in run time.
- advancement in identification techniques with quantitative and validation efficiencies.

However, there is no commonly established analytical technique yet for pesticides analysis in milk and milk-based foods. Since varieties of pesticides from different chemical classes are present in the matrix, which would require different procedures for analysis and thereby making it difficult to establish a single solution applicable to all. Single method to evaluate multipesticide residues in milk and dairy products is therefore still required. The need for such development is based upon the fact that the pesticides used today are more polar and are environment friendly. In coming years, it seems that the use of such pesticides is likely to increase, making it necessary to upgrade the existing methods.

KEYWORDS

- **gas chromatography**
- **maximum residue limits**
- **organochlorides**
- **persistent organic pollutants**
- **pesticide residues**

REFERENCES

1. Abd-Rabo, F. H. R.; El-Salamony, H.; Sakr, S. S. Reduction of Pesticide Residues in Egyptian Buffalo Milk by Some Processing Treatment. *Int. J. Dairy Sci.* **2016,** *11*, 75–80.

2. Abhilash, P. C.; Jamil, S.; Singh, N. Matrix Solid-phase Dispersion Extraction Versus Solid-phase Extraction in the Analysis of Combined Residues of Hexachlorocyclohexane Isomers in Plant Matrices. *J. Chromatogr. A* **2007,** *1176*, 43–47.
3. Abou-Arab, A. A. K. Effects of Processing and Storage of Dairy Products on Lindane Residues and Metabolites. *Food Chem.* **1999,** *64*, 467–473.
4. Agnihotri, N. P. Pesticide Safety Evaluation and Monitoring. In *Proceedings of All India Coordinated Research Project on Pesticide Residues*; Indian Agricultural Research Institute (IARI): New Delhi, India, 1999; unpublished paper pages 10.
5. Ali, A. A.; Metwally, M. M.; Ahmed, N. S.; Anb, A. K. Pesticide Residues in Milk and Their Distribution among Dairy Products. *Egypt. J. Dairy Sci.* **1993,** *22*, 291–299.
6. Anastasio, A.; Mercogliano, R.; Vollano, L.; Pepe, T.; Cortesi, M. L. Levels of Benzo[*a*] pyrene (BaP) in Mozzarlla di Bufala Campana Cheese Smoked According to Different Procedures. *J. Agric. Food Chem.* **2004,** *52* (14), 4452–4455.
7. Angioni, A.; Dedola, F.; Garau, V. L.; Schirra, M.; Caboni, P. Fate of Iprovalicarb, Indoxacarb, and Boscalid Residues in Grapes and Wine by GC-ITMS Analysis. *J. Agric. Food Chem.* **2011,** 59, 6806–6812.
8. Arif, A. M.; Javed, I.; Ayaz M.; Abdullah, M.; Imran, M.; Shahbaz, M.; Gondal, T. A.; Ali, M.; Iqbal, Z.; Salehi, B.; Sharifi-Rad, J.; Martins, N. Organochlorine Pesticide Residues in Raw Milk Samples Collected from Dairy Farms and Urban Areas of Lahore District, Pakistan. *J. Food Sci. Technol.* **2021,** *58* (1), 129–137.
9. Arthur, C.; Pawlyszin, J. Solid Phase Microextraction with Thermal Desorption Using Fused Silica Optical Fibers. *Anal. Chem.* **1990,** *62*, 2145–2148.
10. Bajwa, U.; Sandhu, K. S. Effect of Handling and Processing on Pesticide Residues in Food—A Review. *J. Food Sci. Technol.* **2014,** *51* (2), 201–220.
11. Banerjee, P.; Utture, K.; Dasgupta, S.; Kandaswamy, S.; Pradhan, C.; Kulkarni, S.; Adsule, S. Multiresidue Determination of 375 Organic Contaminants Including Pesticides, Polychlorinated Biphenyls and Polyaromatic Hydrocarbons in Fruits and Vegetables by Gas Chromatography-Triple Quadrupole Mass Spectrometry with Introduction of Semi-quantification. *J. Chromatogr. A* **2012,** *1270*, 283–295.
12. Barkatina, E. N.; Pertsovsky, A.L.; Murokh, V. I.; Kolomiets, N. D.; Shulyakovskaya, O. V.; Venger, O. N.; Makarevich, V. I. Organochlorine Pesticide Residues Basic Food Products and Diets in the Republic of Belarus. *Bull. Environ. Contam. Toxicol.* **1999,** *63*, 235–242.
13. Barriada-Pereira, M.; Concha-Grana, E.; Gonzalez-Castro, M. J.; Munitategui-Lorenzo, S.; Lopez-Mahia, P.; Prada-Rodriquez, D.; Fernandez-Fernandez, E. Microwave-assisted Extraction Versus Soxhlet Extraction in the Analysis of 21 Organochlorine Pesticides in Plants. *J. Chromatogr. A* **2003,** *1008*, 115–122.
14. Basheer, C.; Lee, H. K. Hollow Fiber Membrane-Protected Solid-phase Microextraction of Triazine Herbicides in Bovine Milk and Sewage Sludge Samples. *J. Chromatogr. A* **2004,** *1047* (2), 189–194.
15. Battu, R. S.; Singh, B.; Kang, B. K. Contamination of Liquid Milk and Butter with Pesticide Residues in the Ludhiana District of Punjab, India. *Ecotoxicol. Environ. Saf.* **2004,** *59* (3), 324–331.
16. Bocio, A.; Domingo, J. L. Daily Intake of Polychlorinated Dibenzo-*P*-Dioxins/Polychlorinated Dibenzofurans (PCDD/Pcdfs) in Foodstuffs Consumed in Tarragona, Spain: A Review of Recent Studies (2001–2003) on Human PCDD/PCDF Exposure through the Diet. *Environ. Res.* **2005,** *97* (1), 1–9.

17. Bogialli, S.; Curini, R.; Di Corcia, A.; Lagana, A. Development of a Multiresidue Method for Analyzing Herbicide and Fungicide Residues in Bovine Milk Based on Solid-phase Extraction and Liquid Chromatography-Tandem Mass Spectrometry. *J. Chromatogr. A* **2006,** *1102*, 1–10.
18. Bogialli, S.; Curini, R.; Di Cortia, A.; Laganà, A.; Nazzari, M.; Tonci, M. Simple and Rapid Assay for Analyzing Residues of Carbamate Insecticides in Bovine Milk: Hot Water Extraction Followed by Liquid Chromatography-Mass Spectrometry. *J. Chromatogr. A* **2004,** *1054*, 351–357.
19. Boyacı, E.; Rodríguez-Lafuente, Á.; Gorynski, K.; Mirnaghi, F.; Souza-Silva, É. A.; Hein, D. Sample Preparation with Solid Phase Microextraction and Exhaustive Extraction Approaches: Comparison for Challenging Cases. *Anal. Chim. Acta* **2015,** *873*, 14–30.
20. Bulgaroni, V.; Lombardoa, P.; Rivero-Osimanib, V.; Verab, B.; Dulgeriand, L.; Cerbánd, F.; Riverod, V.; Magnarellib, G.; Guiñazú, N. Environmental Pesticide Exposure Modulates Cytokines, Arginase and Ornithine Decarboxylase Expression in Human Placenta. *Reprod. Toxicol.* **2013,** *39*, 23–32.
21. Burgi, H.; Schaffner, T.; Seiler, J. P. The Toxicology of Iodate: A Literature Survey in View of its Use in Iodized Salt. In *Comprehensive Handbook of Iodine*; Academic Press: San Diego, CA, 2009; pp 909–918.
22. Caboni, P.; Sarais, G.; Angioni, A.; Garau, V. L.; Cabras, P. Fast and Versatile Multiresidue Method for the Analysis of Botanical Insecticides on Fruits and Vegetables by HPLC/DAD/MS. *J. Agric. Food Chem.* **2005,** *53*, 8644–8649.
23. Cai, L. S.; Gong, S. L.; Chen, M.; Wu, C. Y. Vinyl Crown Ether as a Novel Radical Crosslinked Sol-Gel SPME Fiber for Determination of Organophosphorus Pesticides in Food Samples. *Anal. Chim. Acta* **2006,** *559*, 89–96.
24. Coat, S.; Monti, D.; Legendre, P.; Bouchon, C.; Massat, F.; Lepoint, G. Organochlorine Pollution in Tropical Rivers (Guadeloupe): Role of Ecological Factors in Food Web Bioaccumulation. *Environ. Pollut.* **2011,** *159* (6), 1692–1701.
25. Dabiza, N. M. A.; El-Senaity, M. H.; El, Z.; Dein, M. M. Pesticide Residues in Some Market Dairy Products and Distribution of Malathion in Butter, Ghee and Domiati Cheese During Manufacture. *Egypt. J. Dairy Sci.* **1999,** *27*, 345–357.
26. Darko, G.; Acquaah, S. O. Levels of Organochlorine Pesticides Residues in Dairy Products in Kumasi, Ghana. *Chemosphere* **2008,** *71*, 294–298.
27. Dasgupta, S.; Banerjee, K.; Dhumal, K. N.; Adsule, P. G. Optimization of Detection Conditions and Single-laboratory Validation of a Multiresidue Method for the Determination of 135 Pesticides and 25 Organic Pollutants in Grapes and Wine by Gas Chromatography Time-of-Flight Mass Spectrometry. *J. AOAC Int.* **2011,** *94*, 273–285.
28. Dasgupta, S.; Banerjee, K.; Patil, S. H.; Ghaste, M.; Dhumal, K. N.; Adsule, P. G. Optimization of Two-dimensional Gas Chromatography Time-of-flight Mass Spectrometry for Separation and Estimation of the Residue of 160 Pesticides and 25 Persistent Organic Pollutants in Grape and Wine. *J. Chromatogr. A* **2010,** *1217*, 3881–3889.
29. De Lourdes, Z.; Paes, C. M. D. Analysis of Organophosphorus Pesticides in Whole Milk by Solid Phase Microextraction Gas Chromatography Method. *J. Environ. Sci. Health* **2006,** *41* (4), 369–375.
30. Deti, H.; Hymete, A.; Bekhit, A. A.; Mohamed, A. M. I.; Bekhit, A. E. D. A. Persistent Organochlorine Pesticides Residues in Cow and Goat Milks Collected from Different Regions of Ethiopia. *Chemosphere* **2014,** *106*, 70–74.

31. Di Muccio, A.; Pelosi, P.; Barbini, D. A.; Generali, T.; Ausili, A.; Vergori, F. Selective Extraction of Pyrethroid Pesticide Residues from Milk by Solid-Matrix Dispersion. *J. Chromatogr. A* **1997,** *765* (1), 51–60.
32. Diagne, R. G.; Foster, R. D.; Khan, S.U. Comparison of Soxhlet and Microwave-assisted Extractions for the Determination of Fenitrothion Residues in Beans. *J. Agric. Food Chem.* **2002,** *50*, 3204–3207.
33. Dong, J.; Pan, Y. X.; Lv, J. X.; Sun, J.; Gong, X. M. Multiresidue Method for the Determination of Pesticides in Fruits and Vegetables Using Gas Chromatography–Negative Chemical Ionization–Triple Quadrupole Tandem Mass Spectrometry. *J. Agric. Food Chem.* **2011,** *59*, 109–119.
34. Dordevic, T.; Durovic-Pejcev, R. Food Processing as Means for Pesticide Residue Dissipation. *J. Pesticides Phytomed.* **2016,** *31* (3–4), 89–105.
35. EC (European Communities). Commission Directive 199/50/EC of 25 May 199 Amending Directive 91/321/EEC on Infant Formulae and Follow-on Formulae. *Official Journal of European Communities*, **1999,** *139*, 29–31.
36. Eskilsson, C. S.; Björklund, E. Analytical-Scale Microwave-assisted Extraction. *J. Chromatogr. A* **2001,** *902* (1), 227–250.
37. Farrington, J. W.; Takada, H. Persistent Organic Pollutants (POPs), Polycyclic Aromatic Hydrocarbons (PAHs), and Plastics: Examples of the Status, Trend, and Cycling of Organic Chemicals of Environmental Concern in the Ocean. *Oceanography* **2014,** *27* (1), 196–213.
38. Focant, J. F.; Eppe, G.; Pirard, C.; Massart, A. C.; André, J. E.; De Pauw, E. Levels and Congener Distributions of PCDDs, PCDFs and Non-ortho PCBs in Belgian Foodstuff. Assessment of Dietary Intake. *Chemosphere* **2002,** *48* (2), 167–179.
39. Gill, J. P. S.; Bedi, J. S.; Singh, R.; Fairoze, M. N.; Hazarika, R. A.; Gaurav, A.; Satpathy, S. K.; Chauhan, A. S.; Lindahl, J.; Grace, D.; Kumar, A.; Kakkar, M. Pesticide Residues in Peri-urban Bovine Milk from India and Risk Assessment: A Multicenter Study. *Sci. Rep.* 2020, *10* (1), E-article 8054.
40. Gomez, L.; Tiwari, B.; Garcia-Vaquero, M. Emerging Extraction Techniques: Microwave-assisted Extraction. In *Advances in Green Chemistry, Sustainable Seaweed Technologies*; Elsevier: Amsterdam, 2020; pp 207–224.
41. Grimalt, S.; Dehouck, P. Review of Analytical Methods for the Determination of Pesticide Residues in Grapes. *J. Chromatogr. A* **2016,** *1433*, 1–23.
42. Grimalt, S.; Pozo, Ó. J.; Sancho, J. V.; Hernández, F. Use of Liquid Chromatography Coupled to Quadrupole Time-of-flight Mass Spectrometry to Investigate Pesticide Residues in Fruits. *Anal. Chem.* **2007,** *79*, 2833–2843.
43. Grimalt, S.; Sancho, J. V.; Pozoa, Ó. J.; Hernándeza, F. E. Quantification, Confirmation, and Screening Capability of UHPLC Coupled to Triple Quadrupole and Hybrid Quadrupole Time-of-flight Mass Spectrometry in Pesticide Residue Analysis. *J. Mass Spectr.* **2010,** *45*, 421–436.
44. Haug, A.; Hostmark, A. T.; Harstad, O. M. Bovine Milk in Human Nutrition—A Review. *Lipids Health Dis.* **2007,** *6*, 1–16.
45. Hayward, D. G. Determination of Polychlorinated Dibenzo-*P*-Dioxins and Dibenzo-furans Background in Milk and Cheese by Quadrupole Ion Storage Collision Induced Dissociation MS/MS. *Chemosphere* **1991,** *34*, 929–939.
46. Helfrich, L. A. *Aquatic Animals: A Guide to Reducing Impacts on Aquatic Systems*; 2009. http//publc.ext.vt.edu/420/420-013/420-013.htm (accessed on June 3, 2011).

47. Ho, W. H.; Yu, L. J. Determination of Organochlorine Pesticides in Milk by Electron Capture Capillary Gas Chromatography. *J. Chin. Chem. Soc.* **2013,** *47* (3), 469–473.
48. Hopper, M. L. Multivessel Supercritical Fluid Extraction of Food Items in Total Diet Study. *J. AOAC Int.* **1995,** *78* (4), 1072–1079.
49. Jadhav, V. J.; Waskar, V. S. Public Health Implications of Pesticide Residues in Meat. *Vet. World* **2011,** *4* (4), 178–182.
50. Jansson, C.; Pihlström, T.; Österdahl, B. G.; Markides, K. E. A New Multi-residue Method for Analysis of Pesticide Residues in Fruit and Vegetables Using Liquid Chromatography with Tandem Mass Spectrometric Detection. *J. Chromatogr. A* **2004,** *1023*, 93–104.
51. John, P. J.; Bakore, N.; Bhatnagar, P. Assessment of Organochlorine Pesticide Residue Levels in Dairy Milk and Buffalo Milk from Jaipur in Rajasthan, India. *Environ. Int.* **2001,** *26* (4), 231–236.
52. Jordral, M.; Martinez, P.; Angulo, R.; Gallego, M. C.; Pozo, R. Effect of Milk Pasteurization on DDT and Its Metabolites. *Alimentaria* **1995,** *266*, 25–28.
53. Kalantzi, O. I.; Alcock, R. E.; Johnston, P. A.; Santillo, D.; Stringer, R. L.; Thomas, G. O.; Jones, K. C. The Global Distribution of PCBs and Organochlorine Pesticides in Butter. *Environ. Sci. Technol.* **2001,** *35* (6), 1013–1018.
54. Kaushik, G.; Satya, S.; Naik, S. N. Food Processing a Tool to Pesticide Residue Dissipation—A Review. *Food Res. Int.* **2009,** *42* (1), 26–40.
55. Keikotlhaile, B. M.; Spanoghe, P.; Steurbaut, W. Effects of Food Processing on Pesticide Residues in Fruits and Vegetables: A Meta-analysis Approach. *Food Chem. Toxicol.* **2010,** *48* (1), 1–6.
56. Kowalski, C. H.; da Silva, G. A.; Poppi, R. J.; Godoy, H. T.; Augusto, F. Neuro-genetic Multi-optimization of the Determination of Polychlorinated Biphenyl Congeners in Human Milk by Headspace Solid Phase Microextraction Coupled to Gas Chromatography with Electron Capture Detection. *Anal. Chim. Acta* **2007,** *585* (1), 66–75.
57. Lagana, A.; Marino, A.; Fago, G. Evaluation of Double Solid-phase Extraction System for Determining Triazine Herbicides in Milk. *Chrom. Graph.* **1995,** *41*, 178–182.
58. Le Doux, M. Analytical Methods Applied to the Determination of Pesticide Residues in Foods of Animal Origin. A Review of the Past Two Decades. *J. Chromatogr. A* **2011,** *1218* (8), 1021–1036.
59. Lesueur, C.; Knittl, P.; Gartner, M.; Mentler, A.; Fuerhacker, M. Analysis of 140 Pesticides from Conventional Farming Foodstuff Samples After Extraction with the Modified QuECheRS Method. *Food Cntrl.* **2008,** *19*, 906–914.
60. Li, C.; Bradley, R. L.; Schultz, L. H. Fate of Organochlorine Pesticides during Processing of Milk & Dairy Products. *J. Assoc. Off. Agric. Chem.* **1970,** *53*, 127–139.
61. Liska, B. J. Effects of Processing on Pesticide Residues in Milk. *J. Anim. Sci.* **1968,** *27* (3), 827–830.
62. Madan, V.K.; Kathpal, T. S. Decontamination of Pesticide Residues in Milk by Different Kitchen Processes. In *International Conference on Pesticide, Environment, Food Security*; IARI, New Delhi, 2001; pp 267–268.
63. Muller, A.; Bjorklund, E.; von Holst, C. On-line Cleanup of Pressurized Liquid Extracts for the Determination of Polychlorinated Biphenyls in Food Stuffs and Food Matrices Using Gas Chromatography–Mass Spectrometry. *J. Chromatogr. A* **2001,** *925*, 197–205.
64. Nida-Salem, M.; Ahmad, R.; Estaitieh, H. Organochlorine Pesticide Residues in Dairy Products in Jordan. *Chemosphere* **2009,** *77* (5), 673–678.

65. NPIC (National Pesticide Information Center). http//npic.orst.edu/envir/efate.html (accessed June 3, 2011).
66. Ortelli, D.; Edder, P.; Corvi, C. Multiresidue Analysis of 74 Pesticides in Fruits and Vegetables by Liquid Chromatography–Electrospray–Tandem Mass Spectrometry. *Anal. Chim. Acta* **2004,** *520*, 33–45.
67. Pang, G. F.; Fan, C. L.; Liu, Y. M.; Cao, Y. Z.; Zhang, J. J.; Li, X. M.; Li, Z. Y.; Wu, Y. P.; Guo, T. T. Determination of Residues of 446 Pesticides in Fruits and Vegetables by Three-cartridge Solid-phase Extraction–Gas Chromatography–Mass Spectrometry and Liquid Chromatography–Tandem Mass Spectrometry. *J. AOAC Int.* **2006,** *89*, 740–771.
68. Pardez-Lopez, O.; Gonzales-Casteneda, J.; Carabenz-Trejo, A. J. Influence of Solid Substrate Fermentation on the Chemical Composition. *J. Ferment. Bioeng.* **1991,** *71*, 58–62.
69. Payá, P.; Anastassiades, M.; Mack, D.; Sigalova, I.; Tasdelen, B.; Oliva, J.; Barba, A. Analysis of Pesticide Residues Using the Quick Easy Cheap Effective Rugged and Safe (QuEChERS) Pesticide Multiresidue Method in Combination with Gas and Liquid Chromatography and Tandem Mass Spectrometric Detection. *Anal. Bioanal. Chem.* **2007,** *389*, 1697–1714.
70. Pereira, P. C. Milk Nutritional Composition and Its Role in Human Health. *Nutrition* **2014,** *30*, 619–627.
71. Pico, Y.; Viana, E.; Font, G.; Manes, J. Determination of Organochlorine Pesticide Content in Human Milk and Infant Formulas using Solid Phase Extraction and Capillary Gas Chromatography. *J. Agric. Food Chem.* **1995,** *43* (6), 1610–1615.
72. Pietrino, D. Pesticide Residues and Milk Processing. *Ind. Latte* **1991,** *27*, 3–32.
73. Pihlström, T.; Blomkvist, G.; Friman, P.; Pagard, U.; Osterdahl, B. G. Analysis of Pesticide Residues in Fruit and Vegetables with Ethyl Acetate Extraction Using Gas and Liquid Chromatography with Tandem Mass Spectrometric Detection. *Anal. Bioanal. Chem.* **2007,** *389*, 1773–1789.
74. Piri-Moghadam, A.; Ahmadi, F.; Pawliszyn, J. Critical Review of Solid Phase Microextraction for Analysis of Water Samples. *Trends Anal. Chem.* **2016,** *85*, 133–143.
75. Rajashekar, K.; Kondal, K.; Narasimha, K.; Sudhakar, K. Effect of Processing of Milk into Products on the Residue Levels of Certain Pesticides. *J. Food Sci. Technol.* **2007,** *44*, 551–552.
76. Ramos, L.; Eljarrat, E.; Hernández, L. M.; Rivera, J.; González, M. J. Comparative Study of Methodologies for the Analysis of PCDDs and PDCFs in Powdered Full-fat Milk. PCB, PCDD and PCDF Levels in Commercial Samples from Spain. *Chemosphere* **1999,** *38* (11), 2577–2589.
77. Ramos, L.; Hernández, L. M.; González, M. J. Study of the Distribution of the Polychlorinated Biphenyls in the Milk Fat Globule by Supercritical Fluid Extraction. *Chemosphere* **2000,** *41* (6), 881–888.
78. Ramos, J. J.; González, L.; Ramos, L. Comparison of Gas Chromatography-based Approaches after Fast Miniaturized Sample Preparation for the Monitoring of Selected Pesticide Classes in Fruits. *J. Chromatogr. A* **2009,** *1216*, 7307–7313.
79. Raslan, A. A.; Elbadry, S.; Darwish, W. S. Estimation and Human Health Risk Assessment of Organochlorine Pesticides in Raw Milk Marketed in Zagazig City. *Egypt. J. Toxicol.* **2018,** *2018*, 1–8.
80. Ravelo-Pérez, L. M.; Hernández-Borges, J.; Herrera-Herrera, A. V.; Rodríguez-Delgado, M. Pesticide Extraction from Table Grapes and Plums using Ionic Liquid Based Dispersive Liquid–Liquid Microextraction. *Anal. Bioanal. Chem.* **2009,** *395*, 2387–2395.

81. Raza, N.; Kim, K. H. Quantification Techniques for Important Environmental Contaminants in Milk and Dairy Products. *Trends Anal. Chem.* **2018,** *98*, 79–94.
82. Rêgo, I. C. V.; Viana dos Santos, G. N.; Ribeiro, J. S.; Lopez, R. B.; Baia dos Santos, S. B., de Sousa, A.; de Alcântara Mendes, R.; Taketomi, A. T. F.; Vasconcelos, A. A.; Taube, P. S. Organochlorine Pesticides Residues in Commercial Milk: A Systematic Review. *Acta Agronóm.* **2019,** *68* (2), 99–107.
83. Regueiro, J.; López-Fernández, O.; Rial-Otero, R.; Cancho-Grande, B.; Simal-Gándara, J. Review on the Fermentation of Foods and the Residues of Pesticides—Biotransformation of Pesticides and Effects on Fermentation and Food Quality. *Crit. Rev. Food Sci. Nutr.* **2015,** *55* (6), 839–863.
84. Sajid, M. W. Pesticides Residues and Aflatoxins Milk and Their Dissipation during Processing. *PhD (Food Technology) Thesis*; National Institute of Food Science and Technology Faculty of Food, Nutrition and Home Sciences University of Agriculture, Faisalabad, Pakistan, 2015; 158 pp. https://core.ac.uk/download/pdf/132561473.pdf (accessed August 31, 2021).
85. Salas, J. H.; Gonzalez, M. M.; Noa, M.; Perez, N. A.; Dıaz, G.; Gutierrez, R.; Zazueta, R. H.; Osuna, I. Organophophorus Pesticide Residues in Mexican Commercial Pasteurized Milk. *J. Agric. Food Chem.* **2003,** *51*, 4468–4471.
86. Santos, J. S.; Schwanz, T. G.; Coelho, A. N.; Heck-Marques, M. C.; Mexia, M. M.; Emanuelli, T. Estimated Daily Intake of Organochlorine Pesticides from Dairy Products in Brazil. *Food Cntrl.* **2015,** *53*, 23–28.
87. Sharif, Z.; Man, Y. B. C.; Hamid, N. S. A.; Chin, C. K. Determination of Organochlorine and Pyrethroid Pesticides in Fruit and Vegetables using Solid Phase Extraction Cleanup Cartridges. *J. Chromatogr. A* **2006,** *1127*, 254–261.
88. Shimelis, O.; Yang, Y.; Stenerson, K.; Kaneko, T.; Ye, M. Evaluation of a Solid-Phase Extraction Dual-Layer Carbon/Primary Secondary Amine for Cleanup of Fatty Acid Matrix Components from Food Extracts in Multiresidue Pesticide Analysis. *J. Chromatogr. A* **2007,** *1165*, 18–25.
89. Singh, M. (Ed.) *Production and Utilization of Pesticides in Asia and Pacific*; Asian Productivity Organization: Japan, Tokyo, 2002; pp 190–202.
90. Singh, S.; Nelapati, K. Effect of Food Processing on Degradation of Hexachlorocyclohexane and Its Isomers in Milk. *Vet. World* **2017,** *10* (3), 270–275.
91. Smith, R. M. Before the Injection—Modern Methods of Sample Preparation for Separation Techniques. *J. Chromatogr. A* **2003,** *1000*, 3–27.
92. Smith, R. M. Extractions with Superheated Water—A Review. *J. Chromatogr. A* **2002,** *975*, 31–46.
93. Souza-Silva, E. A.; Lopez-Avila, V.; Pawliszyn, J. Fast Direct Immersion Solid Phase Microextraction Coupled with Gas Chromatography–Time-of-flight Mass Spectrometry Method Employing a Matrix Compatible Fiber for Determination of Triazole Fungicides in Fruits. *J. Chromatogr. A* **2013,** *1313*, 139–146.
94. Sporring, S.; Bøwadt, S.; Svensmark, B.; Björklund, E. Comprehensive Comparison of Classic Soxhlet Extraction, Ultrasonication Extraction, Supercritical Fluid Extraction, Microwave Assisted Extraction and Accelerated Solvent Extraction for the Determination of Polychlorinated Biphenyls in Soil. *J. Chromatogr. A* **2005,** *1090*, 1–9.
95. Tan, G. H.; Chai, M. K. Sample Preparation in the Analysis of Pesticides Residue in Food by Chromatographic Techniques. *Pesticides—Strat. Pesticides Anal.* **2011,** *2011*, 27–58.

96. Taylor, M. J.; Hunter, K.; Hunter, K. B.; Lindsay, D.; Le Bouhellec, S. Multi-residue Method for Rapid Screening and Confirmation of Pesticides in Crude Extracts of Fruits and Vegetables Using Isocratic Liquid Chromatography with Electrospray Tandem Mass Spectrometry. *J. Chromatogr. A* **2002,** *982*, 225–236.
97. Thomas, G. O.; Sweetman, A. J.; Parker; C.A.; Kreibich, K. H.; Jones, C. Development and Validation of Methods for the Trace Determination of PCBs in Biological Matrices. *Chemosphere* **1998,** *36* (11), 2447–2459.
98. Thurman, E. M.; Mills, M. S. *Solid-phase Extraction—Principle and Practice*; New York: John Wiley & Sons, Inc., 1998; 372 pp.
99. Tsiropoulos, N. G.; Miliadis, G. E.; Likas, D. T.; Liapis, K. Residues of Spiro-Xaminein Grapes Following Field Application and Their Fate from Vine to Wine. *J. Agric. Food Chem.* **2005,** *53*, 10091–10096.
100. van der Lee, M. K.; van der Weg, G.; Traag, W. A.; Mol, H. G. J. Qualitative Screening and Quantitative Determination of Pesticides and Contaminants in Animal Feed Using Animal Feed Using Comprehensive Two-dimensional Gas Chromatography with Time-of-flight Mass Spectrometry. *J. Chromatogr. A* **2008**, *1186*, 325–339.
101. Wang, J.; Chow, W.; Leung, D.; Chang, J. Application of Ultra-high-performance Liquid Chromatography and Electrospray Ionization Quadrupole Orbitrap High-resolution Mass Spectrometry for Determination of 166 Pesticides in Fruits and Vegetables. *J. Agric. Food Chem.* **2012,** *60*, 12088–12104.
102. Wang, P.; Zhang, Q.; Wang, Y.; Li, X.; Ding, L.; Jiang, G. Evaluation of Soxhlet Extraction, Accelerated Solvent Extraction, and Microwave Assisted Extraction for the Determination of Polychlorinated Biphenyls and Polybrominated Diphenyl Ethers in Soil and Fish Samples. *Anal. Chim. Acta* **2010,** *663* (1), 43–48.
103. Xu, X. M.; Yu, S.; Li, R.; Fan, J.; Chen, S. H.; Shen, H. T. Distribution, and Migration Study of Pesticides between Peel and Pulp in Grape by Online Gel Permeation Chromatography–Gas Chromatography/Mass Spectrometry. *Food Chem.* **2012,** *16*, 161–169.
104. Zhang, H.; Chai, Z. F.; Sun, H. B.; Zhang, J. L. Survey of Extractable Persistent Organochlorine Pollutants in Chinese Commercial Yogurt. *J. Dairy Sci.* **2006,** *88* (5), 1413–1419.
105. Zheng, G.; Han, C.; Liu, Y.; Wang, J.; Zhu, M.; Wang, C.; Shen, Y. Multiresidue Analysis of 30 Organochlorine Pesticides in Milk and Milk Powder by Gel Permeation Chromatography–Solid Phase Extraction–Gas Chromatography–Tandem Mass Spectrometry. *J. Dairy Sci.* **2014,** *97* (10), 6016–6026.
106. Zidan, Z. H.; Fayed, A. E.; Abou-Arab, A. A.; El-Nockrashy, S. A. Monitoring Some Contaminants during the Manufacture of Some Conventional Dairy Products. *Egypt. J. Dairy Sci.* **1994,** *22*, 19–29.

INDEX

C

E

F

L

M

N

O

P

Q

R